Atoms, Corpuscles and Minima in the Renaissance

Medieval and Early Modern Philosophy and Science

VOLUME 36

The titles published in this series are listed at *brill.com/memps*

Atoms, Corpuscles and Minima in the Renaissance

Edited by

Christoph Lüthy
Elena Nicoli

BRILL

LEIDEN | BOSTON

CENTER FOR THE HISTORY OF PHILOSOPHY AND SCIENCE
Radboud University Nijmegen

This book is the result of a project entitled "Mysteries of Living Corpuscles: The Humanist Revival of Presocratic Philosophy and Atomism in Philosophy, Science and Medicine" of the Dutch Research Council (NWO), no. 360-20-320. Extra financing for the research that went into producing this volume was offered by the Faculty of Philosophy, Theology, and Religious Studies of Radboud University, Nijmegen.

Cover illustration: Giordano Bruno, *De triplici minimo et mensura* (Frankfurt, 1591), 50: "Area Democriti" (also included in the chapter by Leen Spruit).

The Library of Congress Cataloging-in-Publication Data is available online at https://catalog.loc.gov
LC record available at https://lccn.loc.gov/2022042182

Typeface for the Latin, Greek, and Cyrillic scripts: "Brill". See and download: brill.com/brill-typeface.

ISSN 2468-6808
ISBN 978-90-04-52891-8 (hardback)
ISBN 978-90-04-52892-5 (e-book)

This book is printed on acid-free paper and produced in a sustainable manner.

Contents

List of Illustrations VII
Notes on Contributors VIII

1 Atoms, Corpuscles, and Minima in the Renaissance: An Overview 1
Christoph Lüthy and Elena Nicoli

2 Atomism in Sixteenth-Century Italian Commentaries on Lucretius 33
Elena Nicoli

3 Galenic Medicine and the Atomist Revival: Elements, Particles, and Minima in Late Renaissance Physiology 56
Elisabeth Moreau

4 Pores, Parts, and Powers in Sixteenth-Century Commentaries on *Meteorologica* IV 87
Craig Martin

5 Atoms, Corpuscles, and Minima in the Renaissance: The Case of Nicolaus Biesius (1516–1573) 113
Christoph Lüthy

6 Mechanical Arts and Biological Development on the Sixteenth-Century World Stage: The Paracelsian Mechanical Philosophy of Petrus Severinus 146
Jole Shackelford

7 Democritus in Francesco Patrizi and Giordano Bruno 176
Leen Spruit

8 Nicholas Hill, an English Atomist 193
Sandra Plastina

9 Finite God and Infinite Space: Conrad Vorstius and David Gorlaeus 227
Kuni Sakamoto

10 Atomism, Mechanism, and Chymistry in the Natural Philosophy of Walter Warner 247
Stephen Clucas

11 Isaac Newton's Atomist Sources: The Case of Bernhard Varenius 268
William R. Newman

Bibliography 285
Index of Names 317

Illustrations

5.1 Copernican diagram. Nicolaus Biesius, *De universitate libri tres* (Antwerp, 1556), 131 116

7.1 "Area Democriti." Giordano Bruno, *De triplici minimo et mensura* (Frankfurt, 1591), 50 189

10.1 Spherical packing of atomic particles. Walter Warner, British Library, Add. MS 4395, f. 71v. Courtesy of the British Library 249

10.2 Refraction through a medium as a series of internal reflexions. Thomas Harriot, British Library, Add. MS 6789, f. 210r. CC0 1.0 Universal, Courtesy of the British Library 257

10.3 "The passage of rays through a medium." Thomas Harriot, British Library, Add. MS 6789, f. 328r. CC0 1.0 Universal, Courtesy of the British Library 258

Notes on Contributors

Stephen Clucas
is Reader in Early Modern Intellectual History at Birkbeck, University of London. He has published widely in the history of early modern philosophy and science, including a number of articles specifically devoted to the history of corpuscular matter theory. He is currently co-editing Thomas Hobbes' *De corpore* (with Timothy J. Raylor) for the Clarendon edition of the Works of Thomas Hobbes.

Christoph Lüthy
is Professor in the History of Philosophy and Science at Radboud University, Nijmegen (The Netherlands). He has published extensively on the history of matter theories and such neo-atomists as Giordano Bruno, David Gorlaeus, Sébastien Basson and Daniel Sennert, as well as on the evolution of scientific (or 'epistemic') imagery. He is also general editor of the journal *Early Science and Medicine*.

Craig Martin
(PhD 2002) is associate professor of the History of Science and Technology at Ca' Foscari University, Venice. He researches the history of early modern medicine and natural philosophy, with special attention given to Aristotelianism and the history of meteorology. His most recent book is a translation of Girolamo Mercuriale's *On Pestilence* (University of Pennsylvania Press, 2022).

Elisabeth Moreau
is an FNRS postdoctoral researcher at the Université libre de Bruxelles (Brussels, Belgium). Trained in history and philosophy of science, she has worked on medicine and matter theories in late Renaissance Europe. Her current postdoctoral project, "From the Alembic to the Stomach," is centred on the medical and alchemical conceptions of digestion and metabolism.

William R. Newman
is professor in the Department of History and Philosophy of Science and Medicine at Indiana University. His numerous publications on premodern natural philosophy, matter theory, and alchemy include *Promethean Ambitions* (Chicago, 2004), *Atoms and Alchemy* (Chicago, 2006), and *Newton the Alchemist* (Princeton, 2019), as well as other monographs, edited volumes, and articles. He is also general editor of the online *Chymistry of Isaac Newton* project (www.chymistry.org).

Elena Nicoli
(PhD 2017) is associate researcher at the Center for the History of Philosophy and Science at Radboud University, Nijmegen (The Netherlands). Her research interests concern the reception of Epicureanism and Lucretius, especially in the Renaissance. She has published articles on the Renaissance commentaries on *De rerum natura* and the influence of Lucretius on philosophers such as Marsilio Ficino and Giordano Bruno.

Sandra Plastina
is associate professor in the History of Philosophy at University of Calabria (Italy). Her research interests focus on the Renaissance and early modern philosophy. A significant part of her research work is dedicated to the thought of early modern women writers and philosophers. Her latest publication is "Italian Women Philosophers in the Sixteenth Century" in *The Routledge Handbook of Women and Early Modern European Philosophy* (London, 2022).

Kuni Sakamoto
(PhD 2012) is associate professor at the School of Arts and Letters, Meiji University (Japan). The main focus of his research is on early modern Aristotelianism and Cartesianism. His works include *Julius Caesar Scaliger, Renaissance Reformer of Aristotelianism: A Study of His* Exotericae exercitationes (Leiden, 2016).

Jole Shackelford
(PhD 1989) is associate professor in the History of Medicine at the University of Minnesota. His research interests fall within the history of science and medicine in the European Renaissance and early modern periods, with an emerging fascination for the history of biological rhythms research, coming to light in his forthcoming trilogy An *Introduction to the History of Chronobiology* (University of Pittsburgh, 2022).

Leen Spruit
is professor of Early Modern Intellectual History at Radboud University, Nijmegen (The Netherlands). His research interests regard history of cognitive psychology, censorship of science and natural philosophy, and the reception of atomism in early modern Italy. His last published book is *Raimondo di Sangro. Cronaca di vita e opere* (with F. Masucci, Naples, 2020).

CHAPTER 1

Atoms, Corpuscles, and Minima in the Renaissance: An Overview

Christoph Lüthy and Elena Nicoli

1 Introduction: One Challenge and Two Questions

We are accustomed to the notion that there is such a thing as 'atomism' and that it has a very long history – we think of titles such as Léopold Mabilleau's *Histoire de la philosophie atomistique* (1895), Andrew van Melsen's *From Atomos to Atom: The History of the Concept of the Atom* (1953) or Andrew Pyle's *Atomism and Its Critics: From Democritus to Newton* (1998). They all suppose not only that in Greek Antiquity there had been coined a word for the ultimate, literally 'uncut' or 'uncuttable' units of matter (*atomoi*), but that this notion defines a historical school of thought, namely 'atomism.' According to this presupposition, the history of this school spans roughly 2,600 years; from the period of their first proponents, Leucippus and Democritus (fifth century BCE), up to the modern age, with the ultimate triumphs being achieved around 1900, when the existence of atoms was first experimentally proven, and in the 1980s, when they were first visualized by means of the newly invented scanning tunneling microscope. When writing *The Atom in the History of Human Thought* in 1998, for example, Bernard Pullman had no hesitation in connecting, at a stroke, Democritus to modern nanophysics as if they were both part of the same unbroken historical trajectory. Thus, he could praise Francis Bacon for predicting, in 1620, that the microscope "would some day reveal Democritus' invisible atoms. He was proven right, of course, three and a half centuries later, with the advent of the scanning tunneling microscope."[1] From Democritus via Bacon to contemporary physics, according to this view, there was one shared intuition, which enjoyed its ultimate verification in contemporary physics.

According to this narrative, atomism first developed in ancient Greece; having found adherents in ancient Rome, most famously in Lucretius, it was then buried for centuries under the weight of Aristotle's counterarguments; finally, it arose like a phoenix from the ashes in the Renaissance, thereby ushering in

1 Pullman, *The Atom*, 224. Cf. Mabilleau, *Histoire de la philosophie atomistique*; van Melsen, *From Atomos to Atom*; Pyle, *Atomism and Its Critics*.

DOI:10.1163/9789004528925_002

a modern worldview that allowed for scientific quantification and the demise of teleological explanations. One of most rapturous versions of this view has been formulated by Stephen Greenblatt, according to whom the recovery of Lucretius' *De rerum natura* was so momentous that "the world swerved in a new direction": the Renaissance was born and "the world became modern," as the subtitle of his best-selling book *The Swerve* declares.[2]

Many questions can be raised against this continuist story, of which the following two seem the most relevant.

To begin with, there is the issue of terminology. Does an author have to use the word 'atom' to be counted among the 'atomists'? This is by no means a marginal issue, given that the most complete document of what is called "ancient atomism," Lucretius' *De rerum natura*, never uses that word, preferring to describe the minuscule bodies falling through the empty universe as *semina rerum*, *materies*, *genitalia corpora*, or *primordia rerum*. The implication of this choice of terms is addressed further below in this Introduction, and much more extensively in Elena Nicoli's contribution to this volume. By the same token, when Plato in the *Timaeus* proposes that the five elements are composed of triangles that add up to regular solids, nowhere does he call them 'atoms.' Still, there have been no qualms in the historiography when it comes to listing both Lucretius and Plato as 'atomists.'[3]

Secondly, there is the question of ontology. Democritus' famous utterance – possibly the only genuine extant fragment of his thought – repeats as follows: "by convention sweet and by convention bitter, by convention hot, by convention cold, by convention color; but in reality atoms and void."[4] Now, if we define 'atomism' as the belief that the cosmos contains nothing else but atoms and void, then we must face the problem that after Antiquity, we will find no single philosopher or scientist who satisfies that criterion. After the recovery of Lucretius' *De rerum natura* in 1417, none of the authors who explicitly championed atoms would make do with Democritus' barren ontology. As Leen Spruit explains in his contribution to this volume, in his *De triplici minimo et mensura* of 1591, Giordano Bruno provided his readers with woodcuts of atoms, which he himself had engraved, and some of which he dedicated to Democritus and to Leucippus. And yet, he insisted that "the atoms and the void don't suffice us."[5] In fact, his cosmos is everywhere animated, thanks to a Neoplatonist unfolding

2 Greenblatt, *The Swerve*, 11, and the subtitle: *How the World Became Modern*.

3 The only exception is Nail, *Lucretius I*, 11, who denies that Lucretius was an atomist.

4 Diels and Kranz, *Fragmente der Vorsokratiker* (6th ed.), fragment 68B9, translated in Taylor, *The Atomists*.

5 Bruno, *De triplici minimo et mensura*, 10: "[…] nobis vero vacuum simpliciter cum atomis non sufficit […]."

of the divinity into all kinds of agencies and thanks to a vivifying ether. Or take Walter Charleton, who, some sixty years later in his *Darknes of Atheism* of 1652, embraced a Christian type of atomism while rejecting the Epicurean notions "that the World was not constituted by God," "that the World was not governed by God," and "that the soul of man doth not survive the funeral of his body."[6] The situation is even more complex with authors who denied the existence of atoms but were generally considered 'atomists' by their contemporaries. The most famous of these was René Descartes, who insisted that his philosophy differed from Democritus' as much as it did from that of Aristotle and hammered resolutely on the indefinite divisibility of his *res extensa*.[7] Still, his richly illustrated explanations of the interaction of corpuscles were widely interpreted as representations of a Democritean world. "He avoids the word 'atom' in vain," Bernhard Varenius concluded in 1647 when reporting to his teacher Joachim Jungius about the debates Descartes had triggered in the Netherlands.[8] In other words, then, neither among those who explicitly endorsed atoms nor among those who sported non-atomic particles do we find anyone who was satisfied with Democritus' limited explanatory apparatus of "atoms and void."

These two issues build up to the following question: if the available histories of atomism include persons who didn't use the word "atom" and if their inclusion is not even based on an ontological correspondence with a Democritean conception of nature, what then are we left with? What is it that can possibly be designated by the expression "atomism and its history"?

2 The Emergence of the Term 'Atomism'

A particularly brutal answer to the above questions would be to dismiss out of hand the term 'atomism' altogether and laugh out of court the idea that it had an age-old pedigree. To begin with, one might point out that the atom of contemporary physics is anything but 'uncuttable': while 'atomic fission', on paper, is an oxymoron, it yet generates energy for good and evil purposes. Indeed, with its nucleus, its halo of electrons and its various forces, it has nothing in common with Democritus' chunklets of universal matter or Epicurus' shower of particles descending through the cosmic void.

6 Charleton, *Darknes of Atheism*, 40–41.

7 Descartes, *Principia*, IV § 202, marginal caption: "Democriti Philosophiam non minus differre a nostra, quam a vulgari.".

8 Varenius, letter to Jungius of 24 December 1647, cited from Elsner and Rothkegel, eds., *Der Briefwechsel des Joachim Jungius*, 686–687: "[…] frustra Cartesio vacabulum atomi in scriptis suis vitante […]."

The only reason that pleads against such a brutal rejection of the term is that, in the course of the past two centuries, philosophers and scientists have routinely invoked the term 'atomism.' While they have done so to describe divergent models of reality, they did in fact believe that they were referring to a doctrinal body that possessed a long prehistory. Even if this assumption had been either erroneous or opportunistic, we would still have to accept it as an actor's category and make sense of it.

But while this is true for more recent times, for the purposes of this volume, it is essential to recognize that the Renaissance did not use the term 'atomism'. In fact, the present book is about a period in which atoms, minima, and particles could be invoked in philosophical reasoning without any automatic allegiance to an atomistic school of thought. Our own histories of atomism in fact respond to the needs of two later centuries, the seventeenth and the nineteenth.

On the subject of the nineteenth century, we may be brief. It will suffice to mention here that the scientific community, notably in the field of chemistry, was at the time deeply divided over the "question of the atom" (*Atomenfrage*).[9] Until 1900 or thereabouts, French chemists by and large rejected atomistic conceptions of matter, while these were particularly favored by English chemists. In 1826, the influential chemist Jean-Baptiste Dumas made his position clear by saying that "if I were master of the situation, I would efface the word 'atom' from science."[10] The 1860 Karlsruhe Congress saw 140 chemists argue over the question of the existence of atoms, molecules, radicals, and equivalents, and the Académie des Sciences repeated the same exercise seventeen years later. But the French remained for the most part unpersuaded. "Who has ever seen a gas molecule or an atom?" went Marcelin Berthelot's polemical demand.[11] This debate subsided only after 1900, when the existence of the atom was experimentally proven, although that was, strictly speaking, a Pyrrhic victory, given that soon afterwards, the composite nature of the atom was also demonstrated, which falsified the notion that the atom was the ultimate and indivisible component of nature.

What is relevant about this nineteenth-century controversy from a historiographical perspective is that, in those decades in which chemists and physicists battled over the postulate of atoms, numerous philosophers and scientists published books and essays in which they offered arguments of all sorts

9 Fechner, *Über die Atomenlehre*, 4.

10 "Si j'en étais le maître j'effacerais le mot atome de la science." Cited from Caullery, *La science française*, 6.

11 "Qui a jamais vu une molécule gazeuse, ou un atome?" Cited from Nye, *Molecular Reality*, 7.

in favor of atomism, including historical arguments – a bit like Renaissance or early modern philosophers rallying for the truth of a given concept because it had a venerable pedigree, being Hermetic or Mosaic or indeed Democritean. Some nineteenth-century authors resorted to the "historicocritical method," as did Kurd Lasswitz in his two-volume *Geschichte der Atomistik vom Mittelalter bis Newton* of 1890, assuming that history could provide a heuristic for deciding on a contemporary issue. Other well-known examples of the attempt to tell of a *longue-durée* prehistory to the current debate were Mabilleau's *Histoire de la philosophie atomistique*, cited previously, or Arthur Hannequin's *Essai critique sur l'hypothèse des atomes dans la science contemporaine* (1899). On the other side of the divide, one finds someone like the neo-scholastic chemist and historian of science Pierre Duhem, who as late as 1902 tried in his *Le mixte et la combinaison chimique: essai sur l'évolution d'une idée* to defend an Aristotelian theory of mixture against the allegedly Democritean matter theory championed by the English.[12]

In other words, the battle of the chemists led to attempts to justify – or reject – the concept of the atom, by recourse to its historical pedigree. The periods that were most closely examined were Antiquity, on the one hand, and on the other the alleged atomistic revival in the seventeenth century. In his *Geschichte der Atomistik*, Lasswitz in fact dedicates hundreds of pages to the seventeenth century. In the process, he – like everyone else who in the nineteenth century traced the history of the concept 'atom' – hit upon the early modern discussions over the existence of ultimate particles, which therefore seemed to historiographers like an anticipation of the contemporary debate. In fact, in the seventeenth century, atoms did move center stage, becoming a contested term and a litmus test in a Republic of Letters that was riven not only along confessional lines, but also in terms of its allegiance to the traditional school philosophy and the rival proposals of various *novatores*.

And so, the nineteenth century's particular obsessions take us to the seventeenth's. But like the nineteenth, the seventeenth century may not be taken as a yardstick by which to measure earlier developments. To be sure, as we will see in this volume, in the fifteenth and sixteenth centuries, an ever-increasing number of writers – timidly at first but then with increasing assertiveness – had invoked atoms, corpuscles, particles, and minima in their writings. But no one at the time took much offense, and certainly no one celebrated or condemned them by viewing them as proponents of 'atomism.' They had moreover characterized these entities in so many different ways, and in fact in

12 Lasswitz, *Geschichte der Atomistik*; Mabilleau, *Histoire de la philosophie atomistique*; Hannequin, *Essai critique sur l'hypothèse des atomes*; Duhem, *Le mixte et la combinaison*.

mutually exclusive ways, that it would have been completely implausible to view them as belonging to the same current of thought or doctrinal category. The reasons moreover why they came up with their particles were decidedly disparate. Next to mathematical, physical, medical, or alchemical reasons, one also finds metaphysical, theological, antiquarian, historiographical or numerological motivations. Open to question also is the utility of pitting atoms and other small particles against Aristotelianism. As Craig Martin has repeatedly argued and as he demonstrates once again in his contribution to this volume, there is an atomistic current in sixteenth-century Aristotelianism that takes its cue from book IV of the *Meteorology*.[13]

Robert Pasnau has made the startling claim that for centuries, the issue of atomism was "a thoroughly peripheral issue, inasmuch as very little turns on whether one thinks the material realm is or is not infinitely divisible."[14] As many of the chapters in this volume attest, this statement is quite correct for much of the Renaissance – although Nicolaus Biesius, discussed in Christoph Lüthy's contribution, may represent a significant exception to this rule. While Pasnau's claim does much to back up our call for the dismissal of the notion of 'Renaissance atomism' altogether, his claim is plainly untrue, however, for the concluding period covered in his book, *Metaphysical Themes 1274–1671*. By the middle of the seventeenth century at the latest, it was impossible to invoke atoms without thereby being taken to reject Aristotle's philosophy, and notably his notion of substantial forms. By that time, explicitly atomistic theories had been developed by anti-Aristotelians such as Giordano Bruno, Sébastien Basson, David Gorlaeus, Jean-Chrysostôme Magnen, Etienne De Clave, or Pierre Gassendi, so that the term 'atomism' and equivalent names were coined to describe what had, by then, seemed to have become a movement.

In the case of the seventeenth century, religion must also be taken into consideration when exploring these theories. However pious the work of an individual philosopher might have looked otherwise, it was the adopted stance vis-à-vis Aristotelian hylemorphism that defined, at least in the eyes of the central organs of the Catholic Church, the orthodoxy or heterodoxy of their work. The 1641 *Querelle d'Utrecht*, so troublesome for René Descartes and his friend and disciple Henricus Regius, show that the rejection of substantial forms in favor of geometrical particles that were pushed about mechanically could trigger censorship also in the Protestant lands.[15]

13 On the commentary tradition of book IV of Aristotle's *Meteorology*, see Martin, *Renaissance Meteorology*.

14 Pasnau, *Metaphysical Themes*, 88.

15 Verbeek, ed., *Querelle d'Utrecht*.

It is therefore important to develop a sense of how the reception of any given term or doctrine could mutate with the passage of time. It is well known how Nicolaus Copernicus could dedicate his *De revolutionibus orbium coelestium* to the pope in 1543, while seventy years later, in 1616, the same book was placed on the *Index of Forbidden Books* "until corrected" (*donec corrigatur*). The fate of atoms is analogous: their use in natural-philosophical and medical explanations only became a problem when they were explicitly sported as the rivals of substantial forms. This development occurred in the Counter-Reformational context of tightening doctrinal rules, and more specifically as a consequence of the decision of the Council of Trent to insist on transubstantiation as the only correct way of interpreting the Eucharist, wherein transubstantiation was defined in the terminology of substance and accident, of matter and substantial forms. This changing context allowed Pietro Redondi to make his famous but ultimately unconvincing claim that Galileo was not condemned for his heliocentrism, but for his atomism.[16]

Redondi quotes the following lines from an "Exercise on substantial forms and physical qualities" from the later seventeenth century, which under the title "Nothing comes from atoms" argues as follows:

> All the bodies of the world shine with the beauty of their forms.
> Without these the globe would only be an immense chaos.
> In the beginning God made all things, so that they might generate something.
> Consider to be nothing that from which nothing can come.
> You, o Democritus, form nothing different starting from atoms.
> Atoms produce nothing; therefore, atoms are nothing.[17]

Atoms *versus* forms: in the course of the seventeenth century, this pair of opposites gradually hardened into a rigid dichotomy. Our historiography tends to define the opposition even more sharply, in terms of 'atomism *versus* hylemorphism.' The addition of this suffix '-ism,' has the effect of turning an opposition of concepts into an opposition of ideologies or worldviews. But is this correct? As Gideon Manning has documented, 'hylemorphism' is an even more recent term than 'atomism.' It was coined only at the end of the nineteenth century and, at first, described a certain type of materialism. Only in the twentieth did it come to refer to the Peripatetic theory of matter and form.[18]

---

16 Redondi, *Galileo Heretic.*

17 Ibid., 340: "*Exercitatio de formis substantialibus et de qualitatibus physicis*, anonymous."

18 Manning, "Three Biased Reminders," 3.

Admittedly, the term 'atomism' is somewhat older.[19] The earliest reference that we have been able to find was in Theodor Gazes's *Antirrheticon* of ca. 1470. However, there, the Greek word *atomismós* means 'indivisibility' and figures as the antonym of the term *merismós*, 'divisibility.'[20] When the term 'atomism' was used in the fifteenth century, whether as a neologism or otherwise, it was therefore not at all intended to designate a school of thought.

After the recovery of Lucretius' poem by Poggio Bracciolini in 1417 and the translation from the Greek of various Epicurean writings, a range of Renaissance philosophers did indeed come to regard Leucippus, Democritus, Epicurus, and Lucretius as standing in the same tradition. However, as Elena Nicoli's contribution to the current volume demonstrates, they did so not primarily because of their use of the term 'atom' (which, as mentioned earlier, Lucretius in fact avoided), but because of such other common doctrines as the mortality of the soul, their denial of divine providence, or their ethics. While references were made in the Middle Ages to '*Democritei*' or '*Epicurei*' as representatives of certain positions on indivisibles, it would seem that the group name 'atomists' came into use only in the seventeenth century, and at first only as a reference to the ancient philosophers. Nicolaus Taurellus, for example, writes in his *Kosmologia* of 1603: "What in the doctrine of the atomists is said about the atoms, is of no importance."[21] And in the vernacular, John Healey writes in his 1610 commentary on Augustine's *City of God*: "Of the Atomists, some confound all, making bodies of coherent remaynders."[22] But, like everyone else who used the term 'atomists' at this time, he was referring to an ancient philosophical standpoint, just as one would speak of 'the Stoics' or 'the Platonists.' It did not yet refer to a current to which a living person could adhere.

Jean-Chrystostôme Magnen takes us one step closer to the modern usage. In the preface to his *Democritus reviviscens* of 1644, he explained that he wished to revive "the philosophy of the atom," that "first-born of all schools of philosophy." It is curious to note, incidentally, that his list of members of this school includes Moschus, the Phoenician, that legendary inventor of atomism, and also Plato and his disciples, but it does not mention Epicurus![23]

---

19 The following paragraphs follow Lüthy, "Atomism in the Renaissance," often verbatim.

20 Cf. Estienne's *Thesaurus Graecae linguae* of 1572. Gazes's passage reads: "ὡς οὐδέν ἐστιν ἄτομον ἄνευ ὕλης, οὐδ' οἷόν τε ἀτομισθῆναι καὶ μερισθῆναι οὐδὲν πλὴν κατὰ τὴν ὕλην. ἀρχὴ γὰρ αὕτη ἀτομισμοῦ τε καὶ μερισμοῦ."

21 Taurellus, *Kosmologia*, 142: "Quod de atomis dicitur ex atomistarum sententia, nullius est momenti."

22 Healey, *St. Augustine of the Citie of God*, 438.

23 Magnen, *Democritus reviviscens*, "Prolegomena," s.p.: "Cum primogenitam inter omnes sapientium sectas Atomorum Philosophiam restituere designaverim [...]."

The abstract name 'atomism' (or 'atomicism') as a school name emerged only in the last quarter of the seventeenth century, and even then, it was used at best fleetingly and in competition with many other '-isms.' Interestingly, moreover, it did not mean what we take it to mean today. For Ralph Cudworth – the person who may in fact have coined the term – it designated specifically the perverted, materialist and atheist variant of a true, ancient "corpuscular philosophy."[24] In his eyes, the original school was that of 'corpuscularism' (which in the modern age had been revived by Descartes), of which 'atomism' was a wickedly deviant form. His contemporary Robert Boyle also distinguished between an 'atomistic' and a Cartesian-style 'corpuscular' philosophy, but unlike Cudworth, he found them both to be valuable and moreover so similar that they could be merged together with "the mechanistic philosophy."[25] That latter expression – "mechanistic philosophy" – had the advantage of clustering various types of philosophy under one umbrella term, irrespective of whether they allowed for indivisible atoms (like Gassendi) or didn't (like Descartes).

Our modern historiography has accepted Boyle's claim that "the atomical philosophy invented or brought into request by Democritus, Leucippus, Epicurus, & their contemporaries" had, by his own century, been "so luckily revived [...] by the learned pens of Gassendus, Magnenus, Descartes, & his disciples."[26] We have often too readily embraced the idea that the seventeenth century's "atomist revival" ultimately produced an overarching "mechanical philosophy" – a philosophy that stands at the cradle of modern science.[27]

But this view overlooks the fact that Boyle's labels have had more success in twentieth-century historiography than in their own day and age. If one examines the natural philosophical literature of ca. 1700, one finds that there existed no terminological consensus over how to demarcate the labels 'atomistic,' 'corpuscular' and 'mechanistic,' not to mention other terms that are now entirely forgotten, such as 'somatistic.' As for the eighteenth century, the term 'atomism' was then used only infrequently. In fact, it was only in the above-mentioned context of nineteenth-century controversies over the explanatory value of atoms in chemistry and physics that the term 'atomism' became an integral component of the scientific jargon. As mentioned above, authors favorable to the notion of the 'atom' began to spread the idea that there was a coherent history of atomism to be told that ran from the Greeks through the Renaissance

24 Cudworth, *True Intellectual System*, ch. 2.

25 Boyle, *Some Specimens of an Attempt*, "Preface."

26 Boyle, *Works*, 13: 227.

27 See, for example, Boas Hall, "Establishment of the Mechanical Philosophy," and Westfall, *Construction of Modern Science*.

up to the chemical atoms postulated after such pioneers as William Hyde Wollaston and John Dalton.

3 Atoms, Corpuscles, and Minima in the Renaissance: Some Preconditions

Now that we have put to bed the idea that there was a notion of 'atomism' before the second half of the seventeenth century, and that the idea of a continuous history of atomism dates to the nineteenth century, we may look with new eyes to the Renaissance. This is what this book tries to achieve. As the reader will notice, in this volume we will still occasionally use the term 'atomism,' even though we are aware that the word was coined after most of the texts under review were written. We don't use the term to refer to a school as such, but rather to a set of theories, based on the idea of small bodies, every time interpreted differently by each author treated in this book.

In Renaissance authors, we find an increasing recourse to the explanation of natural things in terms of small bodies, whether these were called 'atoms,' 'minima,' 'corpuscles' or 'monads.' We may therefore ask the questions: what motivated this increased reliance on invisibly small bodies, and what philosophical allegiance did this recourse to small bodies betray?

In order to answer these questions, we must recall that before the massive introduction of Aristotle's philosophy in the twelfth and early thirteen centuries, atomistic conceptions of matter had not, by any means, been regarded with suspicion. In the twelfth century, for example, William of Conches stated in his *Dialogue on Philosophy* (I, 6) that "when the Epicureans said that the earth consists of atoms, they were correct." By contrast, he rejected the Epicurean cosmogonies as a "fable" that was incompatible with Christian faith and revelation. This medieval combination of an atomistic conception of matter with an insistence on divine creation would in time disappear, however, under the influence of Aristotle.[28]

Because of the dominance of Aristotelian natural philosophy in the universities, from the thirteenth century onwards, Aristotle's view that all extended magnitudes – matter, time, space, and motion – were continuous by nature and could be divided indefinitely, became dominant in natural philosophy and metaphysics. Let us recall, however, that Aristotle's objections were not exclusively aimed at a specific matter theory, but applied to all extended magnitudes. For Aristotle, as much as for his later detractors, to postulate material

28 Robert, "Atomism," 123.

atoms necessarily also meant to postulate indivisibles of space and time. This underlying assumption of isomorphism survived intact into the seventeenth century.[29] This forced someone like Jean-Chrysostôme Magnen, who with his *Democritus reviviscens* strove to replace Aristotle's matter theory with that of Democritus, to conclude that "motion, time and space are analogous, that is to say, proportional."[30] If matter had a discontinuous structure, then the same had to be true also of motion, time, and space.

This isomorphism is powerfully visible in those ninth- to twelfth-century Muslim theologians of the *kalām* (the so-called *mutakallimūn*), who emphatically embraced an overall atomistic conception of time, space, and matter.[31] Their view implied what we would nowadays characterize using the terms 'determinism' and 'occasionalism,' as, according to their model, God re-created the world in each instant, with all secondary causality gone – a view that is obviously quite distant from the doctrines of either Democritus or Epicurus. Reports of their theories entered the Latin West through Avicenna, Averroes, and Maimonides, and were discussed by medieval and Renaissance authors, although, this radically discontinuist view of the world, in which God is the only active cause, found no supporters before the early seventeenth century. Sébastien Basson may have been the first author to overtly flirt with this idea, even while doing so only *en passant*.[32] Given the dictates of isomorphism, however, Basson, too, was forced to conclude that time, space, and motion had to be discontinuous.

In the Latin Middle Ages, the growing acquaintance with Aristotle's arguments against atoms led to questions that were not theological, as the *mutakallimûn*'s had been, but natural philosophical or logical. Natural philosophers in the fourteenth century were, for example, obsessed with the question of mixture: what happens to the qualities of the elements when they are blended into a new, continuous mixture, which according to Aristotle had to be once again a continuum? How did the original ingredients bring about the new "form of the mixt" (*forma mixti*), and how was it possible that sometimes, the original elements could be recovered from the mixt?[33] While these questions remained unresolved in the context of the university disputations, they were very much of concrete relevance to medieval alchemists, whose understanding of mixture and other chemical processes often relied on the assumptions that

29 Palmerino, "Isomorphism of Space, Time and Matter."
30 Magnen, *Democritus reviviscens*, 226.
31 Dhanani, *Physical Theory of Kalām*.
32 Basson, *Philosophia naturalis*, 247 [misnumbered as 227].
33 Maier, "Struktur der materiellen Substanz."

the ultimate carriers of qualities were individual units. William Newman has shown how, from the anonymous thirteenth-century pseudo-Geber (Paul of Taranto) who in his *Summa perfectionis* spoke of the combination of elementary *minimae partes* that together form a *fortis compositio*, there has existed a continuous alchemical tradition that explained chemical operations such as sublimation or mixture in terms of particles. A particularly strong empirical argument in favor of an atomical structure of compounds was constituted by the operation known as *reductio ad pristinum statum*, in which the original ingredients of a mixture were fully recovered from the mixt.[34]

Another set of scholastic authors, beginning with the fourteenth-century Oxford philosopher Henry of Harclay, challenged Aristotle's concept of the continuum and insisted that it was possible to construct a continuum from a finite number (according to some) or an infinite number (according to others) of indivisibles. These positions responded to problems that arose from Aristotle's view of continuous magnitudes and had once again nothing to do with any sympathy for the doctrines of Democritus or Epicurus.[35] In accordance with the same principle of isomorphism mentioned earlier, the debate between indivisibilists and continuists was about the nature of extensions of all sorts, and involved above all geometrical arguments. A popular anti-divisibilist argument, which had already been used by Arabic philosophers, claimed that according to the divisibilists, the diagonal of a square would no longer be incommensurable with its sides, but could be shown to be composed of a precise number of points, each corresponding to a point on the sides.[36] The divisibilists' answer was that this solution clashed not only with common sense, but also violated the principles of geometry. Their rebuttal remained a powerful weapon through the centuries, even into the seventeenth century, where we still find it extensively rehearsed in Libert Froidmont's *Labyrinthus sive de compositione continui* of 1631. It is surprising to the modern reader to see that most indivisibilists felt that they had to accommodate their theories to this objection. Sébastien Basson, for example, simply accepted that diagonals were in reality not lines, but steps.[37] Much more thorough and systematic was the response by Giordano Bruno, who elaborated a veritable geometry of indivisible globules, which explicitly excluded the possibility of diagonal lines. This approach to mathematics, which Bruno shared with Francesco Patrizi, has at times been described as a "realist obstacle" to an understanding of geometrical

34 Newman, *Summa Perfectionis*; idem, "Alchemical Sources"; idem, *Atoms and Alchemy*.
35 Murdoch, "Naissance et développement"; Pabst, *Atomentheorien*; Robert, "Atomism."
36 Murdoch, "Mathesis in philosophiam scolasticam introducta"; Robert, "Atomism."
37 Basson, *Philosophia naturalis*, 420.

magnitudes.[38] In the seventeenth century, several authors tried – in the wake of Andreas Libavius's *Alchymia triumphans* (1607) or Daniel Sennert's *De chymicorum cum Aristotelicis et Galenicis consensu ac dissensu* (1619/1629) – to overcome this difficulty by distinguishing between the mathematical divisibility *ad infinitum* of all extended magnitudes and the physical indivisibility of material particles.

The growing late-Renaissance interest in material indivisibles also has some late medieval antecedents. A small but prominent handful of fourteenth- and fifteenth-century philosophers, less interested in mathematical indivisibles than in physical concerns, had put atoms to work in explaining generation and corruption or condensation and rarefaction. One of the most prominent representatives of this physicalist approach was Nicholas of Autrecourt.[39] The suggestions made by these thinkers did not, however, generate any strong resonances, not least because several of them were condemned for holding heretical views. Nicholas of Autrecourt himself was condemned in 1346. John Wyclif, several of whose views attracted censure in 1377 and 1382, is another notorious case in this respect.[40]

One of the primary reasons why such proposals were frequently suppressed was that, well into the modern age, the Catholic Church considered such atomistic conceptions to be incompatible with its own dogmatic understanding of the Eucharist in terms of the distinction into substance and accident. Although for Autrecourt, for example, atoms were not in conflict with substantial forms, still, there were periods in which the medieval Church took offense at the very use of the term. This was less true for the Renaissance, as various contributions to this volume document. But when the battle against atomistic explanations was once more ignited in the seventeenth century, Wyclif's condemnation could be taken – by friends and foes alike – as the first episode in a perennial opposition between a Christian and a pagan natural philosophy. As late as 1663, this perceived opposition and the question of the Eucharist got René Descartes's works into the *Index of Forbidden Books*, even though he had adamantly rejected the concept of the atom.

But again, as we have suggested earlier, there was no transhistorical doctrinal line followed by the Church. In the Renaissance, Cardinal Cusanus (Nicholas of Cusa) could blend theological, logical, and physical doctrines in the most unorthodox ways, and yet managed not only to avoid all censorship, but to enjoy an untarnished reputation. In several of his writing, Cusanus made recourse to

38 Védrine, "Obstacle réaliste."
39 Grellard, "Autrecourt's Atomistic Physics."
40 Robson, *Wyclif and the Oxford Schools.*

mathematical arguments to strengthen his theological and metaphysical reasoning. In his *De docta ignorantia* (1440), he develops his well-known doctrine of the "coincidence of opposites" (*coincidentia oppositorum*). According to this doctrine, in the field of geometry, the largest and the smallest may be shown to coincide formally, as the infinitely small point and the infinitely large circle share a number of features. By the same token, in the field of metaphysics, God as the all-embracing measure of all things, and its contrary, the invisibly small unit-point, coincide in various formal respects. In his *De ludo globi* (ca. 1462), Cusanus furthermore describes the geometrical point as a dynamic starting point, which is capable of "evolving itself" into atom, line, and body, and which can therefore be present in all things.[41] The unfolding of the point into the All, and its opposite, namely the contraction of the divine All into a point (which is everywhere and anywhere), are routinely illustrated in geometrical terms.

One of the clearest testimonies to Cusanus's intellectual influence is constituted by Giordano Bruno, who repeatedly expressed his admiration for the "divine Cusanus" and developed the latter's doctrine of the "coincidence of opposites" into a complex theory of the "triple minimum."[42] The word 'triple' refers to the fact that Bruno's 'minimum' designates the mathematical point, the monad, and the atom. As for the physical atom, Bruno regarded it (in Cusanian fashion) as the dynamically expansive and ubiquitous minimum in an infinite universe. As Leen Spruit argues in his contribution to this book, it is no coincidence that the atom, which for Bruno is always spherical (since it is the projection of the mathematical point into the third dimension) has the same shape as stars and solar systems (*mondi*), as "we may attribute the same figure to the maximum and the minimum."[43] In fact, not least because of his adherence to heliocentrism and the doctrine of the many worlds, Bruno became ever more Democritean in the course of his philosophical development. The dynamic expansion of the geometrical points into physical atoms – from the atoms into solar systems (*mondi*, in the Italian dialogues), and from the *mondi* into the infinite universe: this, in Bruno's mind, is the path that the original *centra* take through a series of ever larger *circuli* until they reach the ultimate *circumferentia*. The curvature of this ultimate, infinite, and divine circumference was so stretched-out as to coincide with the straight line, its logical opposite. And when, in the opposite movement, the ultimate expansion contracts down to the level of the original unit of expansion, the indivisible

41 Cusanus, *Dialogus de ludo globi*, 578.

42 See, above all, Bruno, *De triplici minimo et mensura*.

43 Ibid., 49: "Eandem ergo figuram maximo tribuimus atque minimo."

point, all geometrical opposites fall together once more into a point that is also the smallest line and the tiniest body.

Giordano Bruno tried to visualize this doctrine in numerous woodcuts of his own making, some of which are inspired by Cusanus's own diagrams.[44] Several of these woodcuts carry names. A key illustration, and one that probably constitutes the very first image of a constellation of physical atoms ever drawn, is called "Area Democriti." It is contained in *De triplici minimo et mensura*, a didactic poem that imitates the style of Lucretius' *De rerum natura*, and is reproduced both on the book cover and in the chapter by Leen Spruit contained in this volume.[45] Here, then, we finally encounter an acknowledged debt to Democritus (who, like Bruno, had postulated infinite worlds), as well as to Lucretius (whose preferred terminology for 'atom,' *semina rerum*, Bruno took as a sign of Lucretius' vitalistic understanding of these particles). And yet, it would be wrong to view Bruno simply as a neo-atomist. As we have mentioned before, Bruno declares that "the atoms and the void don't suffice us," and he fills his cosmos with a vivifying ether.[46] In fact, no single allegedly 'atomistic' author of the early modern period felt that "atoms and the void" were sufficient. To begin with the most important additional element, none wished to exclude a divine Creator. That Creator, however, invariably turned Democritus' 'necessity' and Epicurus' swerving 'chance' into their respective contrary: God either programmed the atoms *ab initio*, or He providentially made sure that they behaved according to His plan. Furthermore, Renaissance and early modern atoms were routinely endowed with souls, innate forces, sympathies and antipathies, and appetites, or else they were vivified by an ether or guided by God. Aggregate titles such as Nicholas Hill's *De philosophia Epicurea, Democritiana, Theophrastica, proposita simpliciter, non edocta* (1601) – which is analyzed in detail in Sandra Plastina's contribution to this book – or Walter Charleton's later *Physiologia Epicuro-Gassendo-Charletoniana: or a Fabric of Science Natural, upon the Hypothesis of Atoms* (1654) are indicative of the combinatorial and syncretistic nature of the alleged revival of ancient atomist ideas.

As Sandra Plastina shows in her chapter, the triple qualifier of Hill's philosophy – *Epicurea, Democritiana, Theophrastica* – is a clear indication of its syncretistic nature. While Democritus and Epicurus can be reconciled with one another without much violence, the introduction into the equation of Theophrasus of Hohenheim – called Paracelsus – creates tensions. In fact, we

44 For a comparison of the woodcuts, see Lüthy, "Centre, Circle, Circumference."

45 Bruno, *De triplici minimo*, 51.

46 Ibid., 10.

here encounter a cluster of ideas that Jole Shackelford discusses in his contribution to this volume on the Paracelsian Petrus Severinus (see below), that is to say, the notion of material particles as seeds and centers of vital forces. But, as Plastina shows, Hill cast his net even more widely. His sources include the two authors discussed in Leen Spruit's chapter, Francesco Patrizi and Giordano Bruno, as well as authors such as Girolamo Fracastoro. The result is a "philosophy neither new nor old" (*philosphia nec nova nec vetus*), which basically follows "the Epicureans and Democriteans" in "approaching the explanation of nature excellently by beginning from the inclination of unstable atoms," but subsequently endows these atoms with all kinds of non-Democritean and non-Epicurean properties.[47] Among these contaminations, we may list the ability of matter to assume forms (matter as *formicapax*, according to Hill's neologism); the ability of the *semina* to imprint signatures (*signatura*) on certain objects; the *minima vis* that inhabits each atom, or the idea, taken from Nicolaus Cusanus via Giordano Bruno, that the center amplifies into a sphere, growing out of the minimum. Hill's decision to express his ideas in aphorisms allowed him, to some extent, to sidestep this conceptual mess and not to have to clean up afterwards. However bizarre the book may be, it is, after Giordano Bruno, the first published attempt to explicitly merge Democritus and Epicurus with contemporary developments, with Paracelsianism, Copernicanism, natural magic, and theological speculation. For contemporaries, Hill's name was thus co-extensive with that of a kind of *Democritus reviviscens*. As Ben Jonson quipped in *Epigram* 134:

> [...] those Atomi ridiculous,
> Whereof old Democrite and Hill Nicholis,
> One said, the other swore, the world consists.

As the case of Hill documents – and an examination of comparable books yields the same result – the new combinations between the atoms of the ancients and modern doctrines were often contradictory, and when compared with one another, they were as different as the natures of the ultimate particles that they each proposed. It is enough to compare Giordano Bruno's ensouled spherical monads with Pierre Gassendi's or Giovanni Alfonso Borelli's highly complex corpuscular shapes, with the unextended point-atoms (*atomi non*

47 Hill, *Philosophia Epicurea*, ed. Plastina, Aph. 138, 103–104: "[...] immo ut audacius, et confidentius loquar, Epicurei, et Democritiani optime naturae explicationem aggressi sunt ab atomorum instabilium inclinatione exorsi, si modo physicorum dubiorum solutionem illinc petent, et derivent apertam."

quanti) of Galileo Galilei's *Two New Sciences* of 1638, with Daniel Sennert's elemental atoms that carried substantial forms, or with René Descartes's curiously shaped particles that were brought about by the variegated movements of the universal *res extensa*. The properties ascribed to the minimal material bodies in each of these models are entirely irreconcilable and also look entirely different when drawn on paper, if they can be drawn at all. What this means is that where the term 'atom' was used, it possessed neither, to use Gottlob Frege's useful terms, a fixed reference nor a stable meaning.

In fact, even those first visual representations of atoms in Bruno's woodcuts are not merely representations of Democritus' atoms, despite their names. They combine Cusanus's theological-geometrical reasoning, Pythagorean monadological and numerological ideas, Lullian combinatorics, Ficinian Neoplatonism, and Copernicus's heliocentrism with Lucretian ideas. It is little wonder, then, that the fortune of these images proved to be so uneven. As for Bruno's belief in a structural similarity between clusters of ensouled spherical atoms and solar systems, it had little influence on later thinkers. By contrast, his geometrical reflections on the ways in which spherical atoms can be stacked had, thanks to Kepler's reflections on the hexagonal shapes of snowflakes, an important influence on seventeenth-century modeling in crystallography and optics.[48] In fact, from Bruno to latter-day illustrations, atoms are predominantly depicted as solid spherules.

4 Reasons for the Increasing Popularity of Atoms, Particles, and Minima

If we examine the Renaissance period between 1400 and 1650, the period between the birth of Nicholas of Cusa and the death of Descartes, we find a slow but evident increase in the reliance on all sorts of atoms, particles, minima, and corpuscles for the explanations of natural phenomena. As has been explained above, there was no overall program behind this proliferation of small particles, nor can any coherence be discerned between the mathematical, metaphysical, medical, numerological, theological, alchemical and physical arguments that were produced in their favor. However, what most of these entities had in common was that they possessed a certain ontological stability, which was then employed to explain upper-level phenomena. Besides that, they shared no other common properties. Some were described as purely material bodies, others as animated entities; some were said to be simple (such

48 Kepler, *Strena seu de nive sexangula*; Lüthy, "Invention of Atomist Iconography."

as Bruno's globules), but most others possessed complex shapes – having hooks and eyes, protrusions and pores. Sometimes, they were said to be made up of one universal matter, but at other times they were distinguished according to different elements or substances. Nor was there any agreement about how many types of them existed, nor how they interacted with other atoms or with the higher levels of material organization. For a range of authors – from Isaac Beeckman through Sébastien Basson to Pierre Gassendi (and probably already Nicolaus Biesius in the 1570s, as Lüthy suggests in this volume), and for Nicholas Hill in 1601 (as Plastina shows in her contribution) – atoms clustered into higher-order molecules. These possessed what we would nowadays refer to as 'emergent properties,' such as being the carriers of life-giving seeds.[49] For René Descartes, by contrast, all physical and chemical properties could be directly reduced to the ultimate level of the corpuscles.

Among the various reasons for the increasing popularity of minute particle theories, several have already been mentioned. If we had to draw up a list of the main factors, one might certainly have to include the following: the increased attention given to book IV of Aristotle's *Meteorology*, which led to the paradoxical phenomenon of an 'Aristotelian atomism'; the recovery of ancient sources, most importantly of Plato's *Timaeus*, Diogenes Laertius' doxographical work, and of Lucretius' *De rerum natura*; the assumption of a *prisca sapientia*, and the increased attention this assumption demanded for Presocratic philosophy, which was sometimes even taken to include the biblical book of Genesis, Moses, and the legendary Phoenician Moschus, with whom Moses was sometimes conflated; the revival of monadological and numerological theories; alchemical theories and experimental evidence; the Galenic revival and the development of physiology as a theory of bodily processes; and finally, from about 1620 onwards, the development of microscopy and the feverish hope for ocular inspection of the ultimate material particles. Although several influences were usually identifiable in any given author or text, they remained distinct enough to lead to a situation of overall conceptual incoherence.

As for the phenomenon of 'Aristotelian atomism,' this emerged out of the attention paid by Italian natural philosophers to book IV of Aristotle's *Meteorology*. In that work, as Craig Martin shows in his contribution to this book, Aristotle seems to abandon his usual four-element theory, for he speaks of two elements only: namely, earth (as the principle of dryness) and water (as the principle of wetness), which mix under the influence of celestial heat. "All recognizable substances in our world contain these two elements," states Aristotle here, "and are to be assigned to one or the other according to the

49 See Joy, *Gassendi the Atomist*, ch. 5.

proportion in which they contain earth or water."[50] In the same work, Aristotle also offers various physical explanations in terms of particles passing through the pores of bodies, as when he speaks of the "water-particles" slipping through the "pores" of earthen bodies.[51] To Renaissance commentators, it seemed that Aristotle had here abandoned his earlier hostility toward the concept of atoms, possibly having recognized his earlier error. Pietro Pomponazzi reported on readers' impressions that, in this particular treatise, "Aristotle therefore Democritizes."[52]

From the late fifteenth century onwards, the *Meteorology* and particularly book IV attracted much attention in Northern Italy. Many commentaries sided with Alexander of Aphrodisias in stating that book IV was, in reality, an independent treatise on perfect mixtures. Agostino Nifo suggested calling this work *Liber de mixtis*; Pietro Pomponazzi proposed *Liber de mixtione*.[53] As Craig Martin stresses in his present chapter, *Meteorology* IV, whatever separate name one chose to give it, was viewed as a treatise whose subject matter lay between natural philosophy and medicine. His insistence on the importance of *Meteorology* IV for the medical theories of Pietro Pomponazzi, Lodovico Boccadiferro, Francesco Vimercato, and Francisco Vallés in fact corroborates the evidence provided in Elisabeth Moreau's chapter in this volume, on the close association between physiological questions and the development of early modern matter theories.

This attention to *Meteorology* IV had a number of interesting effects. Within the Aristotelian university tradition, it led to the emergence of what might indeed be termed an 'Aristotelian atomism,' with all the caution – as discussed above – that such a term requires. Francesco Patrizi mentions a disputation in which a professor confused his opponent by citing passages from "the so-called fourth book of Meteorology," deriving anti-Aristotelian conclusions from it.[54] The intense discussion of the implications of *Meteorology* IV led to a revival of explanations of mixture in which the 'minimal parts' of elements were said to move until they touched, but did not merge and instead merely exchanged their qualities so as to form a common 'temperament' – an explanation that – at least according to Alessandro Achillini, Pietro Pomponazzi or Agostino Nifo – managed to reconcile Aristotle's theory of mixture with that

50 Düring, *Aristotle's Chemical Treatise*, 41, translating Aristotle, *Meteorology* 328a5.

51 See, for example, Aristotle, *Meteorologica*, book IV, ch. ix.

52 Pomponazzi, *Dubitationes*, 43v: "[…] ergo Aristoteles Democrizat […]." Ibid., 47v: "Videtur ibidem Aristoteles Democrizare, nam incidit in questionem Democriti. Qui voluit quod actio fiat per poros." On these passages, see also Lüthy, "Aristotelian Watchdog," 546.

53 See Martin, *Renaissance Meteorology*.

54 Patrizi, *Discussiones peripateticae* (1571), 113v.

of Galen.[55] Here, we have a further witness to the terminological convergence between the authors treated by Moreau and Martin. We should not fail to recognize, however, that similar notions of mixture had a long prehistory. As Danielle Jacquart has documented, we find almost identical definitions of mixture already in medieval physiological treatises, for example in Bartholomew of Salerno.[56]

Outside of the university context, *Meteorology* IV eventually led to explicitly anti-Aristotelian theories. The Padua-trained philosopher-physician Girolamo Cardano developed the two-element theory of *Meteorology* IV into a veritable cosmology. He defined elements as those bodies that could enter into mixtures and form more complex bodies. This definition did not admit fire, which Cardano altogether excluded from the list of elements, while he considered air to be a real substance, but not an element, as it could not mix with the real elements. The function of air was instead to carry celestial heat down to the terrestrial realm.[57]

Julius Caesar Scaliger attacked Cardano's theory in his idiosyncratic but highly influential *Exercitationum exotericarum liber XV. de subtilitate* (1557). Like Cardano, Scaliger was a physician-natural philosopher trained in Northern Italy. In his rebuttal of the former, Scaliger proposed a definition of mixture that was to become famous: "Mixture is the movement of minimal bodies towards mutual contact, so that a union occurs," adding, "For our corpuscles do not touch as Epicurean atoms do, but so that a continuous and single body is produced."[58] In other words, whereas Epicurean atoms remain merely juxtaposed, Scaliger's corpuscles touch and thereby bring about a continuous body defined by common properties. Although Scaliger officially rejected atomistic explanations, such *novatores* as Sébastien Basson, David Gorlaeus, and Daniel Sennert in the early 1600s avidly cited this definition in defense of an atomistic conception of mixture.[59] The same may be said about Cardano's two-element theory, which was opposed by Scaliger: it, too, could easily be 'atomized,' by defining earth and water as material building blocks that heat up under the influence of celestial warmth. An explicitly atomistic version of Cardano's two-element theory is in fact found in David Gorlaeus's *Exercitationes philosophicae* and his *Idea physicae*, both of ca. 1610.[60]

55 See Piccolomini, "De mixtione," 2: 98r–112v; Martin, *Renaissance Meteorology*.

56 Jacquart, "Minima in Twelfth-Century Medical Texts."

57 Ingegno, *Saggio sulla filosofia di Cardano*.

58 Scaliger, *Exercitationes exotericae*, ex. 110.

59 Subow, "Zur Geschichte des Kampfes."

60 On this, see Lüthy, *David Gorlaeus*, 86–88.

In the sixteenth century, however, burgeoning curiosity in the explanatory possibilities offered by the postulate of a structured physical micro-level cannot solely, or even mainly, be attributed to the Aristotelian tradition. More direct influences are to be sought in the recovery of a set of ancient descriptions of atomism, most importantly in Plato's *Timaeus*, Lucretius' *De rerum natura*, and Diogenes Laertius' *Lives of Eminent Philosophers*, as well as in some other genuine or pseudonymous works.

Some fragments of Diogenes Laertius' doxographical account of the views of ancient philosophers had been known before Ambrose Traversari translated into Latin its ten books between 1424 and 1433, but it was this translation and its subsequent publication that started the widespread fifteenth-century diffusion of this text.[61] Diogenes' doxography includes entries on Leucippus, Democritus, and Epicurus. On the first two philosophers, Diogenes adds much to what is said by Aristotle, who remains otherwise our main source. Notably, it depicts Democritus of Abdera as a sage who had extensively traveled to the Orient, had been bestowed with many honors, had written 70 works on a great variety of themes, and had died, almost a saint, at the age of 109. Diogenes furthermore dedicates the last of the ten books of his *Lives* to Epicurus, whose doctrines had previously been known chiefly through Cicero's even-handed discussion, as well as the invectives hurled at him by Lactantius and other Church Fathers. Diogenes Laertius, who seems to have felt philosophically close to Epicurus, not only described the latter's life, but also reproduced three long letters that he attributes to Epicurus and which he takes to explain his doctrines. These three letters cast a new light on the founder of Epicureanism, whom early Christian authors had accused of offering an ethics worthy of swine. It is no coincidence that it was from the tenth book of Diogenes Laertius' *Vitae* that Pierre Gassendi would start his grandiose project of reconstructing Epicurus' philosophy.[62]

A decidedly more important, but usually underestimated source of Renaissance and early modern atomistic thought was Plato's *Timaeus*. This dialogue had exercised a certain influence throughout the Middle Ages, but the concluding parts, in which Plato's atomist ideas are presented, became available only after Marsilio Ficino translated and published the entire dialogue (1484) and also supplied a commentary entitled *Compendium in Timaeum* (1496). Readers encountered in the extensive last part of this late work of Plato's a theory that explained the formation of all four elements out of triangles of

61 Dorandi, "Introduction," 10.

62 Gassendi, *Animadversiones in decimum librum Diogenis Laertii*; see Joy, *Gassendi the Atomist*.

“space” (*chōra*), with each element being made up of a certain number of triangles and defined by a regular atomic shape (the so-called Platonic solids). These shapes were taken to explain the sensory elemental properties (named “secondary qualities” from John Locke onwards). To cite the most convincing example, fire particles burn our skin or produce a sharp taste on our tongue when we eat ‘hot’ foods because they have a sharp, pyramidal shape, which can pierce or lacerate other surfaces. Particularly influential in this account was that it also attempted to explain a range of physiological processes by means of the interaction between our bodily machine and these particles.

Since Antiquity, Plato had often been depicted as a more Christian philosopher than Aristotle; and the first part of the *Timaeus*, in which the demiurge constructs the world, had been known since the early Middle Ages. If atomistic notions needed a respectable face, then it could have been found in Plato. Still, the impact of the *Timaeus* on the evolution of matter theory is complex. On the one hand, its pyramidal fire particle is frequently found even in otherwise non-atomic theories, such as Julius Caesar Scaliger’s *Exercitiones exotericae* (1557). Nicolaus Biesius’s *De natura* (1573), discussed by Lüthy in this volume, provides another case: its interest in an atomistic interpretation of *minima* clearly relies on the *Timaeus*. In Galileo Galilei’s *Saggiatore* (1623), to cite a particularly important later example, one encounters several Timaean concepts, including the notion of the divine mathematician, who has defined all bodies geometrically, and the distinction between the true geometrical quality of the ultimate bodies and the merely sensory, secondary qualities these provoke in us.

Plato’s *Timaeus* plays a curious role in the ever-greater sympathies that Nicolaus Biesius, professor of Galenic medicine at the University of Louvain and imperial physician, came to harbor for atomistic explanations. As Christoph Lüthy explains in his chapter in the present book, Biesius was initially worried about the confessional strife and discord he witnessed all around, and he suspected – somewhat Neoplatonically – that the fragmentation of the divine unity into ever divisible material bodies was the ultimate reason for why the world looked so disjointed. While Biesius’s analysis seems bizarre, Lüthy shows that in the course of his intellectual development, the Louvain professor came to engage with growing conviction a conceptual toolbox of atoms, corpuscles, and *minima*. His strategy was intended not just to stop the fragmentation of matter into nothingness, but ultimately also to explain the properties of organic and inorganic bodies. In the process, he claimed “to follow the Platonists and the Aristotelians” – as if these two groups had postulated one common doctrine! – and this allowed him to merge Aristotle’s notion of the *minimum naturale* with the geometrically defined atomical elements of Plato’s *Timaeus*.

Lucretius' *De rerum natura*, rediscovered by Poggio Bracciolini in an unknown monastery close to Lake Constance in 1417, is another important text that must be discussed here. Next to Plato's *Timaeus*, this was the most extensive ancient source in which atoms played a crucial explanatory role. It is, of course, correct to say that Lucretius' text had never been completely lost, and that earlier in the Middle Ages, it had been copied, used, and cited. In the twelfth century, for example, William of Conches quoted Lucretian verses in his *Dragmaticon philosophiae*.[63] Still, it is obvious that for Bracciolini and his contemporaries, Lucretius' didactic poem came as a surprise. Once brought to Italy, the work spread first in manuscript form, and then more rapidly and widely after its *editio princeps* of 1473, going through at least 79 editions by 1625. It was a text that was much appreciated for its poetic elegance, but at the same time dreaded for its impious character. In fact, the Roman Inquisition made sure that no translation into the vernacular was published.[64] Marsilio Ficino, who may have been the first Renaissance philosopher to compose a commentary on *De rerum natura*, later burnt it.[65] As Elena Nicoli explains in her present chapter, the earliest published reactions to Lucretius' poem – Raphael Francus's *Paraphrasis cum appendice de animi immortalitate* (1504) and Giovanni Battista Pio's *In Carum Lucretium poetam commentarii* (1511), both published in Bologna – clearly indicate that the initial philosophical engagement with this work took place within the context of the heated contemporary debates over the possibility of proving the immortality of the soul on philosophical grounds. It was Lucretius' materialism (which contradicted divine providence, creation, the immortal soul, and other central elements of Christian faith) that initially attracted attention, but not his specific theories concerning the ultimate building blocks of matter.[66]

Subsequently in the sixteenth century, however, we see an increasing number of references to Lucretian matter theory, and not just among such so-called Lucretian *imitatores*. Interestingly, physicians seem to have been particularly interested in Lucretius' theory – in the sixteenth century, in fact, we encounter atoms above all in medical works, and the aforementioned Nicolaus Biesius, who, like Lucretius, wrote one of his works in hexameters, is no exception here. Even though Lucretius invoked "the sacred doctrine of Democritus" and extensively praised Epicurus as his philosophical teacher, his own atomism reflected also Roman medical theories. It has been suggested that his particular interest

63 See Robert, "Atomism."

64 Gordon, *Bibliography of Lucretius.*

65 Nicoli, "Ficino."

66 See Nicoli, *Earliest Renaissance Commentaries on Lucretius*; Prosperi, *Di soavi licor.*

in an atomistic explanation of diseases may have been influenced by one of his acquaintances, the physician Asclepiades.[67] As Nicoli's chapter in this volume shows, irrespective of Lucretius' own beliefs, his avoidance of the term 'atom,' and his preference for the expression "seeds of things" (*semina rerum*) could be interpreted by his readers as evidence that Lucretius' atoms were animated, dynamic units, or at least compatible with substantial forms.[68] Such a 'biological' understanding of Lucretius' atoms is found already in the work of the sixteenth-century physician Girolamo Fracastoro, who in his *De sympathia et antipathia rerum* (1545) explained various physical and medical effects through "effluvia of atoms," but we find it even in the thought of the late seventeenth-century physician Nathaniel Highmore, who spoke of "seminal Atomes."[69]

The Renaissance turn towards corpuscular and atomistic explanations occurred predominantly in the hands of physicians – in fact, many of the authors examined in this book were medical doctors. One of the main reasons for this phenomenon is to be sought in the development of the discipline of physiology. Just as Aristotelians commenting on *Meteorology* IV could – as Craig Martin documents in the present book – end up espousing explanations in terms of pores and particles, so Galenic medical doctors trying to define the functions of elements, mixtures, and temperaments in physiology could end up with a corpuscular terminology, as Elisabeth Moreau argues in her own chapter. Moreau documents that although Galen had rejected atoms and corpuscles, late-Renaissance and early modern physicians working in the Galenic tradition nevertheless moved ever more clearly towards atomistic explanations. Her examination of the physiological writings of Jean Fernel (1497–1558), Andreas Libavius (ca. 1550–1616), and Daniel Sennert (1572–1637) shows how the four elements of Aristotelico-Galenic extraction became recast as ever more decidedly autonomous particles, contiguous *minima* or indivisible atoms, which even when entering into apparently homogeneous mixtures or creating an overall temperament would not lose their identity. All of this happened without necessarily forsaking the notion of substantial forms. In fact, Sennert is remembered for his positing of indivisible atoms as the carriers of such forms.

While the line of influence in Moreau's chapter takes us from Fernel through Libavius to Sennert's atoms-cum-substantial-forms, Jole Shackelford's chapter takes us from Fernel and Paracelsus to Petrus Severinus, and notably to the

67 Stückelberger, "Lucretius reviviscens."

68 See Gemelli, *Aspetti dell'atomismo classico*; and Hirai, *Le concept de semence*.

69 Clericuzio, *Elements, Principles and Corpuscles*, 89.

latter's highly influential *Idea medicinae philosophicae* (1571). Once again, medicine is the driving force behind conceptual innovation, but this time, it takes us – paradoxically – to a mechanical worldview that is yet entirely vitalistic. How is that possible? As Shackelford documents, the process of regular organic development was described as a "mechanical" process by both Paracelsus and Severinus. In what is the most puzzling expression that Shackelford analyzes, that process is described as a "mechanical lithurgy" (*lithurgia mechanica*). This neologism combines the words 'liturgy' and the Greek work 'lithos' ("stone"), although it is not evident whether the latter was intentional; at any rate, with "lithurgy," Severinus refers to organic beings "stepping forth onto this world stage at determined times," in a making manifest of what had previously been hidden.[70] It is a world in which "mechanical spirits" carry out their tasks by transforming the seeds (*semina*) into the organic bodies with which we are familiar. Shackelford convincingly concludes that it was not the imposition of mechanism as such – in what Dijksterhuis has famously called the "mechanisation of the world picture" – that led to a world that opposed dead matter to an Aristotelian world of vital forms and forces. Rather, it was what Shackelford calls the "deadening of the meaning of mechanism, depriving it of spirit, soul, or other internal agency and leaving behind a dead, inorganic machinery." This process had clearly not yet been initiated in the 1601 *Philosophia* of the medical doctor Nicholas Hill, which is analyzed in Plastina's chapter, and which makes use of the vitalist opportunities that Severinus offered for its own atomism.

In fact, physicians preferred to postulate animated matter, in whatever form it came. Many chose to depict elements as living particles, minima or atoms, on the basis perhaps of one additional motivation or legitimation. There exists a body of pseudo-Hippocratic letters, which Rinuccio Aretino had translated in the mid-fifteenth century and which contains a number of letters that describe how Hippocrates had been called to Abdera to heal Democritus, who was said to have gone insane, and who had removed himself to the forest, where he was busy cutting up animals. When Hippocrates visits his prospective patient, it turns out that Democritus is busy in an exercise in comparative anatomy. In his desire to understand why his fellow-Abderites spend their lives in futile endeavors, he is trying to locate the physiological origin of black bile. After that initial meeting, Hippocrates, now with an understanding that Democritus is a more

70 Severinus, *Idea medicinae*, 441. Shackelford's chapter discusses the entire, puzzling passage: "Generatio, progressio est seminum, in qua, ex fontibus, Abyssis & vitalibus Principiis, ordinata corporum explicatione, in hanc mundanam scenam, definitis temporibus, progredientia, Individuorum renovatione, Specierum perpetuitatem custodiant: fiuntque in hac lithurgia, ex invisibilibus visibilia, ex incorporeis corporea, potestate vitalis immortalisque Scientiae, in universa Natura vigentis."

advanced physician than himself, exchanges a number of letters from which he learns further medical secrets from his new master. This body of letters was in many quarters taken as a sign that Democritus' theory of matter was more compatible with medical doctrine than were Aristotle's or even Galen's theory of matter. The most striking product of this tale is Marco Aurelio Severino's *Zootomia democritea* of 1645, which depicts the encounter of Hippocrates and Democritus on the frontispiece, and advocates a new method of comparative anatomy, even so far as to interpret the term 'anatomy' as "an-atomy" (which Severino took to mean "cutting down to the level of atoms").[71]

An altogether different factor in the re-emergence of Democritus' matter theory after the many centuries of Peripatetic refutation was the emerging Renaissance belief in the existence of a primordial *prisca philosophia* or *prisca sapientia*, which had allegedly sprung up in the temporal vicinity of the divine breath of creation. This belief obviously reversed the time axis to the detriment of Aristotle's authority. No longer the victorious systematizer of rudimentary Presocratic intuitions, he could now be seen as a distant afterglow of the original philosophical splendor. The Renaissance's ever-improving textual base appeared to support such a view, because a comparison between Aristotle's judgment of his predecessors and their rediscovered actual or presumed doctrines encouraged humanist historiographers to charge the philosopher with deliberately misrepresenting the views he reported, an accusation that seems to have first been formulated in 1520 by Francesco Pico della Mirandola. Although Democritus was not the main beneficiary of this reversal of fate – *pace* Ernst Cassirer – the perceived overlap between his theory of matter and that expounded in Plato's *Timaeus* suggested that an atomistic conception not only predated Aristotle (which had always been clear), but that the earlier matter theory in some sense represented a venerable *opinio communis*.[72]

Franceso Patrizi, who regularly features on the lists of *novatores*, did little to hide his irritation at Aristotle's physics in general, and at the latter's description of the alleged ineptitude of his predecessors more in particular. As Leen Spruit shows in his contribution to this book, Patrizi in his *Discussiones peripateticae* (1581) applied the same apologetic strategies to Democritus that he also applied to other authors attacked by Aristotle. First of all, he demonstrates that Aristotle's doxographical reports are unreliable. For example, how could Aristotle in one passage maintain that Democritus proposed two principles, the void and the full, while in another maintaining that Democritus had postulated an infinity of principles? On a more constructive note, Patrizi reported

71 Lüthy, "The Fourfold Democritus," 469.

72 Cassirer, *Das Erkenntnisproblem*, 1: 166..

that Democritus was generally regarded as the inventor of physics.[73] But he himself was not particularly interested in atomistic conceptions. That Patrizi's ideas got embroiled in the history of atomism is due to the involvement of characters like Nicholas Hill – discussed in Plastina's chapter – who merged Patrizi's physics with Bruno's monadic atomism and the ancient views of Democritus and Epicurus.

As far as the re-evaluation of Aristotle's position in the history of philosophy is concerned, the most extreme doxographical version must have been that by Sébastien Basson, who in the preface to his *Philosophia naturalis* of 1621 described Aristotle as a solitary rebel against a holy ancient consensus, which involved the *prisci philosophi* – Plato, Hippocrates, Democritus, and Empedocles included.[74] In the late sixteenth century, the figure of "Mo[s]chus, the Phoenician" – the true inventor of atomism according to Strabo, who in turn had taken it from Posidonius – was furthermore identified by some with the biblical Moses. This implausible idea, which never became mainstream, represented the culmination of a veritable sanctification of an atomistic matter theory and in some quarters was integrated into what was proposed as a Mosaic physics – a physics that took its cue no longer from Aristotle, but from *Genesis*.[75]

Another version of this quest for primeval wisdom involved numerology, which could assume a range of Pythagorean, Cabbalistic and Neoplatonic overtones. As mentioned earlier, Giordano Bruno's own type of atomism was partially indebted to this tradition.[76] Combinations of Pythagoreanism and monadic conceptions of reality persisted well into the eighteenth-century, however, receiving an additional boost from Gottfried Wilhelm Leibniz.[77]

Up to now, we have stressed the contributions made by physicians to the development of corpuscular and atomistic reasoning. But time and again, we have encountered metaphysical and theological themes – for example, in Nicolaus Cusanus, Nicolaus Biesius, Francesco Patrizi, or Giordano Bruno. In fact, as Kuni Sakamoto demonstrates in his contribution to this volume, theological work also contributed to the development of atomistic matter theories. Sakamoto analyzes the case of Conrad Vorstius, the German theologian who was appointed as Jacob Arminius's successor as professor of theology at Leiden but because of his Socinian sympathies was never allowed to teach there.

73 Patrizi, *Discussiones peripateticae* (1581), 3: 293.

74 Basson, *Philosophia naturalis*, "Ad lectorem."

75 On this movement, see Sailor, "Moses and Atomism"; Walker, *Ancient Theology*; Blair, "Mosaic Physics."

76 On this, see Yates, *Giordano Bruno*.

77 Neumann, *Monaden im Diskurs*.

After his condemnation at the Synod of Dort in 1619, he was expelled from the Netherlands. In responding to questions regarding the attributes of God, Vorstius was wondering what the ubiquity of God implied for his presence in all things, and whether denying that He possessed a circumscribed quantity would not imply that He was "substantially whole even in the tiniest things like the burrow of a beetle or the tip of a spear."[78] Sakamoto explains how Vorstius, balking at that implication, instead imposed a limited quantity and hence spatial finitude to God. His detractors protested that he thereby physicalized God. David Gorlaeus followed Vorstius in subjecting God to a general ontology: like all beings (*entia*), God must possess unity, truth, goodness, existence, locality, and durability. In this extended sense, God, likewise, could be seen as an atom: He too was indivisible, one, and enduring, just like physical atoms. But contrary to Vorstius, Gorlaeus did not deny that God's essence was infinite. According to his ontology, material atoms occupy finite spaces within this world, while God fills the infinite space within and beyond this world.

5 Empirical Arguments for the Existence of Atoms

Most of the arguments discussed up to now in favor of the existence of atoms, or otherwise stable corpuscles or minima, have been of a speculative nature. However, in the later Renaissance, and possibly beginning with Fracastoro's above-mentioned *De sympathia et antipathia rerum* (1545), empirical arguments became ever more prominent, even if Christoph Meinel has convincingly shown that "in atomism, there was no experimental proof possible, although most corpuscular theories of the seventeenth century explicitly claimed to be based upon experience."[79] Those who thought that dust motes or water vapors were atoms and therefore accessible to ocular inspection, were sorely disappointed by Jean-Chrysostôme Magnen's calculation that an atom of frankincense had to be at least 7.7×10^{19} smaller than a grain of the same substance.[80] The empirical proof therefore had to be indirect. But none of them was decisive. After all, the magnet could be explained by atomic effluvia – but also in terms of occult sympathies between bodies; epidemics could likewise by explained by seminal particles – but also by corrupt air; condensation and rarefaction

78 Vorstius, *Tractatus de Deo*, notes to disputation III, 240: "Et sane mirum videtur esse paradoxon, deum sic omnis quantitatis expertem esse, ut totus substantialiter sit in re qualibet, etiam nimima, vi.g. in antro scarabei, aut in apice unius pili, etc."

79 Meinel, "Early Seventeenth-Century Atomism," 68. See also Kangro, "Erklärungswert und Schwierigkeite der Atomhypothesen."

80 Magnen, *Democritus reviviscens*.

could be explained through particles and the void – but Aristotelians had of course for centuries provided non-atomistic explanations for these very same phenomena. As mentioned above, one of the best proofs, and one that was invoked from Paul of Taranto up to the seventeenth century, was the *reductio ad pristinum statum*, a chemical operation in which the original ingredients of a mixture are recovered – for example, silver previously dissolved in nitric acid. Such a recovery is indeed hard to reconcile with the Aristotelian theory of mixture.[81] However, while this specific chemical operation favors an understanding of chemical substances in terms of stable atoms and molecules, most other operations do not; and, in fact, the *reductio ad pristinum statum* seemed, to many, to be a tedious exception upon which no coherent new theory of matter could be built.

In the manuscripts of Walter Warner from the 1620s, which Stephen Clucas analyzes in his chapter, we encounter a practitioner who sought mechanical and atomistic explanations for chymical and physical processes. As is typical for a practitioner, he was not overly concerned with the names and the metaphysical status of the units he invoked to do their explanatory work, calling them at one time *particulae*, and then again *atomi, minima, singula, individua*, or *simpla*. As Clucas points out, this allowed him to blur the boundaries between the scholastics' *minima naturalia* and the ancient atomists' rigid units. But however atomistic the terminology may at times have been, and however much we may want to insert Warner into the history of an ever more mechanical approach to chymistry, Clucas shows how much of the explanatory work was still done by spirits and forces. Furthermore, in Warner, we encounter once more the influence of Daniel Sennert's chymical thinking, which combined the idea of atoms as carriers of properties with traditional notions of substantial forms.

The complexity of the genealogy of atomistic conceptions in the domain of chymistry is also evident in William R. Newman's chapter, which focuses on one of the hitherto unstudied sources of Isaac Newton's chymical ideas. That source, Bernhard Varenius, has been mentioned in this Introduction once before, namely as the former student of the Hamburg pedagogue Joachim Jungius, who in 1647 reported that everyone in the Netherlands was discussing Descartes and that Descartes avoided the term 'atom' in vain, given that everyone understood the true nature of his philosophy. As Newman demonstrates, Varenius was strongly influenced by the atomism of Jungius's so-called "syndiacritical" analysis of natural processes, which in turn was once more inspired by Sennert's writings – Jungius had published an abridgement of Sennert's

81 Newman, *Atoms and Alchemy*.

De chymicorum cum Aristotelicis et Galenicis consensu ac dissensu. But the genealogical line Sennert – Jungius – Varenius – Newton gets complicated by Varenius's debt to Descartes's corpuscular philosophy, which he came to admire during his sojourn in the Netherlands. Newton, who edited Varenius's *Geographia*, therefore encountered the following passage, which Newman quotes from the 1682 English translation:

> Many of the Ancients, as *Democritus* and *Leucippus*, determined that the whole World consisted of very little solid pieces, which differ only in their various *figures*, *shapes*, and *magnitude*: and them many of the later *Philosophers* do follow; and of late *Cartesians* endeavoured by such an *hypothesis* to declare all natural appearances.[82]

In Newman's chapter, the threads of medieval chymical sources come together with a plethora of other sources, including Sennert's analysis of elements, Jungius's syndiacritical understanding, Varenius's theory of corporeal and spiritous waters consisting of fixed and unfixed ingredients, and Descartes's ideas concerning the agency of shaped corpuscles. In the face of so many converging influences, Newman explains how, for Varenius, the perennially used term 'atomus' came to mean any kind of small, semi-permanent corpuscle endowed with what we would call secondary qualities. Where Nicolaus Biesius had used four different terms and Walter Warner six, Varesius had settled for one term, 'atomus,' but that terminological reduction did not in the least diminish the range of its properties, and effects.

Interestingly, for a certain period, it was hoped that the microscope might settle all of these issues, by granting ocular access to the atom's hidden secrets, and in fact, for a few decades, this new instrument appeared – also rhetorically – to provide the most forceful evidence for an atomistic conception of matter. After all, one of Aristotle's most successful epistemological objections to an atomistic conception of matter had been that if Democritus were correct, then a person with perfect eyesight (such as the mythological Lynceus) would perceive a very different world than we do, namely a discontinuous, perceptibly atomized world.[83] After the emergence of microscopes as a 'philosophical instrument' around 1620, it turned out that magnified objects did indeed look different than they appeared under inspection by the naked eye. Although this did not itself prove the existence of atoms, let alone allow us to determine

82 Varenius [and Sanson], *Cosmolography and Geography*, 30. The Latin is found on p. 42 of Newton's 1672 edition.

83 Aristotle, *De generatione et corruptione*, book II, ch. 10.

what are their shapes, it did invalidate Aristotle's belief in the adequacy of our senses. The first philosopher to recognize this was Francis Bacon, who was probably also one of the first people to peer through Cornelis Drebbel's newly invented compound microscope. When speaking of this instrument in his *Novum organum*, he remarked: "if Democritus had seen this, he would have rejoiced a lot, believing a method to have been discovered to see the atoms."[84] Although Francis Bacon held Democritus in great esteem and also postulated material micro-structures and schemata, he was however not persuaded by an atomistic explanation of natural phenomena.[85] Still, his aphorism about the rejoicing Democritus was handed from one author of microscopy to the next. At the height of the mania for microscopy, Nathaniel Highmore wrote that he had seen Descartes's magnetic effluvia through a microscope, while Henry Power and Robert Hooke were persuaded that further improvements in the design of optical glasses was going to be sufficient to render atoms visible.[86] In the event, they didn't, and the credulousness of the seventeenth-century microscopists was to become an object of derision for the more worldly wits of the eighteenth-century.

Despite the best efforts of the Catholic Church, and in some minor ways also certain Protestant currents, to try to contain what could be described as a veritable atomist hype, the seventeenth century ended with much of Europe believing that Aristotle's description of bodies in terms of matter and form was unhelpful or outright wrong, and that there were indeed material structures to be found beneath the visible realm. However, no consensus had been reached about how many structural levels existed below what was visible; whether there would ever be an ultimate level to be found; and if so, of how many types of particles would that ultimate level be constituted, and by what characteristics would these types be defined.

And so, whilst in the heyday of the mechanical philosophy in the second half of the seventeenth century, the Republic of Letter had tended to reason along atomist lines of thought, the eighteenth century saw a return to a more agnostic stance. In a way, the eighteenth-century reliance on atoms, particles, and corpuscles was as unprogrammatic and ontologically dissonant as had been the case for any parallel theories in the Renaissance. It was, above all,

84 Bacon, *Instauratio magna* [=*Novum organum*], aphorism II§39: "[…] quale perspicillum si vidisset Democritus, exiluisset forte, et modum videndi atomum (quem ille invisibilem omnino affirmavit) inventum fuisse putasset."

85 Rees, "Atomism and Subtlety in Bacon"; Manzo, "Bacon and Atomism."

86 Highmore, *History of Generation*, 117; cf. Power, *Experimental Philosophy*; Hooke, *Micrographia*.

from the mid-nineteenth century onwards that atoms returned to the fray and atomism became an ideology.

Acknowledgments

This book is the result of a project entitled "Mysteries of Living Corpuscles: The Humanist Revival of Presocratic Philosophy and Atomism in Philosophy, Science and Medicine" of the Dutch Research Council (NWO), no. 360-20-320, which was carried out at the Center for the History of Philosophy and Science (CHPS) at Radboud University, Nijmegen, The Netherlands. This project was steered by Christoph Lüthy and Hiro Hirai from 2013 to 2018 and formally involved also Elena Nicoli and Kuni Sakomoto. A range of further researchers, including Frederik Bakker, Delphine Bellis, Davide Cellamare, Elisabeth Moreau, Carla Rita Palmerino, and Leen Spruit also had some informal associations with the project during their stay at the CHPS.

Extra financing for the research that went into producing this volume was offered by the Faculty of Philosophy, Theology, and Religious Studies of Radboud University, Nijmegen.

Finally, we wish to express our gratitude to Brian Clarke for language-editing this entire volume with the utmost care and patience.

CHAPTER 2

Atomism in Sixteenth-Century Italian Commentaries on Lucretius

Elena Nicoli

The story of the return of Lucretius' *De rerum natura* to Renaissance Italy has been examined from different perspectives in the last decade, but it can still easily be misinterpreted. According to a widespread account, the rediscovery of a manuscript of *De rerum natura* in 1417 and the consequent diffusion of Lucretius' Epicurean ideas directly impacted the alleged 'Renaissance revival of atomism' in the late sixteenth and seventeenth centuries. According to this view, in challenging Aristotelian hylemorphism, this revival would, in turn, have paved the way for the Scientific Revolution. This sequence of contingencies has given some scholars the illusion that the rediscovery of Lucretius might have been a revolutionary event that triggered the Renaissance or even modernity *tout court*. It is, therefore, necessary to abandon the grand, conventional narrative, in which some crucial details are lost, in favor of small, local narratives, which allow us to grasp how sixteenth-century scholars read *De rerum natura* and understood Lucretius' ideas.

To this end, in this chapter, I shall examine some crucial passages taken from the extant exegetical works published in Italy on Lucretius in the sixteenth century: Raphael Francus's *Paraphrasis cum appendice de animi immortalitate* (1504), Johannes Baptista Pius's commentary (1511), and Girolamo Frachetta's *Spositione* (1589).[1] I will draw from these sources a few topics that are central to the early modern reception of Lucretius' atomism to answer the following central question: How was Lucretius' atomic matter theory explained in Renaissance commentaries on *De rerum natura*? An answer to this question will clarify how Renaissance scholars interpreted the Epicurean matter theory and will contribute to a better understanding of the actual role of the rediscovery of Lucretius' text in the so-called revival of atomism in the late sixteenth and seventeenth centuries.

---

1 On the lost fifteenth-century "*commentariola*" by Ficino, see Nicoli, "Ficino," 330–361. A complete overview of the Italian commentaries on Lucretius from the sixteenth up to the twenty-first century can be found in the book chapter by Milanese, "Italian Commentaries," 195–215.

 | DOI:10.1163/9789004528925_003

1 Raphael Francus's *Paraphrase* (1504)

Printed in Bologna in 1504, Raphael Francus's *Paraphrasis in Lucretium cum appendice de animi immortalitate* is the first extant printed attempt at interpreting Lucretius' *De rerum natura* after its rediscovery in 1417. Francus's *Paraphrasis* is a particularly relevant case study because, besides providing us with the earliest extant Renaissance explanation of *De rerum natura*, it is also a pertinent first example of the Renaissance interest in Lucretius' physics. According to some scholars, the first commentaries on Lucretius' *De rerum natura* were chiefly philological, with little attention given to its philosophical and scientific contents.[2] But this assessment seems misplaced for Francus's *Paraphrasis*, which focuses exclusively on the doctrinal contents of the poem, with no attention given to its philological and poetic aspects.

Raphael Francus was probably born around 1477 in Florence. He studied both in Florence and Bologna, where, in 1504, he published his *Paraphrasis in Lucretium*, which was perhaps his first work.[3] Also in Bologna, seven years later, Johannes Baptista Pius was to write the first systematic commentary on Lucretius (see below), which was to be published by Hieronymus Baptista de Benedictis, son of Joannes Antonius, who had edited Francus's work. This direct Bologna connection between the first two Lucretian publications is certainly conspicuous: since Francus was a student in the city when Pius was engaged there as a teacher, it is possible that they met and each discussed with the other their projects, that is to say, respectively a paraphrase and a commentary on Lucretius. Once Francus was back in Florence, he began to teach, first at the Florentine *Studio*. He was subsequently a professor in Pisa, where he taught logic, moral philosophy, and natural philosophy. The date of his death is uncertain.

Beyond what the title would suggest, Francus's work is an exegesis, more than a mere paraphrase. While his *Paraphrasis* does summarize the contents of Books I–III of *De rerum natura*, it also contains several digressions concerning the most challenging passages of the poem. Although his original project would almost certainly have included at least the paraphrase of *De rerum*

2 Wilson, *Epicureanism*, 17.

3 Information about Francus's life is lacking and sometimes contradictory: there are many variants of his name, which makes it difficult at times to piece together his life and to identify his works. In some Latin texts, he is called *Raphael Francus*, in some others, *Raphael Franciscus*, while Italian variants of his name are *Raffaele Franco*, *Raffaele Franchi*, or *Raffaele Franceschi*. He was also known as "il Celatone"; see Rhodes, "Raphael Franciscus," 79–81; Pizzani, "Dimensione cristiana," 313–333.

natura's fourth book, he stopped short after the third.[4] There may be several reasons why he took this decision: 1) Francus was mainly interested in general physics, as it appears from the topics of his other publications;[5] 2) He wanted to create for himself an opportunity to discuss the nature of the human soul, the status and nature of which, at that time, stood at the center of a major controversy (see below); 3) Finally, he may simply have run out of time, patience, or money.

1.1 *Francus's 'Appendix on the Immortality of the Soul'*

In the fourteenth century, Dante had portrayed the Epicureans as denying the soul's immortality; for this reason, they were the only philosophers to be included in the *Inferno* among the heretics: "Within this region is the cemetery | of Epicurus and his followers, | all those who say the soul dies with the body."[6] This passage in Dante's *Divina Commedia* provides an insightful testimony that the soul's mortality was perceived as one of the most outrageous consequences of Epicurus' philosophy.

That Francus was aware of the heterodox content of Lucretius' Book III with respect to the nature of the soul is shown by his addition of a separate *Appendix de animi immortalitate*. There, he first presents the Platonic theory of the soul, then Aristotle's theory of the same, and finally, he replies to Lucretius' "quibbles" (*cavilla*).

In keeping with the traditional condemnation of Epicurus' doctrine on the soul, right at the outset of the *Appendix*, Francus writes: "It seems that we will give value to our work if we reject the harmful and weak opinion of the Epicureans about the soul."[7] Even though this statement seems to suggest that the *Appendix* was an open condemnation of Lucretius' theories, there is clearly

---

4 In two passages (*Paraphrasis*, fol. 10v and 18r–v), Francus promises to discuss the *simulacra* in what would have been the paraphrase of the fourth book of the *De rerum natura*; ultimately, this paraphrase was never written.

5 Francus also published a *Verificatio universalis in regulas Aristotelis de motu non recedens a communi mathematicorum doctrina*; this work was published in Florence, in 1516, by Bernardo Zucchetta together with another work by the same Francus entitled *Solutio obiectorum contra suam positionem: quae est velocitatem in motu attendi penes excessum proportionum moventium supra mobilia*.

6 Dante, *Inferno*, X.13–15: "Suo cimitero da questa parte hanno/ con Epicuro tutti suoi seguaci/ che l'anima col corpo morta fanno." I use Mandelbaum's translation. Interestingly, these souls can see the future but not the present. The *contrapasso* is evident: those who were so materialistically attached to the present can, in hell, only know events that are remote in time; on this passage, see Robert, *Epicure aux Enfers*.

7 Francus, *Paraphrasis*, fol. 27r: "Operae praecium facturi videmur si nocuam et becillam (sic) Epicureorum de anima sententiam repudiaverimus."

a disproportionality between the first two parts of the *Appendix*, which contain a detailed analysis of Plato's and Aristotle's doctrines, and the last one, in which Francus hastily confutes Lucretius' theories. Given the brevity of the latter, the *Appendix* should perhaps be regarded as a precautionary measure taken by Francus to avoid censorship rather than as a genuine attempt to disprove Lucretius' doctrine.

Moreover, by comparing what Francus writes in his paraphrase with the few historical documents concerning his own life, it would seem that he had in fact a reputation as a *philosophus Lucretianus*.[8] After Francus's death in 1524, Giovambattista Pelotti, notary of the Florentine *Studio*, writing to its director, Francesco Del Nero, commented as follows: "without doubt, we received a very complicated task [when charged] to put on the right track that restless mind [*sc.* Francus], which none could heal."[9] In the same letter, Pelotti mused that Francus, after his death, might have been received at the gates of Tartarus by Pluto, who was very glad of his coming, and that Epicurus and Lucretius, among other philosophers, greeted him, saying: "At last, you have come!"[10]

Indeed, Francus was a bit of an *enfant terrible*: he had caused trouble in Pisa, where, according to his own testimony, he had fueled an intense debate on the soul's immortality within the context of the great sixteenth-century Italian controversy over this subject. In a letter of 1517 to Francesco Del Nero, Francus writes:

> This year, I have been involved in numerous disputes and fights. For this, I have clarified some mysterious and hidden issues about the nature of the soul, but I still have to explain some of them [...]. But I dared to defend Themistius and Alexander of Aphrodisias against Averroes, and, with no small risk, I protected them from the false accusations and the unfavorable judgments, now in a private dispute, where frequently it

8 See Pizzani, "Dimensione cristiana," 331–333, and Prosperi, "Lucretius in the Italian Renaissance," 215.

9 "Procul dubio provinciam difficillimam accepimus ad dirigendam illam inquietam mentem quam nullus sanare potuit." Quoted in Verde, "Il secondo periodo," 118.

10 "Audivimus quam primum eius anima se contulit ad tartareas ianuas, Pluto ille princeps inferum de eius adventu multo gaudio affectus fuit. Epicurus, Lucretius et plurimi filosophi (sic) obviaverunt ei dicentes: venisti tandem [...]." Quoted in ibid., 128 n. 35; see also Brown, "The Return," 102 and 171 n. 43. There is a discrepancy between the date reported in Verde's chapter (2 September 1524) and that contained in the online inventory of the Carte Strozziane, kept in the "Archivio di Stato di Firenze" (N/190, series I, part 2, string CXXXIV), where this letter is dated 25 December 1524; see <www.archiviodistato.firenze.it/asfi/fileadmin/risorse/allegati_inventari_on_line/n190_inventario_2.pdf>, last accessed 12 February 2022 (unfortunately, I could not check the document myself).

> was necessary to fight violently with the most illustrious men of our Gymnasium, and then again in a public disputation, led as I was by the respect for ancient Peripatetics, certainly with a lot of effort, but perhaps with no less success. [...] At the Gymnasium, things would be going well if only a Perettus [*sc.* Pomponazzi] or a Suessanus [*sc.* Nifo] were here.[11]

This letter reflects a local episode in the great controversy about the immortality of the soul. This controversy, which raged between the fifteenth and the sixteenth century, was kindled by the publication of the Latin translations of Aristotle's commentators, and specifically of Themistius' *Paraphrasis* of Aristotle's *De anima* by Ermolao Barbaro (1481), and of Alexander of Aphrodisias' *Enarratio de anima* by Girolamo Donato (1495).[12]

Francus's claim in this letter that he "dared to defend Themistius and Alexander of Aphrodisias against Averroes," led Alison Brown to conclude that "since Alexander of Aphrodisias could be assimilated to Lucretius, this might suggest that he [*sc.* Francus] was arguing as a Lucretian."[13] Attractive though this idea might sound, since it would support the claim that Francus was indeed a *Lucretianus*, I believe that this passage must be interpreted differently. Francus maintains that he protected Themistius and Alexander "from the false accusations and the unfavorable judgments" brought against them by Averroes. Averroes had described Alexander as a natural-philosophical materialist and presented him as the exemplary exponent of the soul's mortality.[14] As for Themistius, Averroes deferred to him as the direct source of his own doctrine of a single and separate intellect common to all human beings – a doctrine that could also lead to the denial of the soul's immortality. In turn, Aquinas, in the *Tractatus de unitate intellectus*, had discredited Averroes' interpretations of both Themistius and Alexander by showing that their positions

11 "Maximis enim in comptentionibus ac certaminibus hoc anno versati sumus. Ad hoc archana et recondita de animae natura partim a nobis aperta sunt partim deinceps aperienda [...]. Verum ausi sumus Themistii Alexandrique Aphrodissei partes adversus Averroem tueri, et quidem bono periculo tum familiari disceptatione, ubi frequenter cum maximis nostri Gymnasii viris fuit acriter dimicandum, tum vero publica disputatione, observantia veterum peripatheticorum ducti, illos a calumnis atque importunitatibus asseruimus, magno quidem cum labore nec minori fortasse cum fructu. [...] Res Gymnasii optime se haberent modo aliquis Perettus aut Suessanus adforet." Quoted in Verde, *Lo Studio fiorentino*, 376–377. On the sixteenth-century controversy about the immortality of the soul, see Spruit, "The Pomponazzi Affair," 225–246.

12 Francus explicitly refers to the Ermolao Barbaro's translation of Themistius in the *Appendix* (fol. 33r–v).

13 Brown, *The Return*, 77.

14 See Kessler, "Alexander," 10–18.

were much closer to his own, that is, one in favor of the soul's immortality.[15] In Francus's time, scholars could finally read Themistius' *Paraphrasis* of *De anima* and Alexander's *Enarratio de anima* in Latin translation, and could decide who had given the correct interpretation, Averroes or Aquinas.

For this reason, the fact that Francus claims to have defended Alexander against Averroes does not necessarily mean that he supported the mortality of the soul following Lucretius, but rather that he defended the Greek commentator against a false accusation. Interestingly, in the part of the *Appendix* devoted to the *Perypatheticae rationes*, Francus often refers to Themistius and Alexander. He claims that "even Alexander of Aphrodisias admitted that part of our soul is eternal, incapable of suffering, and separable."[16]

All in all, Francus's attitude towards Lucretius is not easy to determine: even though he was reputed to be a *Lucretianus*, he did not publicly endorse Lucretius' doctrines. At the same time, however, his attempt in the *Appendix* to disprove Lucretius' theory of the soul's mortality – whether by design or not – seems weak. For this reason, the *Appendix* has been viewed by some scholars as a precautionary step taken by Francus to steer clear of censorship. Still, it is also plausible that he may have added it simply because he wanted to create an opportunity to clarify his position on the controversial topic of the immortality of the soul.

1.2 *Francus on Lucretius' Matter Theory*

When explaining *De rerum natura*, Francus unexpectedly passes over the more poetic sections, such as the Hymn to Venus, to deal immediately with

15 According to Aquinas, both the agent intellect and the passive intellect must be conceived as faculties of the human soul, while the Averroists maintained that the agent intellect is not a faculty of the human soul, but, rather, a separate entity, thanks to which human beings engage in what we call thinking. The proof for immortality and incorruptibility, which would result from an activity that does not employ a corporeal organ, would therefore apply only to this separate entity, not to each human soul. Contrary to Aquinas, Latin Averroists consequently denied the immortality of the human soul; see Thomas Aquinas, *Tractatus de unitate intellectus contra Averroistas*, 2.55: "Quod autem Alexander intellectum possibilem posuerit esse formam corporis, etiam ipse Averroes confitetur, quamvis, ut arbitror, perverse verba Alexandri acceperit, sicut et verba Themistii praeter eius intellectum assumit. Nam quod dicit, Alexandrum dixisse intellectum possibilem non esse aliud quam praeparationem, quae est in natura humana, ad intellectum agentem et ad intelligibilia: hanc praeparationem nihil aliud intellexit, quam potentiam intellectivam quae est in anima ad intelligibilia. Et ideo dixit eam non esse virtutem in corpore, quia talis potentia non habet organum corporale, et non ex ea ratione, ut Averroes impugnat, secundum quod nulla praeparatio est virtus in corpore."

16 Francus, *Paraphrasis*, fol. 34r: "Namque et Alexander Aphrodiseus nostrae animae partem sempiternam, impatibilem et separabilem concessit."

Lucretius' ontology, physics, and psychology. This choice is particularly significant; indeed, as mentioned before, Francus's paraphrase involves only the philosophical contents of the *De rerum natura*, in a procedure that gives the lie to the assumption that the first commentaries on Lucretius were chiefly of a philological nature, with little attention given to its philosophical dimension.

1.2.1 The Infinite (in)divisibility of Matter

As is well known, Epicurus and Lucretius refused to accept matter as infinitely divisible, and therefore postulated the existence of indivisible parts, namely atoms. One of Lucretius' arguments in favor of the atomic theory is that things cannot be divided into infinity. If infinite division were possible, a small thing would be equal to the sum of all things, since they would both consist of infinite parts (Lucr. 1.615–635). Francus's commentary on this passage is as follows:

> And I additionally construe this in the following manner: for a shorter and more exact reasoning, let the biggest and highest thing be called A; while the tiniest and smallest thing be called B, then we proceed like this: what contains infinite parts that are capable of existing separately, this, without doubt, is considered infinitely large *in act*. But then, both A and B contain infinitely many parts of that kind; indeed, each [of them] is proven to be infinitely large. But if by chance anyone should object to our argument and put forward as the reason for the inequality the fact that, clearly, the parts of which A is composed are bigger than those of which B [is composed], even though whichever set of parts is infinite, certainly reality at once overturns even this [argument]. For, we will take at least one from the small parts of B, thence we will add another one and, in this way, we will produce a body, thereafter we will add another and yet another in succession to infinity. So, it will happen that B is proven to be of infinite magnitude because it always increases through addition, however small, and because all its infinitely many parts, each of which is of some quantity, are present at the same time. For the rest, this opinion [i.e., Lucretius' opinion] only opposes and disagrees with Anaxagoras and those who compose things of infinite parts which exist separately. But according to the Peripatetics, natural bodies are made up of a finite number of parts, capable of existing separately *in act*, but of an infinite number of parts *in potency* or non-existent.[17]

17 Ibid., fols. 9v–10r: "Quod ita adstruimus: vocitetur brevioris exactiorisque ratiocinationis gratia, maxima quaeppiam et suprema res A. B vero minutissima ac minima, tunc ita pergimus. Quod infinitas seorsum idoneas existere partes continet, id nimirum actu

Francus explains here the Lucretian idea that if things were made of infinitely many parts, all things would have to be equally immense. He, therefore, resorts to the Aristotelian distinction between infinite *in potentiality* and *in actuality*. According to Aristotle, no extended body can in actuality be divided infinitely; it is, however, indefinitely divisible, that is, it could potentially be divided *ad infinitum*, but this potentiality will never be wholly actualized.[18] Then, by associating this Aristotelian idea with the Lucretian claim that things cannot be divided to infinity, Francus blends the two theories, choosing to interpret their positions as much more compatible than they really were. At the same time, Francus, besides accommodating Lucretius' view to Aristotle's, applies the former's criticism only to Anaxagoras, who reportedly thought that things were composed of infinite parts (see also Lucretius, *De rerum natura* 1.844).

Interestingly, Aristotle's explanation wherein he denies the actual divisibility of spatial magnitudes *ad infinitum* was to become one of the preferred arguments used by those commentators who wanted to make Lucretius' atomism more acceptable to the eyes of their readers. Indeed, we shall see that a very similar argument in defense of Lucretius' atomic theory was also to be found in Frachetta's *Spositione*.

1.2.2 Francus on the Inadequacy of an Atom-Based Ontology

One of the few instances in which Francus seems very critical of Lucretius is when he comments on the passage concerning the theory of infinite worlds (2.1023–1104). While the influence of Marsilio Ficino is to be detected throughout the *Paraphrasis*, on this occasion, there is an explicit reference to the latter's *Theologia Platonica*:

> [Marsilio Ficino] in his [*Platonic*] *Theology* beautifully showed that every function and movement must be ascribed to some virtue and quality

infinttum (= infinitum) habetur. Quum igitur tam A quam B infinibiles illius modi portiones cohibeat, profecto utrumque infinitum convincitur. Quod si forte aliquis nostro occurrat argumento causamque inaequalitatis in medium afferat, quia scilicet partes quibus A componitur grandiores sint quam illae quibus B, tametsi quaelibet sint infinitae. Verum enimvero vel hoc de subito proruit. Nam minimum quandam ex b portiunculis sumptitabimus, inde alteram adiungemus corpusque aliquod hoc pacto conflabimus, inde aliam atque aliam deinceps in infinitum adiicemus. Ita fiet ut infinibilis b probetur magnitudinis, cum semper adiectione quantulacumque etiam adcrescat, cumque infinitae omnes suae partes, quarum quaelibet aliquanta est, simul adsint. Caeterum haec ratio solummodo Anaxagorae hisque, qui infinitis seorsum existentibus partibus res componunt, adversatur et pugnat. Verum apud Perypateticos finitis actu portionibus seorsum existere idoneis, infinitis vero potentia seu inexistentibus naturalia corpora conficiuntur."

18 See Aristotle, *Physica* III 6 206a 16–18.

> superadded to the body, or superior to the body itself. In fact, by its nature, a body impedes an action more than causing it. The corpuscles that these [*sc.* Democritus, Epicurus, and Lucretius] describe, can therefore by no means be the first beginnings of things. Therefore, they could not build this or other worlds, seeing that they are devoid of motion.[19]

Even though Francus aims to disprove that Lucretius' atoms could be capable of creating infinite worlds, the argument he uses is not primarily against multiple worlds, but against any atom-based cosmogony. Even if there were only one world – according to his argument here – it could not have been formed by atoms.

Francus is most probably referring to the second chapter of the first book of the *Theologia Platonica* (of which the title states "Corpus natura sua nihil agit"), in which Ficino insists on the idea that bodies are not capable by themselves of performing any activity, least of all creating things. In order to act or create, they need something else, namely a virtue or a quality implanted in them. Ficino explicitly refers to the ancient atomists as an example of those who believed the contrary, that bodies acted by their own mass.[20]

Having seemed to have accepted the idea of the existence of atomic particles as plausible, Francus now refutes the concept of an ontology exclusively centered on atoms (and void). He feels the need to add something else, and the objection he raises would be one of the most common points of criticism directed toward the atomists: if everything, including the soul, is made up of void space and inanimate atoms, how is it possible that there exist living things in the world? According to Francus, Lucretius' atom-based ontology provides an insufficient basis for explanations. As we shall see in our next section, it is precisely from the necessity to solve this problem that an innovative interpretation of ancient atomism was soon to arise.

---

19 Francus, *Paraphrasis*, fol. 20r: "[Marsilius Ficinus] sua in Theologia pulcherrime commonstravit functionem omnem ac motum in virtutem quandam et speciem corpori super additam seu corpore ipso praestantiorem referri debere. Nam suopte ingenio corpus actionem potius impedit quam concinnet. Itaque nullatenus, quae corpuscula isti effingunt rerum primordia esse queunt. Igitur non hunc non alios mundos adstruere potuerunt quandoquidem motus expertia sint." On Ficino and Lucretius' atomism, see Nicoli, "Ficino," 330–361, esp. 360.

20 Ficino, *Platonic Theology*, 1.2.4: "Si quid igitur agere corpora videantur, non ex ipsa sui mole, ut Democritii, Cyrenaici, Epicurei putaverunt, sed ex aliqua vi et qualitate illis insita operantur." The inclusion of Cyrenaics among the atomists is puzzling: Ficino must have thought that, since the Cyrenaics were hedonists, like the Epicureans, they must have been atomists as well.

2 Johannes Baptista Pius's *Commentary* (1511)

Printed in Bologna in 1511, Pius's commentary is – properly speaking – the first commentary on Lucretius' poem, as it covers all six books of *De rerum natura*.[21] Compared to Francus's paraphrase, it has moreover a more traditional set-up, as it carries a *lemmata* section in addition to the source text under explication. This commentary provides us with a particularly significant case study, primarily because of Pius's peculiar working method. In fact, it looks as if the commentary had been intended not so much to clarify Lucretius' work as to collect the most relevant and conspicuous information related to the topics covered or expressions employed by Lucretius. The result is that Pius's commentary is not the most reliable source for understanding Lucretius' philosophy, but is an excellent source for knowing what a Renaissance reader might have thought, or might have been interested in, when reading the *De rerum natura*.

Johannes Baptista Pius was probably born in Bologna, where he was a pupil of Filippo Beroaldo the Elder, and studied natural philosophy under Alessandro Achillini. He later became a university professor: he taught rhetoric and poetry in Milan, Bologna, and Rome. Pius also published several editions of classical authors, including Plautus and Lucretius.[22] Pius already knew Lucretius' *De rerum natura* in 1501, when he wrote his eight-folio *Praelectio in Titum Lucretium et Suetonium Tranquillum*, which is probably the written text of a lecture he gave in Bologna. Although, in the *praelectio,* there is only one explicit reference to Lucretius, who is positively described as "the first among the poets," this text proves Pius's precocious interest in the poem.[23]

Even though the reason that led Pius to compose a voluminous commentary on Lucretius is not explicitly stated anywhere in the paratexts, it is still possible to collect information from external sources. Some episodes concerning Pius's life provide, in fact, hints at the motivation that pushed him to undertake this project. One of the motivations may have been that he was a sympathizer of Epicureanism; at least this is how Leonardo Marso, his colleague at the *Sapienza*, presented him in the funeral oration written for Pius around 1542/43:

---

21 Portions of this section have been previously published in my chapter on Pius: Nicoli, "Atoms," 235–250.

22 Information about Pius's biography can be found in Del Nero, "Note sulla vita," 247–263; and the entry by Conti, "Pio."

23 Pius, *Praelectio*, fol. A1v "Primus hic animus deus in humano corpore hospitans ausus autore Manlio oculos alieno immittere caelo, indidemque deus effectus, uti Lucretio poetarum antistiti in uno eo placuisse video."

> But he [*sc.* Pius], not some half-educated man, while deeply engaged with the opinion of the stupid Epicureans, was also very studious of the Christian religion, more than anybody would believe. He fought with heretics with all 'his sails and horses', <as they say>, and not with the incenses of the ancients, but, having confessed his faults to the priest with a living voice, each year, he strove strenuously to purify his soul.[24]

Marso does not specify to what extent Pius was committed to Epicureanism, since his main goal was to show that he was, nevertheless, a good Christian who eventually confessed his faults and repented. However, supposing Marso's testimony can be accepted as fact, then it is reasonable to assume that, among other reasons, Pius might have written his commentary on *De rerum natura* out of a genuine interest in Lucretius' Epicurean philosophy.

Moreover, it can hardly be a coincidence that only a few years earlier, in Bologna, the *Paraphrasis in Lucretium* by Raphael Francus (1504) had been published by the same publisher that also edited Pius's commentary. It is more than possible that this precedent had also encouraged Pius to write his work on Lucretius.

2.1 *Pius on Atoms as Seeds*

Some telling clues as to Pius's understanding of Lucretius' atoms can be found in the comment on the passage in which Lucretius claims that nothing can arise out of nothing (1.159–191). According to Lucretius, a given organism can only grow from appropriate and fixed seeds (*certa semina*); not all things may be born from all, in fact, because, in particular things, there resides a distinct faculty (*secreta facultas*). In this passage, Lucretius establishes the organizing and creative power of his seminal atoms and seems to go beyond a purely materialist view of composition by mere aggregation, by endowing his atoms with almost miraculous generative powers.[25]

Commenting on this Lucretian passage, Pius interprets *secreta facultas* as "*Secreta*. Separate, distinct. *Facultas*. Capacity to generate [*gignendi potentia*]."[26] The second definition is noteworthy: in fact, the noun *potentia* cannot be

---

24 "Sed hic plumbeorum Epicureorum opinioni, nec quivis mediocriter eruditus, inhaerebat Christianae religionis supra quam credibile cuiquam sit studiosissimus erat. Cum haereticis velis, equisque ⟨ut dicitur⟩ decertabat, nec Antiquorum suffimentis, sed viva voce cum sacerdote confessus admissa, singulis annis animam purgare enixissime studebat." Quoted in Novoa, "Leonardo Marso," 250–253.

25 Sedley, *Lucretius*, 197.

26 Pius, *Commentarii*, fol. 19v: "*Secreta. Seiuncta, separata. Facultas. Gignendi potentia.*"

found anywhere in Lucretius but is crucial to the scholastic tradition, where it translates Aristotle's *dynamis* as "potentiality."[27] Analyzing *facultas* in terms of *potentia*, Pius reveals again his Aristotelian bias: he makes clear not only that Lucretius' *semina* are endowed with a generative power – as Lucretius himself seems to say – but also that, in them, resides a potentiality to develop into some actuality. In the same fashion, Pius defines *semina* as "the causes that bring to the things their own essence."[28] He thereby suggests that the seeds are the bearers of the things' essence, not mere chunks of matter that create bodies by their mutual aggregation and disaggregation. Such a conception of *semina* as an informing principle could already be found in Marsilio Ficino (1433–1499), who, differently from Lucretius, conceived however of the seeds as something immaterial, and in this was closer to the Stoic and Plotinian idea of seminal reasons than to Lucretius' atoms.[29]

When Pius arrives at the passage in which Lucretius presents the causes of atomic motion and the swerve (Lucr. 2.216–293), thanks to which the atoms deviate a little from their trajectory, he comments as follows:

> *Quare in seminibus*. Lucretius has already argued that the bodies are moved by an intrinsic force that produces motion. Now, he says that the same thing happens to the principles, which are moved by an internal cause, i.e., their own form, if we trust Aristotle. *In seminibus*. Herein the causes of things and the principles.[30]

In this passage, Pius first suggests that atoms are driven by an intrinsic force inherent to their nature and productive of their motion. Then, he moves from a physical to a metaphysical level, suggesting that an Aristotelian form resides in the atoms and is the cause of their movement. A little further on in the text, he clarifies this idea: "The principles are moved by a hidden cause, in no other way than we are dragged and moved by an innate, internal form."[31]

---

27 Galen's use of δύναμις, translated into the Latin *facultas*, was very common among medieval and Renaissance natural philosophers and physicians.

28 Pius, *Commentarii*, fol. 19v: "Semina rerum. Causae quae rebus esse suum afferunt."

29 See Nicoli, "Ficino," 356.

30 Pius, *Commentarii*, fol. 58v: "Quare in seminibus. Disseruit iam Lucretius a vi intrinseca corpora moveri gignente motum. Nunc idem accidere principiis inquit, quae moventur ab interiori causa hoc est a forma sua si credimus Aristoteli. In seminibus. In causis rerum et principiis expone."

31 Ibid.: "Principia a causa latente moventur non secus ac nos ab innata interiori forma trahimur et agimur."

Pius explains Lucretius' first principles not as merely material and structural elements, as parts of a purely mechanical explanation, but as bearers of forms. Whereas Lucretius, in this passage, simply maintains that the swerve is the cause of *voluntas* in living beings, Pius's interpretation of the text proceeds as if atoms themselves had an internal will, which he identifies with Aristotle's concept of form. If atoms are bearers of form, they must have a causal, formal principle within them, which makes them develop according to their specific nature.

Finally, commenting on a passage in Book II on the possible creation of other worlds elsewhere in the universe (Lucr. 2.1072), Pius writes:

> *Quae*. [Their own] nature and natural impetus move the seeds of things. Lucretius wants to suggest that the seeds are not moved by Nature, that is, God, but by a natural instinct and impulse, that is to say, inherent to the nature of those principles.[32]

Pius supports the idea of a spontaneous action of the atoms as the cause of change, given that they are endowed with a natural impulse that is inherent to them. With this interpretation, Pius enhances a specific aspect of Lucretius' philosophy, namely the organizing and generative power of the seminal atoms. At the same time, he provides a link between the Aristotelizing scholastic doctrine of substantial forms and Lucretius' conception of atoms.

Therefore, it appears evident that Lucretius' atoms, defined as *semina rerum*, are understood by Pius as more than mere Democritean chunks of matter. It must be obvious, however, that the emergence of a vitalistic understanding of atoms in Pius and other Renaissance authors echoed certain passages in Lucretius' own text as well as its description of atoms as the "seeds of things." This explains why such an interpretation is found not only in Pius's own commentary, but also in other Renaissance authors from Ficino onwards.[33] However, more clearly than many other Renaissance commentators, Pius grasped the ambiguity and polyvalence of Lucretius' concept of *semina rerum*, and chose to enhance the generative and vitalistic power of his "seeds," re-contextualizing and adapting them to the demands of Christian, Platonic, and Scholastic traditions.

---

32 Ibid., fol. 81r: "Quae. Natura et naturalis impetus. Mouet semina rerum. Non a natura hoc est a deo vult innuere Lucretius semina moueri sed ab instinctu et impressione naturali hoc est insita naturae illorum principiorum."

33 Nicoli, "Ficino," 352. The concept of *semina* in the Renaissance has been thoroughly examined in Hirai, *Le concept de semence*. For a detailed discussion of Ficino's concept of seeds, see idem, "Concepts of Seeds," 257–284.

3 Girolamo Frachetta's *Spositione* (1589)

The *Breve spositione di tutta l'opera di Lucretio* (Venice, 1589), written by the Italian scholar and political writer Girolamo Frachetta (1558–1619), is remarkable because it is the first printed text in the vernacular that explained the content of *De rerum natura*.[34] There is no extant evidence of the first Italian translations of Lucretius' poem, attributed to Giovan Francesco Muscettola (ca. 1530) and Tito Giovanni Ganzarini, also known as "lo Scandianese," which have never been published.[35] The translation of the celebrated Alessandro Marchetti, completed in 1668, was to be printed posthumously only much later, in 1717, after a long period during which ecclesiastic authorities successfully prevented its publication.

Frachetta's work is neither a translation nor an ordinary commentary, but rather an exegetical work, the goal of which is to explain the contents of *De rerum natura* in Italian to an audience that might have found Lucretius' Latin text inaccessible.

What is also remarkable about Frachetta's work is that it was the first publication on Lucretius after more than seventy years, during which no other edition or commentary on *De rerum natura* had been published in Italy. Most probably, the reason for this lack of publications was that Lucretius' poem had become rather unwelcome to the authorities of the Counter-Reformation due to its heterodox content. In 1559, after the institutionalization of the *Index of Prohibited Books*, Lucretius' text initially escaped the list in its original Latin form. What the censors successfully prevented, however, was the publication of translations into the vernacular languages: they probably calculated that if the poem was available only in Latin, its audience would remain limited. In this hostile context, Frachetta's *Spositione* stands out as the only attempt in that period to publish an explanation of Lucretius' poem, and moreover in the vernacular, which, of course, rendered the contents of *De rerum natura* accessible to a larger audience.

But did the choice of the vernacular make Frachetta's work different from those of his predecessors? I would contend that this choice may have forced Frachetta to be extra cautious in his commentary on Lucretius' philosophy. The *Spositione* hinges entirely on a comparison between Lucretius' doctrine and that of Aristotle, the latter serving as his philosophical benchmark

---

34 The only two studies devoted to Frachetta's *Spositione* are Gambino Longo, "La *spositione* de Lucrèce," 185–200 and Coleman, "Translating Impiety," 55–71.

35 See Prosperi, *Lost in Translation*, and the recent book by the same author, *Il fantasma di Lucrezio*.

throughout the paraphrase. As the full title of the *Spositione* clarifies, Frachetta aims to show the respects in which Lucretius' Epicurean doctrine agreed with the truth and with Aristotle's philosophy, and in which respects it did not.[36]

As will be explained shortly, Frachetta wrote his *Spositione* during his sojourn in Rome (1582–89), but his interest in Lucretius probably has earlier origins and can be dated to his university years (the 1570s–1581), when he was a student in Padua. There is much evidence to show that both Lucretius' work and atomistic ideas were already widely circulating among Paduan scholars during the second half of the sixteenth century.[37]

Girolamo Frachetta was born in 1558 in Rovigo, in the Veneto region.[38] He spent his youth in his hometown, where he studied humanities (*lettere umane*) at the public school under the supervision of Antonio Riccoboni. He later moved to Padua, where he began to study philosophy at the *Studio* and met the famous Aristotelian philosopher Francesco Piccolomini (1523–1607). In 1582, Frachetta moved to Rome, where he offered his services to Cardinal Luigi d'Este. In 1583, he dedicated to this cardinal his *De universo assertiones octigentae*, an encyclopedic collection of theses concerning the corporeal and incorporeal universe. He defended these theses in June of the same year in a public disputation in the church of Santa Maria sopra Minerva. However, Frachetta's enthusiasm for controversial ancient doctrines, including those of Lucretius, provoked the hostile reaction of ecclesiastical censors, who forbade the publication of 112 of the 800 theses, especially those concerning the intellective soul, the cabala, the names of God, and the cabalistic concept of the divine Sephiroth.[39] It is even more surprising, therefore, that his commentary on a text like that of Lucretius was not also forbidden. During these first years in Rome (1582–1586), Frachetta and other intellectuals gathered around the

36 The original title is *Breve Spositione di tutta l'opera di Lucretio, nella quale si disamina la dottrina di Epicuro, e si mostra in che sia conforme col vero e con gl'insegnamenti di Aristotile, e in che differente.*

37 For an in-depth analysis of Lucretius' reception in Padua in the second half of the sixteenth century, see Ceccarelli, "Reading Lucretius in Padua," 219–234.

38 An account of Frachetta's life can be found in Baldini, "Frachetta, Girolamo," and idem, "Girolamo Frachetta," 241–264.

39 See Baldini, "Girolamo Frachetta," 241–264. The fact that *De universo* underwent preliminary censorship in Rome, where it was published, suggests that the Master of the Sacred Palace was in charge of it, since he was responsible for the books published in the so-called "district of Rome." Unfortunately, his archive has disappeared, so we could not retrieve any further documentation on this issue. I would like to thank Leen Spruit for this information.

Cardinal Luigi d'Este and founded the *Accademia degli Incitati*.[40] In the same period, Frachetta decided to start writing commentaries on literary works in the vernacular. The first one was his *Spositione sopra la canzone di Guido Cavalcanti "Donna mi prega etc.,"* which was published in Venice in 1586. Also in Venice, in 1589, he published his *Spositione di tutta l'opera di Lucretio*, which is dedicated to the bishop Scipione Gonzaga, his new protector after the death of Luigi d'Este. The choice of these two texts cannot be a coincidence and show an unequivocal interest in Epicureanism; not only was Lucretius' poem regarded as a manifesto of Epicureanism, but Guido Cavalcanti, Dante's friend, was also reputed to have been an Epicurean. He is described by Boccaccio, in the *Decameron*, as one who "was somewhat inclined to the opinion of the Epicureans."[41]

Later in his life, Frachetta was to become a political informer of Scipione Gonzaga, whereupon he now dedicated himself to politics and abandoned his literary and philosophical activity. Eventually, after an adventurous further career, he ended up in prison in 1619 and died a few months after his release at the age of 61.

Frachetta's *Spositione* is divided into two parts, each consisting of six *lettioni*, which are the texts of actual lectures he pronounced at the *Accademia degli Incitati*. The first six *lettioni*, which correspond to the six books of *De rerum natura*, summarize and paraphrase the most important passages of the poem and comment on them. The second part (from *lettione settima* to *lettione duodecima*) consists of a detailed commentary on the hymn to Venus, which opens Lucretius' work. In the first part, Frachetta's work is concerned exclusively with the philosophical content of Lucretius' poem and with detailed speculations on the meaning of some crucial passages. But in the second part, which is devoted to one of the most celebrated poetical sections of *De rerum natura*, Frachetta is more interested in elucidating the content and in clarifying certain philosophical problems posed by the poem's opening (why, for example, did Lucretius invoke Venus given that he did not believe in divine providence?).

Frachetta introduces his work with a dedicatory epistle to cardinal Scipione Gonzaga and a letter to the reader. A detailed table of contents follows the

---

40 For Frachetta as the co-founder of the *Accademia degli Incitati*, see the epistle to the readers in his *Spositione di tutta l'opera di Lucretio* (n.p.) and the dedication of *Spositione sopra la canzone di Guido Cavalcanti*, IIr. The special connection between some of the sixteenth- and seventeenth-century academies and the *De Rerum Natura* has been noticed by Prosperi, *Il fantasma*, 98.

41 Boccaccio, *Decameron*, 756 (day VI, novel IX): "[…] *egli* [Cavalcanti] *alquanto tenea della oppinione degli epicuri*." At the same time, Dante depicts Cavalcanti's father as consigned to the sixth circle of the *Inferno*, amongst the Epicureans (see *Inferno* X. 52–72).

paratexts. In it, the author lists all the most notable themes, including an extensive entry on atoms, and another – the longest entry in the *Spositione* – on Lucretius' errors.

3.1 *Lucretius' Atoms and Aristotle's Prime Matter*

In *De rerum natura*, Lucretius shows that nothing can arise out of nothing, and nothing can be reduced to nothing (1.149–264). Frachetta claims that both Aristotle and Lucretius maintain that what is corrupted does not dissolve into nothing, but is resolved into the matter out of which it has first been generated. Elaborating on this concept, Frachetta goes so far as to assert that both Aristotle and Lucretius conceived prime matter as eternal and capable of acquiring any form whatsoever; having been deprived of one form, it can be endowed with another.[42] One recognizes Aristotle's theory in this description, but not that of Lucretius, who did not reason in terms of forms. As we have already shown, however, atomism, in this period, coexisted with Aristotelian hylemorphism, and the two theories could be deemed compatible.

Frachetta then introduces the concept of atoms: they are "according to Democritus' opinion, which Epicurus followed, eternal bodies, indivisible and invisible because of their smallness."[43] Interestingly, Frachetta does not deny the existence of atoms; instead, he engages in an explicit defense of Lucretius' atomism. The title of Frachetta's next section declares: "The opinion about the atoms is defended" (*Opinione de gli atomi si difende*). He cites a passage in *De caelo* (III 4 303a) in which Aristotle criticizes Leucippus and Democritus for supporting the atomistic theory. But surprisingly, Frachetta takes the side of those on the receiving end of the attack, claiming that their opinion about atoms is, after all, not so different from Aristotle's prime matter:

> I will just say that, if we want to understand this in depth, we will find that this opinion concerning the atoms, that is, that they are matter, out of which are created and into which are dissolved the things that are generated and corrupted, and that they are corporeal, indivisible, invisible and eternal, is not much different from Aristotle's about prime matter. Because also prime matter, according to Aristotle, is [1] eternal and, if we

42 Frachetta, *Sposizione*, 7: "Ciò fatto si dà a provare Lucretio con quattro ragioni che le cose, che si corrompono, non si risolvono in niente, ma nella materia medesima, onde si generano. La qual propositione è altresì tenuta comunemente per vera da naturali Filosofi, & da Aristot. Il quale pone per ciò la materia eterna, & con potestà di ricevere tutte le forme, come fa anco Lucretio: acciò che spogliandosi d'una, possa riceverne un'altra."

43 Ibid., 7–8 "[…] gli atomi; che sono, per opinion di Democrito; la qual seguitò Epicuro; corpi eterni per picciolezza, non divisibili, & non vedevoli […]."

> trust Simplicius, [2] is corporeal by its very nature. And as it has no form, therefore it is [3] invisible, and it can also be said to be [4] indivisible because none of its parts can be distinguished from any other since it is all of the same kind unless various forms are introduced; for the cause of which, or to the degree to which it has [forms], may be said to be not only one but many.[44]

Frachetta, in this passage, strives to demonstrate that Aristotle's prime matter possesses a series of attributes that are similar to those employed by Lucretius to describe his atoms, that is, [1] eternity, [2] corporeality, [3] invisibility, and [4] indivisibility. While these attributes suit Lucretius' atoms very well, it should be clear that not all of them are suitable as a description of Aristotle's prime matter.[45]

From an Aristotelian perspective, proving the eternity of prime matter should not be complicated. Since Aristotle believed that all things must come into existence from an underlying substratum and that prime matter is the substratum of all things, it must have either created itself or else it must be eternal (*Phys.* I 7 190b 1–10, 25). Obviously, from Aristotle's perspective, a substratum cannot generate itself, and therefore, prime matter must be eternal.

As for corporeality, it is questionable whether this characteristic may be applied to Aristotle's prime matter. In fact, in *Metaphysics*, Aristotle claims:

---

44 Frachetta, *Sposizione*, 8–9: "[D]irò solo, che, se noi vogliamo penetrar ben entro, troveremo cotale opinione de gli atomi, cio è, che siano materia, di cui si compongono, & in cui si disciolgono le cose che si generano, & che si corrompono, & sieno corporei, indivisibili, non vedevoli, & sempiterni: esser poco differente dall'opinione di Aristotile della materia prima, perche etiandio la materia prima, secondo Aristotile, è eterna: & se noi crediamo a Simplicio, è di sua natura corporea; come che non habbia forma niuna; per la qual cosa è invisibile: & si può dire anco indivisibile; imperoche non si distingue alcuna sua parte da altra; essendo tutta d'una medesima guisa; se non con l'introducimento di varie forme; per cagion delle quali, ò per lo riguardo, che v'ha, si può dire, non essere una, ma molte."

45 Not only is it difficult to provide a univocal definition of Aristotle' prime matter, but the very concept of prime matter has often been called into question by scholars who wondered whether Aristotle actually even postulated such a thing. Ancient and early modern commentators, on the other hand, seem to have had no doubt that Aristotle believed in a prime matter. It was possible also to view prime matter as a logical precondition for the analysis of anything in terms of matter and form. All in all, there is substantial disagreement among scholars over what was the nature of Aristotle's prime matter and whether it existed at all; see, among others, Robinson, "Prime Matter," 168–188; Charlton, "Prime Matter," 197–211; Graham, "The Paradox," 475–490; Byrne, "Prime Matter," 197–224; and Sorabji, *The Philosophy*, 253.

> By matter I mean that which in itself is neither a particular thing nor of a certain quantity nor assigned to any other of the categories by which being is determined. For there is something of which each of these is predicated, whose being is different from that of each of the predicates.[46]

From this passage, it appears that prime matter lacks all positive determinations; for this reason, some commentators have assumed that it must have been incorporeal. However, in another passage, Aristotle stated that "when length and breadth and depth are taken away we see nothing left unless there is something that is bounded by these."[47] This statement encouraged other commentators, including Simplicius, to maintain that prime matter is a three-dimensional extension (*diastēma*), separate from determinate dimensions, which would give it a particular magnitude. Frachetta relies and expands upon Simplicius' interpretation when he defines Aristotle's prime matter as extended and corporeal. He declares that "prime matter [...], if we trust Simplicius, is corporeal by its very nature." This specific interpretation allows him to pair Aristotle's prime matter with Lucretius' atoms. Significantly, Frachetta also adds that, according to Simplicius, Aristotle's prime matter is corporeal by its very nature, even though it has no form. According to his interpretation of Simplicius, being corporeal must be an essential disposition of prime matter: it does not become so only by some superimposed form.

According to Frachetta, another quality that both Lucretius' atoms and Aristotle's prime matter have in common is indivisibility. There is, however, a substantial difference between the two: Lucretius' atoms consist of smallest parts inseparably cohering and are therefore undoubtedly indivisible. According to his interpretation of Aristotle, on the other hand, prime matter is indivisible because none of its parts can be distinguished from the others before the introduction of various forms.[48] Frachetta again tackles this topic of the indivisibility of prime matter, a few pages later in the *Spositione*, but now from a different perspective:

> Even though Aristotle would not agree with Lucretius when the latter says that matter consists in minimal bodies or that there are bodies absolutely (*assolutamente*) indivisible, nevertheless he would grant indivisibility to prime matter, in the way we said before. Moreover, he would admit that

46 *Met.* VII 3 1029a 20–23, transl. Ross.

47 Ibid. VII 3 1029a 16–19.

48 Frachetta, *Sposizione*, 9: "non si distingue alcuna sua parte da altra, essendo tutta d'una medesima guisa, se non con l'introducimento di varie forme."

> this kind of matter is in itself most simple, and he would not deny that minimal natural bodies are given, which, albeit not absolutely (*assolutamente*), because as quantity they can be divided into infinity, at least as natural bodies (*come naturali*) are indivisible; that it is so is clear from what he says in the first book of *Physics* in the 36th part and elsewhere.[49]

Frachetta here maintains that, according to Aristotle, prime matter is indivisible, and that there are minimal natural bodies, which can – in principle – be divided to infinity; but, as natural bodies, they are indivisible, just like Lucretius' atoms. To endorse this theory, Frachetta cites Aristotle's *Physics* I 4 187b 14–34, where it is said that the repeated subtraction of parts from a finite natural body must come to an end; the minimum is the limit below which a substantial form cannot be sustained.[50] Frachetta uses this passage as evidence that Aristotle endorsed the existence of minimal physical parts, which cannot be further divided and are therefore consistent with Lucretius' atomism.

In the *Spositione* passage quoted above, Frachetta also emphasized that, according to Aristotle, there cannot be indivisible bodies; unless we consider them as natural bodies, in which case they are indeed indivisible. Here, Frachetta might be interpreting the problem of the divisibility/indivisibility of matter in the light of the Aristotelian distinction between potentiality and actuality: no extended body can in actuality be divided infinitely. Each extended body is, however, indefinitely divisible; that is, it could potentially be divided *ad infinitum*, but this potentiality will never be wholly actualized.[51] Like Lucretius' atoms, some magnitudes cannot be further divided in the natural world. We recall, as mentioned above (subsection *1.2.1.*), that Francus had formulated a similar comparison.

---

49 Ibid., 13: "[…] se ben Aristotile, non converrebbe con Lucretio, in dire, che la materia sia, corpi minimi, ò in dire, che ci sieno corpi assolutamente indivisibili, non pertanto concederebbe la indivisibilità alla materia prima; nel modo, che davanti dicemmo. & ammetterebbe detta materia essere da per se semplicissima, ne negherebbe, che non si dieno corpi naturali minimi, li quali sieno, se non assolutamente; conciosiacosa che come quantità possino dividersi in infinito, almeno come naturali, non divisibili, la qual cosa stare in questa maniera, appare per quello che egli dice nel primo libro della Phisica alla parte 36 & altrove."

50 On the theory of *minima naturalia*, see especially Murdoch, "Minima Naturalia," 91–131.

51 For the distinction involved in the division of *corpora naturalia* into either potentiality or actuality, see Averroes, *Aristotelis de physico auditu commentaria*, I c. 36 (as a clarification of Aristotle, *Physica* I 187b 15–16), to which Frachetta probably refers.

3.2 *Lucretius' Void and Aristotle's Interpretation of Plato's* Chōra *as Matter*

Frachetta's interpretation of Lucretius' theory of void space is quite self-contradictory. At first, commenting on 1.329–397, he explicitly claims that Lucretius' opinion of the void differs from that of Aristotle. Lucretius is furthermore wrong, he explains, because there is nothing that Nature eschews more than void, since where there is void, there is no nature, "given that nature is material and corporeal form."[52] This criticism relies mainly on Aristotle, who had put forward many arguments to discredit adherents to belief in the void in the fourth book of *Physics* (IV 6 213a 12–9 217b 28).

Interestingly, Frachetta later seems to reconsider his opinion on the void, or at least tries to look at it from another perspective. Commenting on the passage in which Lucretius argues in favor of the infinity of space and matter, that is, in favor of atoms (*De rerum natura* 1.958–1051), Frachetta reaffirms that nature does not admit any kind of infinite, if not in potency (*la natura non ammette infinito d'alcuna guisa, se non in potenza*). Moreover, with reference to Aristotle, he argues that Lucretius is wrong when he claims that there is an infinite space beyond the universe. In the end, however, Frachetta tries once more to find a compromise and he adapts Lucretius' opinion to the Platonic and Aristotelian views:

> But, again, in this passage, I do not want to omit to mention that not only do the atoms correspond to prime matter, as we have shown earlier, but also to the void, because the matter is bare and can accommodate forms, and can be defined as the place of the latter – as Plato described it – in the same way as the void is bare, can accommodate bodies, and is the place of these. [Therefore, this matter] is infinite by its nature, that is to say, that it has no boundary or end, given that the limit depends on the forms.[53]

52 Frachetta, *Sposizione*, 9: "[…] percioche non v'è cosa che più schifi la natura che il vacuo, conciosia cosa, che dove è vacuo non sia natura, essendo la natura, forma materiale, & corporea."

53 Ibid., 19: "[…] ma non voglio restar di dire etiandio sopra cotal passo, che la materia prima, con cui hanno corrispondenza, non solamente gli atomi, come avanti mostrammo, ma anco il vacuo; in quanto la materia è nuda, & capace di forme, & puo dirsi luogo di esse: come da Platone è detta: nella guisa che il vacuo è nudo, & capevol de corpi, & luogo di essi: è di sua natura infinita, cioè a dire che non ha termine o fine alcuno. conciosiacosa che la terminazione dipende dalle forme."

In other words, Frachetta here identifies Lucretius' combined conception of both atoms and void with Aristotle's prime matter. He was not unique in Aristotelizing the ancient atomists in this way, as the same tendency can be found also in the handbook of natural philosophy written by the Jesuit Benedict Pereira, *De communibus omnium rerum naturalium principiis* (1576).[54]

Frachetta had already extensively demonstrated that Lucretius' atoms are the same as Aristotle's prime matter, and now he claims that both atoms and the void are equivalent to it, since they are bare; capable of receiving any form, in the case of matter; capable of receiving bodies, in the case of void; they can each be considered the place of forms and bodies, respectively. What is more, being both atoms and void, prime matter may – before the imposition of form – be considered infinite; it is the superimposition of form that provides matter with boundaries. However, it must be pointed out that Frachetta, in this context, illegitimately takes the contrary of finite to be infinite rather than indefinite.

In this discussion, Frachetta refers to Plato's alleged notion of matter as the recipient of forms and as their place. His description would have been more appropriate for Plato's idea of the receptacle of all becoming, which has been variously interpreted by ancient and modern scholars, each time differently, either as matter or as space (*chōra*).[55] The first who arbitrarily identified Plato's space (*chōra*) with matter (*hylē*) was Aristotle himself, in a statement contained in his *Physics*: "hence Plato in the *Timaeus* identifies space and matter."[56] Later, in the Renaissance, this interpretation was accepted by Ficino.[57]

As we have seen before, Frachetta first establishes a connection between Lucretius' atoms and Aristotle's concept of matter. Based on the Aristotelian reinterpretation of Plato's idea of space (*chōra*), which he also holds to be identical with matter, he also identifies Lucretius' void space with the prime matter. Ultimately, his aim is to show that Lucretius' atoms and void can all be assimilated to Aristotle's concept of prime matter, conceived as a permanent substrate capable of receiving forms and bodies. However, it must be evident that this interpretation is entirely implausible: if taken seriously, it would lead

54 Pereira, *De communibus*, 136: "ergo apud hos philosophos [sc. Democritus, Leucippus, Epicurus] inane & atomi locum tenent materiae."

55 The debate over whether Plato's receptacle has to be interpreted as space or matter has divided scholarship ever since the days of Aristotle. See Algra, *Concepts of Space*, 72–118, who argues for a theory of "the receptacle-as-both-space-and-matter."

56 Aristotle, *Physics* IV 2 209b 11–12, transl. Ross. This view, however, is not genuinely Plato's.

57 Ficino, in his translation of Plato's *Timaeus*, provided the following marginal title for *Tim.* 52a 8 – b 1: "that matter is place" (*quod materia est locus*), essentially attributing Aristotle's interpretation directly to Plato; see Ficino, *Diuini Platonis Opera*, 484.

to a kind of Parmenidean monism, in which the totality of all things is just one single chunk of extended matter. Frachetta seems to forsake the advantage of atomism vis-à-vis Parmenides' monism, an advantage which had precisely consisted in a dualism that allows for multiplicity as well as change.

4 Conclusion

Let us go back to our opening question: how was Lucretius' atomic matter theory explained in the earliest Renaissance commentaries on *De rerum natura*? Not as we would have expected it to be explained. First, when the first commentators of *De rerum natura* elucidated the poem, they did not place it in a historical-philosophical perspective, nor did they try to clarify Lucretius' text. On the contrary, they often attached their conceptions to it and remodeled the text according to their own philosophical views. Furthermore, there is no explicit condemnation of Lucretius' atomism as such to be found in these texts, but rather a tendency to adapt and reframe it within other philosophical contexts. What the first exegetes of Lucretius criticized were instead the consequences of Lucretius' materialist philosophy: first and foremost, the denial of the immortality of the soul.

Secondly, more than the presence of atoms and void in Lucretius' natural philosophy, these commentators seemed to resent the absence of something else: an element of soul or energy that could make these particles capable of generating things. Consequently, the most remarkable development of these new interpretations of Lucretius' philosophy is the overlapping of the conception of the atom with that of Aristotelian matter, and the consequent attribution of a form to atoms. A similar development was later to be found in Bruno's vitalist atomism or in Sennert's concept of *atomi-cum-forma*.

After the rediscovery of *De rerum natura*, those who commented on Lucretius' text reinterpreted his matter theory vitalistically and embedded it in a teleological account of nature. This mutation opened up the possibility for the spread of a model based on atoms, gradually replacing Aristotle's natural philosophical system.

CHAPTER 3

Galenic Medicine and the Atomist Revival: Elements, Particles, and Minima in Late Renaissance Physiology

Elisabeth Moreau

1 Introduction

In the course of the past decades, studies on the history of matter theories have shown the heterogeneous nature of Renaissance atomism in physics, mathematics and alchemy.[1] Stimulated by the rediscovery of Lucretius' poem and the Latin translation of Epicurus' epistles, this "atomist revival" was also built upon novel interpretations of Aristotelian natural philosophy as developed since the late Middle Ages.[2] Renaissance atomistic theories, therefore, did not necessarily entail an allegiance to the philosophy of Democritus. They often involved eclectic interpretations that posited a select number of atomistic conceptual ingredients, and above all, the structure of matter into indivisible and discontinuous units. This is particularly evident in the prolific terminology of these theories, which alternately defined atoms as "elements," "particles" and "*minima*."

Whereas historians have emphasized the multidisciplinary nature and conceptual disparateness of the atomist revival, they have not investigated how medicine, and specifically the authority of Galen, have contributed to the emergence of this revival. And yet, despite Galen's rejection of atoms and corpuscles, it is possible to trace atomistic conceptions of the elements in Renaissance Galenic medicine. This helps explain why in the medical texts of the early seventeenth century, the notion of the element was at times merged with that of the atom in order to explain the body's functioning. As the elements were a central aspect of Galenic medicine, the historiographical lacuna on this topic is all the more surprising. The present chapter aims to fill this gap

---

1 See Lüthy et al., eds., *Late Medieval and Early Modern Corpuscular Matter Theories*; Clericuzio, *Elements, Principles and Corpuscles*; Newman, *Atoms and Alchemy*; Meinel, "Early Seventeenth-Century Atomism."

2 On the atomist revival of the Renaissance, see Lüthy, "Atomism in the Renaissance"; Halleux, "Atomisme."

 | DOI:10.1163/9789004528925_004

by considering the atomistic redefinition of the element in Galenic physiology, a medical field that directly intersects with matter theory.[3]

Early modern physiology was rooted in ancient and medieval medicine following Galenic and Aristotelian philosophy.[4] In the late Antiquity, the pseudo-Galenic treatise *Introduction, or The Physician* defined physiology as the part of theoretical medicine that was dedicated to the healthy body.[5] Medieval medical works such as Avicenna's *Canon of Medicine* and the *Articella* further developed physiological knowledge as the study of "natural things" (*res naturales*). This formulation points to the initial meaning of physiology as the investigation of nature (*physis*), with Aristotle's physics serving as the foundation of medical learning. In accordance with this model, the human body and the natural world shared the same composition out of four elements (earth, water, air and fire), which in turn were endowed with the four primary qualities (hot, cold, dry, moist). The balanced or imbalanced "mixture" of the four qualities determined the state of health or "temperament." From the late Middle Ages through to the early modern period, physicians explored the notion of temperament or "complexion" in relation to Galenic and Aristotelian accounts of matter-form, the elements, and mixture.[6] In the process, they developed eclectic views on the nature of the elements and their structure within the living body.

Centered as it was on the notions of elements, mixture, and temperament, Galenic physiology triggered new theories of matter in the late Renaissance. In this chapter, I will consider the interpretations of three important medical figures of that period, each of whom explored a different facet of Galenic physiology and did so in an original way: the first, imbuing it with Platonic philosophy; the second, with chymistry; and the third, with atomism.

I will first examine the Platonic account of the French physician Jean Fernel (1497–1558), whose *Universa Medicina* was a major textbook of Galenic medicine in the early modern period.[7] His *Physiologia*, first published in 1542 as *De naturali parte medicinae* [*On the Natural Part of Medicine*], diffused the

3 For the Latin edition of Galen's works, I will use Galen, *Galeni opera omnia*, ed. Kühn.

4 On early physiology at the intersection of natural philosophy and medicine, see Nutton, "*Physiologia* from Galen to Jacob Bording"; Cunningham, "The Pen and the Sword," Part 1; Bylebyl, "The Medical Meaning of *Physica*."

5 See Galen, *Introductio seu medicus*, 2, vol. XIV, 677–678.

6 On early modern theory of matter-form, see Manning, ed., *Matter and Form in Early Modern Philosophy*; Lüthy and Newman, eds., "The Fate of Hylomorphism: 'Matter' and 'Form' in Early Modern Science."

7 On Jean Fernel, see Kany-Turpin, ed., "Jean Fernel"; Henry and Forrester, "Introduction" to Fernel, *The Physiologia*, transl. Forrester; idem, "Jean Fernel and the Importance," in Fernel, *On the Hidden Causes of Things*, transl. Forrester, 3–66.

term "physiology" as the medical study of the functioning of the healthy body. Next, I will investigate the medical theory of the German physician Andreas Libavius (ca. 1550–1616), who was an important figure in the history of German alchemy at the dawn of the seventeenth century.[8] In the history of early modern science, he is well-known for his endless polemic against Paracelsus and his followers, as well as his *Alchymia* (1606), a sophisticated textbook on alchemical operations and instruments. Interestingly, Libavius also proposed a medical philosophy that drew on medieval alchemical sources. The connection between his medical thinking and his chymical theory of matter still needs further examination in order to elucidate his later atomistic ideas. Finally, and on the basis of Libavius's and Fernel's accounts, I will consider the medical views of the German physician Daniel Sennert (1572–1637), who served as a professor of medicine at the University of Wittenberg.[9] By synthetizing Aristotelian natural philosophy, Galenic medicine and Paracelsian chymistry, Sennert would become established as an important reference for seventeenth-century sciences.

Following their various philosophical inclinations, Fernel, Libavius and Sennert redefined the elements as discontinuous particles, *minima* or atoms, in a medical context that was pervaded by Galenic philosophy.[10] In exploring their views, this chapter aims to show the role of the Galenic tradition in the rise of early modern atomistic explanations. In particular, it stresses the need to include medical authorities such as Galen and Avicenna among the sources of the atomist revival. I will examine this specific aspect in the next section in relation to the elements and mixture in the medical tradition. The ensuing sections on Fernel, Libavius and Sennert will retrace two particular contexts in which the concept of discontinuous matter was developed from a physiological angle, namely the element as the material component of the body, and its first level of organization in "homeomerous" body parts.[11]

8 On Andreas Libavius, see Moran, *Andreas Libavius and the Transformation of Alchemy*; Newman, *Atoms and Alchemy*, 66–84; Hannaway, *The Chemists and the Word*; Gilly, "The 'Fifth Column' within Hermetism."

9 On Daniel Sennert, see Michael, "Sennert's Sea Change"; Newman: Atoms and Alchemy, 85–156; Hirai, Medical Humanism and Natural Philosophy, 151–172; Stolberg, "Particles of the Soul."

10 The term "chymistry" and its derivatives here refer to the post-Paracelsian transition, from the late Renaissance to the eighteenth century, between medieval alchemy and Lavoisier's modern chemistry: see Newman and Principe, "Alchemy vs. Chemistry."

11 On Fernel's concept of temperament and mixture, see Moreau, "Elements, Mixture and Temperament."

2 Elements and Mixture in the Galenic Tradition

The Aristotelian notions of "element" and "mixture" raised important debates in the scholastic philosophy of the Middle Ages. The discussion was centered on alternative interpretations of mixture as developed by Avicenna, Averroes, Thomas Aquinas, and John Duns Scotus, among others.[12] Their respective views have been the subject of several studies, as has their legacy in the Renaissance.[13] In contrast, the reception of the Galenic theory of mixture in the early modern period has received insufficient attention. In this section, I examine Galen's argumentation in *On the Elements According to Hippocrates* and its reception in a major work for the medical tradition, Avicenna's *Canon of Medicine*.[14] The survey of these canonical medical sources will serve as a useful preamble to the appraisal of Jean Fernel's *Physiologia* in the next section.

Galen's account of the elements was based on the humoral medicine of Hippocrates and the physics of Aristotle. Following Hippocrates' *On the Nature of Man* and Aristotle's *On Generation and Corruption,* Galen defined elements as the "smallest parts" of bodies that had the "first and simplest" nature. The elements were necessarily multiple in number and contrary to one another, because they were subject to change. For this reason, the indivisible, discrete and insensible atoms of Democritus and Epicurus had to be dismissed, as they rendered both the alteration of compounds and sensorial phenomena like pain altogether impossible. As for the number of elements, it was determined by four tangible qualities, which were distributed into two pairs of contraries, namely active (hot / cold) and passive (moist / dry). Each element was associated with one pair of these qualities: fire (hot-dry), earth (cold-dry), air (hot-moist) and water (cold-moist).

In accordance with Aristotle's physics, Galen viewed the process of generation and destruction (corruption) as fundamental to natural change. Bodies were composed of elements and underwent physical change through the action and passion of their respective qualities. The same held true also for living beings. As they belonged to the sublunary world, living beings were subject

12 On ancient and medieval interpretations of mixture, see Woods and Weisberg: "Interpreting Aristotle on Mixture"; Maier, *On the Threshold of Exact Science*,124–142; Petrescu: "John Duns Scotus."

13 See Emerton, *The Scientific Reinterpretation*; Deitz, "*Falsissima est ergo haec de triplici substantia Aristotelis doctrina*"; Navarro-Brotons, "Matter and Form in Sixteenth-Century Spain."

14 Galen, *On the Elements According to Hippocrates*, transl. de Lacy; Galen, *De elementis ex Hippocrate*, I, 413–508. Concerning Avicenna's *Canon*, I use Costeo and Mongio's edition of Avicenna, *Canon medicinae* (1595).

to change as a consequence of the mixture of the four elements that defined them. Initially, in the embryonic stage, these elements were provided by the maternal blood, which allowed for the development of the organism. After birth, the body assimilated the elements through the digestion of food, given that edible plants and animals were equally composed of the four elements.

Involved as they were in physiological operations such as nutrition, reproduction and growth, the elements of the body played an essential role in the constitution of the temperament, which in turn defined the state of health. Such a temperament resulted from the union of elements into a balanced "mixture" (*krasis* or *mixis*).[15] Whereas Galen suggested that mixture required a moderate and uniform proportion of elements, he refrained from further commentary on Aristotle's interpretation:

> It is not necessary for physicians to understand how things that are mixed through and through, whether the mixtures are of qualities only, as Aristotle supposed, or of corporeal substances that pass through each other; that is why Hippocrates said nothing about these matters but was content with the mere fact that the elements are mixed in their entirety.[16]

In this passage, Galen alludes to Aristotle's description of mixture in *On Generation and Corruption*.[17] Following the Aristotelian physics, elements become united through the action and passion of their contrary qualities, which result in the formation of homogeneous and moderate compounds. These compounds were brought about and completed by their respective substantial forms, while the elements continued to be present, but merely in potentiality. With this description, Aristotle aimed to establish the equilibrium and homogeneity of compounds in contradistinction to the atomists' spatial juxtaposition of discrete material units. In keeping with this reasoning, Galen rejected atoms and corpuscles, while endorsing the Hippocratic requirement of a balanced ratio between the four qualities.

Despite his reluctance to provide further details on the behaviour and status of the elements during mixture, Galen at one point did align his interpretation of mixture with the model of liquids:

---

15 On Galen's theory of elements and mixture, see Guyomarc'h and Marchand, eds., "Studies on Galen's *De elementis*"; Boudon-Millot, "La notion de mélange."

16 Galen, *On the Elements*, 137 = *De elementis*, I, 489.

17 Aristotle, *On Generation and Corruption*, 1.10, 327a30–328b24.

> In the mixing of wine with water, for example, and in the breaking up of the parts of each into the smallest bits, it happens that each of them acts on the other and is acted on by it, and that they share their qualities with each other the more readily the smaller the bits into which they have been broken. [...] Moreover, it is concordant with this reasoning that the qualities of things that have been mixed more thoroughly and for a long time are more closely united.[18]

Here again, Galen relied on Aristotle's *On Generation and Corruption*, which stated that the ability of bodies to mingle was facilitated by the juxtaposition of their smallest parts, in particular in the case of liquids.[19] Along these lines, Galen suggested that mixture entailed the reduction of bodies to their smallest elemental parts, which closely united in the compound. We encounter this view of the elements as the smallest parts or "minimal particles" (*minimae particulae*) in numerous medical works of the Galenic tradition throughout the centuries, from the medieval *Pantegni* of Haly Abbas, which was translated by Constantinus Africanus in the eleventh century, to Renaissance Latin translations of Galen's *On the Elements According to Hippocrates*.[20]

In the second half of the twelfth century, physicians of the school of Salerno further developed the Galenic approach to the element as a minimal particle.[21] As Danielle Jacquart has shown, they elaborated the definition of the temperament as a process of mixture (*commixtio*) and the resulting "complexion" (*complexio*) in accordance with the *Isagoge* of Johannitius.[22] In this spirit, Bartholomew of Salerno deemed the element a simple particle, and considered mixture a union *per minima*, thereby merging Galen's account with Haly Abbas's *Pantegni* and Plato's presentation of the four elements in the *Timaeus*. The latter treatise was available through Calcidius' partial translation (fourth century), which was further commented upon by the School of Chartres in the twelfth century. In turn, the interpretation of mixture as a union of minimal particles was transmitted in a fundamental work of medieval alchemy: the *Summa Perfectionis* of pseudo-Geber (late thirteenth century).[23] As

18 Galen, *On the Elements*, 139 = *De elementis*, I, 490.

19 Aristotle, *On Generation and Corruption*, 1.10, 328a32–36.

20 See *Liber Pantegni* in Isaac Israeli, *Omnia opera Ysaac*, fol. iv; Galen, *De elementis ex Hippocrate sententia*, transl. Winter von Andernach (Paris, 1528); Idem, *De elementis libri duo*, transl. Niccolò Leoniceno in *Prima classis humani corporis* (1541), 2r–9v; idem, *De elementis libri duo*, transl. Trincavelli (1548).

21 Jacquart, "*Minima* in Twelfth-Century Medical Texts."

22 Eadem, "De *crasis* à *complexio*."

23 Newman, *Atoms and Alchemy*, 31.

William Newman has shown, this treatise was important for the emergence of atomistic theories in seventeenth-century chymistry, particularly in the works of Libavius and Sennert.

Avicenna also played a major role in the discussion as to the nature of elements and mixture, a question to which it is mostly his works on Aristotelian natural philosophy and metaphysics that have attracted the attention of historians.[24] In *On Generation and Corruption* (*Liber tertius naturalium*) and in his *Book of Healing* (*Sufficientia*), Avicenna described the mixture of elements in relation to the action and passion of their qualities. But instead of following Aristotle in postulating a presence of the elements merely in potency, he supposed the persistence of their substances as "firm" or "fixed" forms that were united in the compound.[25] Following a Platonic emanationist interpretation, Avicenna further stated in his *Metaphysics* that the compound was brought about by the introduction of a form of divine origin. This form came from a "giver of forms" (*dator formarum*), which was an emanation of the active intellect.[26] The celestial form of the compound then achieved its well-prepared matter resulting from the mixture of elements.[27]

Importantly, Avicenna's account of mixture echoed in the medical theory of his *Canon*, which was translated into Latin by Gerard of Cremona in the twelfth century. In this work, Avicenna subscribed to Galen's Aristotelian doctrine of matter and form.[28] He adopted Galen's definition of elements as bodies and "first parts" which could not be further divided, and stated that the temperament resulted from the action and passion of qualities. In his view, the formation of the temperament required the reduction of elements into contiguous minute parts.[29] Furthermore, as he suggested in the *Canon*, the "specific form" of bodies was of a different nature from that of the complexion

24 On Avicenna's theory of mixture, see Stone, "Avicenna's Theory of Primary Mixture"; McGinnis, *Avicenna*, 83–89.

25 Avicenna, *Liber tertius naturalium* ed. van Riet and Verbeke, 63; idem, *Sufficientia*, 1.6, in *Avicenne opera* (1508), 17rb; ibid., 1.10, 19rb. For the English translation of *Sufficientia*, see Avicenna, *The Physics of the Healing*, transl. McGinnis, vol. 1, 48 and 68.

26 Avicenna, *Liber de philosophia prima*, 9.5, ed. van Riet and Verbeke, 488–489.

27 Hasse, "Avicenna's 'Giver of Forms'"; Davidson, *Alfarabi, Avicenna and Averroes*, 74–126.

28 On the late medieval and Renaissance reception of the *Canon*, see Siraisi, *Avicenna in Renaissance Italy*; Chandelier, *Avicenne et la médecine en Italie*.

29 Avicenna, *Canon*, 1.1.2.1, vol. 1, 9a: "Elementa sunt corpora, et sunt partes primae corporis humani, et aliorum, quae in corpora diversarum formarum dividi minime possunt." Ibid., 1.1.3.1, vol. 1, 11ab: "Complexio est qualitas, quae ex actione ad invicem et passione contrariarum qualitatum in elementis inventarum: quorum partes ad tantam parvitatem redactae sunt, ut cuiusque earum plurimum contingat, plurimum alterius: provenit."

of the elements.[30] While the complexion was the result of the union of minute elemental parts, the specific form pointed to the presence of a non-elemental, divine nature within the living body. In *De viribus cordis* (also translated as *De medicinis cordialibus*), a treatise on pharmacology attached to the printed edition of the *Canon*, Avicenna elaborated on the celestial origin of this form and its relation to the body's life and spirit.[31]

Avicenna's account of mixture sparked numerous comments in natural philosophy, mainly for its suggestion that the "fixed" forms of elements remained in actuality within the compound. Medieval and Renaissance philosophers debated one another over rivaling interpretations of mixture within an Aristotelian perspective. What was at stake was the persistence or destruction of the elemental forms and their qualities in the "mixt," namely the compound. As Anneliese Maier and Norma Emerton have pointed out, these debates triggered novel ideas concerning material change and propelled the debate towards a corpuscular redefinition of the elements in the early modern period.[32] Most importantly for our purposes, these interpretations of elements and mixture happened within a medical context that transmitted the Galenic description of elements as "minimal particles" (*minimae particulae*). This view continued to circulate in influential medical works such as the *Pantegni* (*editio princeps* 1515) and the *Canon* (*editio princeps* 1507). In the late Renaissance, these Galenic and Avicennian accounts of elements and mixture were widely diffused in a major work of medical philosophy, Jean Fernel's *Physiologia*.

3 The Twofold Constitution of Living Bodies

In early modern medicine, the French physician Jean Fernel was famous above all for his *Universa medicina* (1567), a systematic textbook of Galenic medicine, which was divided into *Physiologia*, *Pathologia* and *Therapeutices*.[33] However, it is mostly his *De abditis rerum causis* [*On the Hidden Causes of Things*] that

30 Ibid., 1.2.2.1.15, vol. 1, 111b: "Et sua quidem operans substantia, est illud, quod operator cum forma suae speciei operatur, quam acquisivit post complexionem, quod quum eius simplicia se commiscuerunt, et ex eis generate fuit res una; praeparavit se ad recipiendum speciem, et formam additam super illud, quod habent simplicia."

31 Avicenna, *De viribus cordis*, 1, in *Canon*, vol. 2, 334a–341a.

32 Maier, *An der Grenze*, 3–143; Emerton: *The Scientific Reinterpretation*, 76–105.

33 Fernel, *Universa medicina* (1567). I use the following editions of the *Physiologia* and *De abditis rerum causis*: Fernel, "The *Physiologia* of Jean Fernel (1567)," transl. Forrester; idem: *De abditis rerum causis*, in Jean Fernel, *On the Hidden Causes of Things*, ed. and transl. Henry and Forrester, 105–736.

has drawn the attention of the historians of science for its Platonic account of medical philosophy and its original explanation of epidemic diseases.[34] Nonetheless, Fernel was also innovative in his *Physiologia*, in which he introduced his Platonic views in a concise and elegant style. This treatise adopted the framework of the seven "natural things" that was developed by medieval medical textbooks such as the *Isagoge* of Johannitius, the *Pantegni*, and the *Canon*. In that context, the *Physiologia* addressed the issue of elements and mixture in order to explain the functioning of the body.[35]

3.1 *Elements as Contiguous Portions*

For Renaissance physicians like Fernel, the elemental composition of the body was a difficult academic subject. This question was related to the body's structure at the scale of its visible and invisible components. According to Aristotelian natural philosophy, living beings were organized into three main levels of body parts: described from the lowest stratum upwards, these were the elements, the homogeneous parts, and the organic parts. Looking from the highest stratum downwards, the living body appeared most readily to be made up of organic or "anhomeomerous" parts such as the limbs and organs. These consisted of homogeneous or "homeomerous" parts, such as flesh, bones, skin, muscles, and tissues, which were in turn composed of the four elements.[36] In addition, each organic part was defined as an instrument performing a physiological operation for the whole body. For instance, the body was the instrument of the soul, the eye that of vision. All organic parts required a good "conformation" in order to carry out all bodily operations. However, they lost their organic character when divided into their homeomerous parts, which were themselves subdivisible into elements. This scheme brought to the fore several questions. How was one to explain the resolution of living beings into elements? Conversely, how did the arrangement of elements result in an organism? Questions such as these validated Fernel's approach in presenting the body's composition starting from its simplest components.[37]

---

34 Hirai, *Medical Humanism*, 46–79; Blank, "Jean Fernel"; Deer Richardson, "The Generation of Disease."

35 See Moreau, "Elements, Mixture," 51–58.

36 On Aristotle's division of the body parts into homeomerous and anhomeomerous, see Aristotle, *History of Animals*, 1, 486a6–9; idem: *Parts of Animals*, 1, 640b; idem, *Generation of Animals*, 1, 715a1–15. On homeomerous parts, see Aristotle, *On Generation and Corruption*, 1, 321b16–25; idem, *Meteorology*, 4, 378b9–390b23; see also Berman: "Aristotle on Like-Partedness."

37 Fernel, *Physiologia*, 2.1, 182–187.

Fernel drew his theory of elements and mixture from Galen's *On the Elements According to Hippocrates*. Along the lines of this treatise, he rejected atomism, made a distinction between elements and qualities, and adopted the theory of humors.[38] He defined elements as the smallest parts of physical division and determined their number through four tangible qualities, organized by contrary pairs. For Fernel, these elements and their primary qualities constituted the human body. If physical pain proved that there was a multiplicity of basic elements, incineration showed that they were defined by the four qualities, while nutrition attested to the elemental nature of the four humors that were contained in blood.[39]

In his account of temperament, Fernel declared that the state of health resulted from the mixture of elements. His argumentation relied on ancient and medieval debates over rivaling models of mixture. Fernel followed Galen and Aristotle in rejecting the atomistic and corpuscular conceptions of mixture as a mere assemblage of discontinuous material units, such as *minima* and corpuscles.[40] In doing so, he explained Aristotle's refutation of juxtaposition as a response to models of mixture that were traditionally associated with atomistic interpretations. Most importantly, Fernel insisted that the compound (*mistum*) was made of elemental substances whose forms remained intact.[41] Whereas this view would imply an Avicennian interpretation of mixture, Fernel, as a Renaissance physician, was reluctant to acknowledge his intellectual debt to "Arabic" physicians, and more generally, to the scholastic tradition. Instead, he debunked competing models of mixture that were discussed in medieval texts. These models included the Thomist view on the destruction of the elemental forms in favor only of qualities, as well as the Averroistic interpretation of "broken" intermediate forms. As will be argued towards the end of this chapter, the Averroistic model was in fact much debated in late-Renaissance medicine.

Surprisingly, Fernel eventually embraced a conceptual mixture of the various models that he had previously refuted, while avoiding their respective aporias. At first, he preserved the Aristotelian definition of mixture. The qualities of the elements were subject to action and passion, while their forms

38 Ibid., 2.4, 194–199.

39 Ibid., 2.5, 199–201.

40 Ibid., 2.8, 210: "Atqui si hoc modo corporum substantiae totis immiscentur, fit saltem illorum ad usque minima divisio, sic ut huius minimum corpusculum, illius minimam particulam contingat: id certe indignum philosophia videri solet, credere aliquid esse in rerum natura minimum, quod dividi nequeat"; see Aristotle: *On Generation and Corruption*, 1.10, 327a30–328b25.

41 Fernel, *Physiologia*, 2.6 and 2.8, 200–204 and 210–211.

remained *in potentia* within a uniform compound.[42] However, Fernel used an ambiguous terminology concerning the status of the elements in the *mistum*. He defined them as "minute yet non-minimal" portions, and described their arrangement by using an oxymoronic formulation, speaking of a "continuous juxtaposition."[43] According to Norma Emerton, this interpretation corresponds to a "minimist" conception of the elements as had been adopted by the Italian physicians Julius Caesar Scaliger (1484–1558) and Girolamo Fracastoro (ca. 1478–1553).[44] In Emerton's analysis, minimist interpretations proposed a corpuscular understanding of mixture following Aristotle's *On Generation and Corruption*. The process of mixture operated a transient moment of union where elemental particles formed a continuum before receiving a form. For Emerton, this description was popular in Renaissance accounts of biological phenomena because it proposed a convincing model of mixture that counterbalanced the atomistic viewpoint.

In this context, it is important to mention that Renaissance physicians frequently viewed elements as particles and *minima* that were involved in the process of mixture. However, these particles should not be confused with the *minima naturalia* expounded in Aristotle's *Physics*. Such natural *minima* usually designated a lower limit of material division beyond which matter could no longer support a given substantial form.[45] As John Murdoch has pointed out, Aristotle's doctrine of mixture in *On Generation and Corruption* did not rely on *minima* (ἐλάχιστα), but merely on small parts. As a consequence, the particles of mixtures and the *minima naturalia* played different roles in the works of Aristotle. Nonetheless, a series of Renaissance Aristotelian philosophers from the school of Padua – including Agostino Nifo (ca. 1473–1538), Julius Caesar Scaliger (1484–1558) and Jacopo Zabarella (1532–1589) – merged

42 Ibid., 2.7, 206–207: "Neque vero potest ut illa vere misceantur, nisi contrariis inter se pugnent qualitatibus, per quas pariter agentia atque patientia mutationes conversionesque subeant [...]. Hae quidem nequaquam dissipatae aut extinctae oblitescunt, sed repressa duntaxat exuperantia ad moderationem mediocritatemque referuntur. Hinc corpus unum prodit consimile, quod re ipsa simpliciter unicum iam est: nam contraria quae in permistionem venerant non energia, id est efficacia vel actu, sed potentia duntaxat supersunt et manent."

43 Ibid., 2.8, 210–212: "Primum enim quatuor illae mundi simplices naturae [...] in exiguas non autem quam minimas portiones se distrahunt, eoque se ordine componunt, ut quaeque tandem alterius diversique generis cuipiam cohaerescat [...]. Hoc positu exiguae portiones, suam formam, qualem ante permistionem, integram quaeque retinent [...]. Quanquam igitur substantiae καθόλου temperari nequeunt, sed duntaxat continua appositione connecti: qualitatum tamen consummata est permistio."

44 Emerton, *The Scientific Reinterpretation*, 103.

45 Aristotle, *Physics* 1.4, 187b35–188a13. See Murdoch, "The Medieval and Renaissance Tradition."

the two concepts. Such an amalgamation of the natural *minima* with the small material chunks involved in mixture stimulated the emergence of Aristotelian corpuscularianism.[46]

As I have argued above, Galenic medicine was also involved in Renaissance conceptions of elements as *minima*. In *On the Elements According to Hippocrates*, Galen departed from Aristotle by defining elements as minimal parts (ἐλάχιστα). Whereas Fernel did not employ these *minima* in his writings, he did emphasize the smallness and contiguity of the elemental parts. In this way, he aligned his thinking with Avicenna's definition of complexion and mixture, most notably, by claiming that the elemental forms remained intact and that only their qualities were altered during mixture. It is for this reason that the German historian of atomism Kurd Lasswitz affiliated Fernel's theory of mixture in *Physiologia* to that of Avicenna and considered the French physician as one of the few atomist philosophers of his time.[47] Interestingly, this verdict echoes Scaliger's account of the Avicennian interpretation of mixture; the heap of discrete entities thus described were comparable with Democritean atoms insofar as the elemental forms remained "fixed" within the compound, and hence in actuality.[48]

Although it would be misleading to consider Fernel an atomist philosopher, his interpretation of mixture was remarkable for its suggestion that the elements were of a discontinuous nature, being contiguous minute portions. This description forms a stark contrast with his initial rejection of mixture as a juxtaposition of discrete *minima* and corpuscles in keeping with Aristotle's objections in *On Generation and Corruption*. This contrast probably explains why Fernel insisted so much on distinguishing the minute portions of mixture from *minima*. And yet, the terminology that he applies to mixture frequently invokes the term "aggregate" (*concretio*) with reference to the union of elements. As this term, in the Latin translation of Galen's *On the Elements According to Hippocrates*, designates the combination of atoms (σύγκριμα), we may legitimately point to a certain ambiguity on Fernel's part vis-à-vis the Galenic tradition.[49]

---

46 Lüthy, "An Aristotelian Watchdog."

47 Lasswitz, *Geschichte der Atomistik*, vol. 1, 452.

48 Scaliger, *Exotericarum exercitationum*, 34v; see Sakamoto, *Julius Caesar Scaliger*, 147.

49 On this point, see Riolan the Elder's criticism in his commentary on *Physiologia*: Riolan, *Commentarius ad Fernelii librum De elementis*, 1.7, in *Opera omnia*, 11: "Fernelius definit mixtionem, diversorum in unum et idem concretionem, quae definitio, ut dicam libere, mihi non placet, quia concretionis nomen in Democrito et Empedocle Aristoteles reprehendit, estque verbum methodicorum proprium authore Galeno, qui atomorum coagmentationem syncrima nuncupabant."

We have just seen that Fernel described mixture as an assemblage of minute portions whose intact forms were preserved in the *mistum*. What remains to be determined is his stance concerning the substantial form of the compound. In Fernel's theory, the *mistum* is in fact achieved by the introduction of a superior form once mixing has occurred.[50] This superior nature, it would seem, had an external origin, and this appears to point once more to Avicenna's conception of mixture. In his natural philosophy, Avicenna had stated that the temperate matter of the compound received a form that was introduced by a celestial instantiation or "giver of form." If Fernel were to follow this view, this would imply that it was the superior form, not the elements, that conferred the homogeneous and continuous character to the body, which was initially composed of contiguous portions. This is why we must turn now to an examination of Fernel's conception of the superior form.

3.2 *Innate Heat and the Substantial Form*

In Fernel's physiology, the status of the substantial form was closely connected to the constitution of the living body. Designated as the temperament of the whole body (*totius temperamentum*), this constitution arose from the mixture of all organic parts (limbs and organs). These were made of homeomerous parts (muscles, veins, arteries and other tissues), which in turn were the result of the successful mixture of the four elements. In theory, this scheme would entail that the temperament of the whole body was equal to the sum of the temperaments of all body parts. However, Fernel asserted that the simple addition of the temperaments of organic parts was equivalent to the temperament of a corpse, which was mainly composed of the element earth, cold and dry.[51] This observation followed from Aristotle's *On Longevity and Shortness of Life* wherein life was defined as hot and moist and death as cold and dry.[52] According to Fernel, the hot living part of the body corresponded to the vital

50 Fernel, *Physiologia*, 2.8, 212–213: "Haec autem forma seu perfectio sui ubique persimilis est adeo, ut non modo particulae quae sub aspectum recidunt, et quae ex tenuibus elementorum portionibus constant, sed et ipsa elementorum fragmenta, quae seorsum non aliis permista substantiis intelligimus, totius speciem gerant. Mistum enim iam est horum unumquodque, totiusque temperamentum accepit: nihil igitur prohibit quo minus composite totius species immigret in omnia."

51 Ibid., 3.7, 238–9: "Quandoquidem propria cuiusque partis temperatura quam quatuor elementorum permistio tulit, non alia inest viventi atque animali iam primum extincto. [...] Unde intelligitur viventis eum calorem qui per corpus diffusus omnia regit ac moderatur, et cuius appulsu vita inest, cuius excessu mors irruit, a partium temperamentis differre, neque ab his proficisci."

52 Aristotle, *On Longevity and Shortness of Life*, 466a18–23.

heat. The temperament of the whole body therefore needed to include vital heat in addition to the body's elemental core.

In Renaissance medicine, the debates on vital heat were centered on an ambiguous passage in Aristotle's *Generation of Animals*. It stated that the vital heat transmitted by the seed was not composed of the element fire but of spirit (*pneuma*), which was "analogous" to the element of the stars.[53] In their commentaries on this passage, Renaissance physicians sought to clarify the non-elemental and quasi-celestial status of the spirit that composed vital heat. Renowned for his Platonic interpretation of this topic, Fernel affirmed the celestial origin of innate heat (*calor innatus*) as a vital principle. Nonetheless, his understanding of this vital principle remained anchored in the Aristotelian and Galenic traditions, as it merged Aristotle's definition of life with Galen's theory of vital heat and spirit (*pneuma*).

Faithful to this Galenic and Aristotelian framework, Fernel defined vital heat as a physiological principle that was related to the soul. According to the medical tradition, this hot entity animated bodies and operated physiological functions such as reproduction, growth and nutrition.[54] Although vital heat was traditionally considered to be composed of fire and air, Fernel pointed to its supra-elemental nature whose origin was "hidden elsewhere." To support his claim, he took the example of plants as qualitatively cold bodies which, nonetheless, were animate and hence possessed of vital heat. For Fernel, the supra-elemental nature of vital heat came from its conveyor, the spirit (*spiritus*), which served as an equivalent of the Galenic notion of *pneuma*.[55] *Spiritus* played a key role in physiological functions, as it assumed the functions of the vegetative or "natural" soul, which was common to all living beings. Being the seat of vital heat and the instrument of the soul, the spirit was essential for animating the living body. Maintained by the inspiration of air, it circulated in the arteries but vanished at the time of death, leaving the arteries empty. As Fernel underlined, this fine entity was attached to the body throughout its lifespan. For its actual origin, one had to look to aether, the fifth celestial element that, according to Aristotle's *On the Heavens*, was to be found in the superlunary world.[56]

As Hiro Hirai has shown, Fernel adopted a Platonic interpretation of the relationship between the celestial spirit and the soul.[57] The celestial spirit was

53 Idem, *Generation of Animals*, 2.3, 736b34–737a2. See Walker, "The Astral Body."

54 Fernel, *Physiologia*, 4.1, 256–261; see Freudenthal, *Aristotle's Theory of Material Substance*; Debru, *Le corps respirant*.

55 Fernel, *Physiologia*, 4.2, 260–261.

56 Aristotle, *On the Heavens*, 1.3, 270b20–25. See Solmsen, "The Vital Heat."

57 See Hirai, "Humanisme, néoplatonisme et *prisca theologia*."

a celestial garment that was pure, immortal and eternal, as it covered the "thick and condensed" part of the mortal body. The material body, in turn, provided an elemental dimension to the spirit. Consequently, the spirit of the living body had a less pure nature than the celestial and divine spirit from which it proceeded. In its terrestrial existence, the spirit was "inserted" (*insitus*) in the living body as if "banished in a black and dark prison," waiting to be released to return to his celestial home at the time of death.[58] In support of this duality of spirit as both celestial and corporeal, Fernel quoted Alexander of Aphrodisias's *Problemata*, declaring that the spirit played the role of a bond (*vinculum*) between the celestial and terrestrial worlds within the human body.[59] As Hirai has pointed out, Fernel borrowed this reference from an important Platonic work, Marsilio Ficino's *Theologia Platonica* (1482).[60] This conception of the *vinculum* buttressed Fernel's description of the spirit as both "incorporeal and corporeal," "immortal and mortal," "divine and terrestrial." With this interpretation, he sought to have clarified Aristotle's description of the spirit as an entity analogous to the element of the stars, while stressing its intermediate status between a divine and terrestrial nature.

In Fernel's medical philosophy, the vital heat and the celestial spirit were closely connected to the superior form of living beings. His explanation of this topic is found in *On the Hidden Causes of Things*, a treatise that he had composed before the first edition of the *Physiologia* (1542) but only published in 1548. There, he asserted that the body's substantial form had a celestial and divine origin and was therefore different from the mixture of the elements.[61] This form was transmitted by the spirit inserted into the body, which carried the vital heat. Proper to all living beings, this innate heat (*calidum innatum*) stemmed from a celestial entity, the world-soul (*anima mundi*). The latter acted as the giver of forms conveyed by the world-spirit (*spiritus mundi*). As Hirai has shown, this ontology relied on Ficino's Platonic philosophy, which Fernel further developed in a medical context by combining it with Galen and Aristotle.

This Platonic account of innate heat and the substantial form further elucidates Fernel's considerations on the temperament of living beings. In his view, this temperament resulted from the introduction of the superior form into

58 Fernel, *Physiologia*, 4.2, 262: "Duobus hisce corporibus iam stipatus animus, in tertium hoc mortale caducumque corpus, seu potius in tetrum et tenebricosum carcerem tanquam exul deiectus, terrarium fit hospes, donec effracto carcere alacer et liber in patriam reversus, municeps fiat et civis deorum."

59 Ibid., 262–265.

60 See Hirai, "Humanisme"; Fernel, *On the Hidden Causes*, 2.7, 480–481.

61 Fernel, *On the Hidden Causes*, 2.18, 698–699; see Hirai, *Medical Humanism*, 46–79; Blank, "Jean Fernel," 9–21.

a temperate body, which, as we have explained, was made up of contiguous elemental portions. Innate heat corresponded to the formal part of the living body, which was of divine origin and which supervised the vital functions of the material organism. Just as the superior form achieved the mixture of elements, innate heat animated the whole body and completed its temperament.

In merging the Ficinian and Avicennian interpretations of the living body and its connection to the heavens, Fernel's account of the temperament was widely discussed in early modern medicine. As we will see in our next sections, Fernel bequeathed two major ideas to medical philosophy: the celestial and divine origin of life, and the composition of matter in discrete elemental portions.

4 Elemental Bodies and the Quintessence

In considering the early modern discussions on the elements of the body, we must turn now to chymical medicine and one of its chief actors, the German physician Andreas Libavius. While promoting chymistry as an academic discipline to be grounded in medicine and natural philosophy, most of his works were conceived in opposition to publications by Paracelsian philosophers. It was in one of these polemical treatises, *Novus de medicina veterum tam Hippocratica quam Hermetica tractatus* [*New Treatise on the Hippocratic and Hermetic Medicine of the Ancients*, 1599] that Libavius expounded his medical theory of the elements, mixture and the temperament.[62] This treatise took aim at the *Apologia chymica* [*Chymical Apology*, 1597] of the Italian physician Giuseppe Micheli or Michelius, a Reformed Paracelsian scholar who had settled in the Dutch Republic.[63] In his *Apologia*, Michelius had criticized some letters that Libavius had published in *Rerum chymicarum epistolica forma* [*Epistolary Form of Chymical Things*, 1595]. Libavius, in turn, replied to this criticism in his *Novus de medicina tractatus*, in which he systematically went about dismantling the argument offered in the *Apologia*.

Apart from its polemical intention, Libavius's treatise aimed to demonstrate the historical compatibility between alchemy and the authority of Aristotle and Galen, authorities harshly criticized by the new Paracelsian philosophy. In keeping with this stance, his *Novus de medicina tractatus* sought to merge the precepts of chymistry with those of learned medicine. Libavius justified

62 Libavius, *Novus de medicina tractatus*. On this treatise, see Moran, *Andreas Libavius*, 43–49; Kahn, *Alchimie et paracelsisme*, 354–357; Moreau, "Reforming the *Prisca Medicina*."

63 Michelius, *Apologia chymica*.

this compromise as an endeavor to improve therapy by means of chymical remedies.[64] On the basis of recipes attributed to late-medieval alchemists such as Ramon Lull and Arnald of Villanova, he expounded upon this therapeutic objective in the second part of the treatise.[65]

In this context, Libavius's theory of the elements and the temperament was primarily a reaction against the Paracelsian criticism of the Galenic tradition. His treatise *Novus de medicina tractatus* targeted a widely-read digest of Paracelsian medicine, the *Idea medicinae philosophicae* [*Idea of Philosophical Medicine*, 1571] by the Danish physician Petrus Severinus (1540–1602).[66] For Severinus, the elements were only the envelopes and "rudiments" of things, whose function it was to foster seeds of celestial origin. As the foundations of nature, these seeds provided the chymical and vital properties of bodies. To support this claim, Severinus took up the Paracelsian distinction between the fertile powers of the celestial seeds (*dynameis*) and the sterile (*relollacea*) qualities of the material elements. Disparaging the qualities traditionally associated with the elements, Severinus objected to the Galenic conception of the temperament. In his view, Galenic medicine was a "corporeal philosophy" for its emphasis on the material components of bodies, that is to say, on the elements.[67] Following this reasoning, Severinus substituted the chymical powers of the three principles (*tria prima*), which originated from the celestial seeds, for the qualities of the four elements.

In response to this Paracelsian criticism of the elements and the primary qualities, Libavius argued that the elements were the first constituents of bodies, from which stemmed the chymical principles. Nonetheless, the relationship between the elements and the chymical properties of bodies remained central for him. Indeed, Libavius strove to reunite the Galenic conception of elements and temperament with the chymical principles. To this purpose, he took the perspective of medieval alchemy and Aristotelian natural philosophy.

4.1 *The First Compounds of Elements*

Libavius's strategy consisted in highlighting discrepancies in the interpretation of the elements and the *tria prima* in the corpus of texts attributed to Paracelsus himself.[68] Paracelsus was of course well known for his account of the *tria prima*

64 Debus, *Science, Medicine and Society*, 151–165.

65 Libavius, *Medicinae hermeticae expositio fidelis*. The treatise consists of a commentary on pseudo-Lull's *Clavicula* and pseudo-Villanova's *Rosarium philosophorum*.

66 Severinus, *Idea medicinae philosophicae*. On Severinus, see Shackelford, *A Philosophical Path for Paracelsian Medicine*, and Shackelford's chapter in the present book.

67 Severinus, *Idea*, 65 and 134–135.

68 For the works of Paracelsus, I use Paracelsus, *Bücher und Schrifften*, ed. Huser.

(Salt, Sulfur and Mercury) as the first components of bodies, from which the four elements were derived. However, as Libavius pointed out, Paracelsus had regularly acknowledged that the elements preceded the *tria prima* in the order of divine creation, for instance in the (apocryphal) treatise *De pestilitate* [*On Pestilence*].[69] For Libavius, the key to solving this inconsistency laid in another treatise by Paracelsus, *De primis tribus essentiis* [*On the First Three Essences*]. Although this text advocated the resolution of elements into the *tria prima*, it distinguished between common and prior elements.[70] This point was crucial for Libavius, because it suggested that Paracelsus conceded the difference between pure elements in their natural places and elements that composed bodies. Rooted in Aristotelian physics, the distinction between pure elements and elemental bodies allowed Libavius to identify the latter with the *tria prima* of the Paracelsian philosophers. Constituting the first compounds of elements, such elemental bodies (*elementata*) possessed a powerful essence, which the chymists attempted to separate from their elemental core.[71]

For his view of the *tria prima* as the first compounds of elements, Libavius could draw upon his sources on late medieval alchemy. These treatises, indeed, established a distinction between elements (*elementa*) and elemental bodies (*elementata*). For instance, the *Rosarium philosophorum* [*Rosary of philosophers*, 1504], a collection of alchemical texts attributed to Arnald of Villanova (fourteenth century), defined the element as the first body subject to composition. Earth, water, air and fire, in their natural state, were not pure and simple elements. They were always encountered as blended, one with another or others, into elemental bodies (*elementata*).[72] Another source for Libavius, the *Correctorium alchemiae*, an alchemical treatise attributed to the English

69 Libavius, *Novus de medicina tractatus*, 123: "Sed chymici veteres omnes mercurio elementa priora faciunt, [...] (quod et Paracelsus in libro de pestilitate repetit)"; see Paracelsus, *De pestilitate*, III, 30.

70 Libavius, *Novus de medicina tractatus*, 123: "Paracelsus vulgaria elementa iterum resolvit in elementa, quod fortasse sic aestimaret ex veterum sententia nullum elementum circa nos in nostro loco esse purum, hoc est, omnia vicissim composita esse ex elementis prioribus. In libro de essentiis tribus, quicquid ex elementis productum est ex tribus esse putat, quae principia vocat"; see Paracelsus, *De primis tribus essentiis/Von den ersten dryen Essentiis*, III, 15–23.

71 Libavius, *Novus de medicina tractatus*, 42: "Elementatum ex elementis tanquam membris constat. Id Paracelsitae, maximeque Severinus iudicant in se comprehendere essentiam ex principiis, (quae tria fingunt, sulphur scilicet, salem, et mercurium) ortam."

72 Anon.: *Rosarium philosophorum* (Frankfurt, 1550), fol. g2v; see Calvet, "Étude d'un texte latin"; Telle, "Remarques sur le *Rosarium philosophorum*."

physician Ricardus Anglicus (fourteenth century), distinguished the common elemental entities from the four elements.[73]

The notion of elemental body (*elementatum*) could also be taken to refer to the medieval Platonic tradition as transmitted by the French philosopher William of Conches in *De philosophia mundi* (ca. 1129). This school of thought defined *elementata* as those elemental bodies that were perceptible in the physical world. By contrast, the *elementa* were purely theoretical entities, only accessible to the mind.[74] At the cosmogonic stage, the formation of the elemental compounds followed upon the biblical chaos. This process caused the creation first of elements through to more complex bodies, which were first elemental and then organic. Remarkably enough, this explanation overlapped with medical considerations on the body parts as they were being developed in the *Pantegni* during the same time period. According to this view, the body consisted of elements (*elementa*), homeomerous parts, which were homologous with the *elementata*, and organic parts.

In accordance with his definition of the *tria prima* as elemental bodies, Libavius took the chymical distinction between impure elements and pure principles to be conventional and related to a practical context. Indeed, it was only by analogy that the *tria prima* were named "principles," since the principles of art differed from those of nature.[75] However, Paracelsus and his disciples had misunderstood the analogical dimension of the terms "principles" and "essences." As Libavius explained, the ancient alchemists had used this terminology in a merely experimental framework. Being chymical essences extracted by distillation, the *tria prima* were elemental compounds and hence subject to the rules of Aristotelian physics, which therefore had to obey the "axioms" of elements and mixture.[76]

For Libavius, then, the chymical principles were composed of elements and characterized by primary qualities. Mercury was cold and moist and composed

---

73 Ricardus Anglicus, *Correctorium alchemiae*, in Geber, *De alchemia*, 302. See Telle, "Ricardus Anglicus."

74 Silverstein, "*Elementatum*"; idem: "Guillaume de Conches"; Jacquart, "*Minimum*"; Elford, "William of Conches."

75 Libavius, *Novus de medicina tractatus*, 119: "Scimus autem artem etiam elementa exhibere, quae non raro cum principiis coincidunt: sed tamen itam cum analogia ad plane externa. […] Est magisterium elementorum in chymia; est principiorum. […] Nec elementa chymica sunt ultima compositionis. Sunt enim duntaxat artis, et ex analogia nomen acceperunt."

76 Ibid., 47–48: "Quod autem attinet ista tria, patior quidem chymicos intra artis suae septa manentes ita symbolice loqui. Sed si universalem physicam spectes et ista succumbent mistionum elementorumque axiomatis. Ita Michelius per mercurium intelligit aquam; per sulphur, oleum; per salem, terram."

of water, Sulfur was hot and composed of air and fire, and Salt was hot and dry and composed of earth.[77] As for the sensory properties of the *tria prima* – for instance, the thickness of Sulfur – they all derived from the secondary qualities of the elements.[78] As Libavius recalled, this had already been understood by early alchemists. Bernardus Trevisanus made mention of four efficacious qualities, while Ricardus Anglicus emphasized the primary qualities in the composition of Sulfur and the formation of diseases.[79]

To uphold the precedence of the elements over the *tria prima* in the physical order of things, Libavius eventually referred to the Scriptures. He specifically recalled that the Book of Genesis mentioned the creation of Adam from a "viscous slime."[80] This primordial material had to be understood as a compound of water and earth, into which God infused a seminal power.[81] However, the divine nature of this power was restricted to the episode of Creation, when it was first introduced into the elements by the divine breath. Subsequently, it became immanent in elemental bodies as first compounds (*prima mista*), which manifested themselves in physiological functions and chymical phenomena.

Consequently, Libavius did not deny the existence of seminal powers but claimed their enclosure in elemental bodies that derived from the primordial slime of water and earth. Since these seminal powers were involved in the formation of elemental, that is, homeomerous bodies, they obeyed the rules of Aristotelian physics. On this point, Aristotle's *Meteorology* was a particularly convenient source for Libavius because it discussed the homeomerous parts of

77 Ibid., 45: "Cum qua [mercurius] necessario est humiditas. Dicitur enim aqua essentialis esse. Sulphur, oleum et aeream igneamque partem nominant. Calidum ergo. Sal terra est, et simul calidus siccus."

78 Ibid., 119: "Oleum, seu sulphur coagulat ex liquido. [...] necesse est, duos humores in eo fuisse, liquidum, et crassum. Crassum vero secundarum qualitatum e mistione est."

79 Ibid., 108 and 44; see Bernardus Trevisanus: *De chymico miraculo*, 27; Ricardus Anglicus: *Correctorium*, 288–294. On Bernardus Trevisanus, see Kahn, "Recherche sur le livre attribué au prétendu Bernard le Trévisan."

80 Libavius, *Novus de medicina tractatus*, 120: "Haec sunt ultima elementa, idque tam evidens est, ut pluribus oraculis crebro inculcetur [...]. Videmus item etiamnum, hodie ex aqua et terra fieri limum viscidum, et hunc conglutinari, aut concrescere in lapidem. Si hic resolvitur in ultima; non in mercurium, sulphur et salem, nisi haec sint elementorum symbola, sed in elementa vulgata solvetur." See Genesis 2:6–7.

81 Libavius, *Novus de medicina tractatus*, 48: "Non fugit nos, autorem creaturae initio miscuisse contemperasseque ista prima et postea efficaciam prolificam seminariamque inspirasse. Sed tamen eius rei vestigia videmus in ruditer compositis, in imperfecte mistis, in resolutionibus, in nutritione et augmentatione, in vita inter elementa. Nemo est, quin intelligat hominis corpus ex semine et sanguine agente interno principio ad similitudinem generantis, post primum Adamum effectum esse."

natural bodies such as flesh, bones and metals.[82] All these materials resulted from the mixture of elements and were predominantly composed of water and earth, which allowed them to coagulate.[83] This Aristotelian subtext buttressed Libavius's understanding of elemental bodies as first compounds enclosing seminal powers from physiological, chymical and cosmogonical viewpoints.[84]

4.2 *Temperament and Seminal Powers*

In response to the Paracelsian denigration of primary qualities as "mortal shadows of things," Libavius sought to restore the concept of temperament as the mixture of qualities. In his *Novus de medicina tractatus*, he explored this theme through a detailed assessment of the Galenic doctrine of *krasis*. Following Galen's *On the Usefulness of the Parts*, Libavius stated that the temperament was related to particles resulting from the mixture of primary qualities.[85] These particles corresponded to the homeomerous parts of the body, such as bones, skin, veins and tissues. Libavius pointed out that Galen, however, had not clarified the relationship between *krasis* and the substantial form of the body, nor had he specified whether mixture related to the corporeal parts of elements. To elucidate these questions, Libavius broached the nature of the substantial form of the body.

Like Fernel, Libavius considered the substantial form according to a twofold conception of temperament. While the material side of *krasis* resulted from the mixture of the elements and the balance of primary qualities, its formal side was related to the supra-elemental form of the *mistum*. For Libavius, this superior form derived from the propagation of the seeds that had been introduced into the elements during the divine act of creation.[86] God had in the very beginning created elemental compounds and inserted seminal powers into them. Through their seeds, living beings engendered similar beings, following

82 See Newman, *Atoms and Alchemy*, 66–84; Martin, *Renaissance Meteorology*.

83 Aristotle, *Meteorology*, 4, 384b24–384b33 and 388a10–388a13.

84 See Crisciani, "Il corpo nella tradizione alchemica." Obrist, "Les rapports d'analogie"; Taub, "Physiological Analogies and Metaphors."

85 Libavius, *Novus de medicina tractatus*, 94–95: "De Galeno, quod temperamentum definiverit formam rerum substantialem, oblivione peccasti. Non enim rerum formam esse crasin scripsit, quam similaribus mistis duntaxat assignavit. [...] Patet ex hoc non nudas qualitates merasque intelligi, sed una comprehendi corpora simplicia, quae ad mistionem concurrunt [...], quanquam [...] nolit determinare utrum totae qualitates totis misceantur tantum, an etiam corporeae partes."

86 Ibid., 38: "Ego soleo dupliciter de crasi loqui. Intelligo enim interdum primarum qualitatum conspirationem, qua aliquid calidum, humidum, frigidum, siccum, etc. dicitur: interdum formam misti, qua mistum est, licet substantiam habeat non ex concursu elementorum, qualis fit, cum ex aqua et terra nascitur lapis, sed ex seminaria propagatione instituta in elementis et ex materia universi a Creatore."

the divine blessing in Genesis, the injunction to be fruitful and multiply.[87] These seminal powers now resided in the chymical essence of any given body, an essence that Libavius identified with the innate heat. In other words, the elemental core of the body encloses a seminal principle of divine origin, which supervises the physiological functions of reproduction and nutrition. Within the body, this seminal power acts in the blood and the seed, guaranteeing the perpetuation of individuals and of their species.

Libavius also understood the superior form of living beings within a chymical framework. The substantial form of the body was, indeed, related to the "quintessence."[88] As posited in Lull's alchemical *Testamentum* (fourteenth century), this quintessence was a spiritual entity that was intermediate between body and soul.[89] John of Rupescissa (ca. 1310–ca. 1370) had subsequently added a pharmacological dimension to this notion in *De consideratione quintae essentiae* [*Consideration on the Fifth Essence*].[90] According to this treatise, the quintessence of bodies could be extracted by distillation for medicinal purposes. Libavius, in turn, viewed quintessence as a celestial entity that was included in the elements and was associated with the vital principle of bodies, that is, the innate heat. On this point, he referred to the distinction in the *Rosarium* between *elementum*, *elementatum* and *quinta essentia*.[91] Although the quintessence was an incorruptible entity distinct from the elements, it could yet be extracted from the elemental bodies. Nonetheless, Libavius recalled that early alchemists had been aware that the notion of quintessence as a product of distillation referred to the practical art. It was only by analogy with the divine creation that it applied to the celestial essence.[92]

---

87 Ibid., 107: "Sane ita Deus ex elementis constituit mista, iisque insevit seminarias rationes, iuxta quas unumquodque produceret suum semen et gigneret simile, sicut in animalibus vox oraculi iubet ea crescere et multiplicari. Illae rationes seminariae in ea parte sunt, quam chymici essentiam vocant, Galeni succum, aut calidum innatum substantiale"; see Genesis 1:28.

88 Libavius, *Novus de medicina tractatus*, 80: "Sed quinta essentia est corpus per se subsistens, differens ab omnibus elementis [...]. Haec [...] Galenici sciunt ab Aristotele, Galeno caeterisque eadem scribi de tota natura, et calidi innati substantia [...]. Aliqui istam naturam coelitus deduxerunt. Sed rectius semineo tribuitur principio, nec tam est aliena ab elementis, quin in eis conservetur, imo initio creationis etiam ex iisdem sit concinnata, accedente divina virtute in eis instituta."

89 Pereira, "Heavens on Earth."

90 Johannes de Rupescissa, *De consideratione quintae essentiae rerum*, 15–21. On Rupescissa, see DeVun, *Prophecy, Alchemy, and the End of Time*; Halleux, "Les ouvrages alchimiques"; Multhauf, "John of Rupescissa."

91 Libavius, *Novus de medicina tractatus*, 80; see s.n., *Rosarium*, fol. g2v.

92 Libavius, *Novus de medicina tractatus*, 43: "Nam et veteres mentionem faciunt mysterii seu arcani, magisterii, quintae essentiae, et similium. Sed in sua arte permanserunt, nec nisi analogia quadam ad explicationes naturales, quatenus arti inservirent suae,

Here again, Libavius attempted to embed the Paracelsian explanation of seminal powers into a traditional framework. With his account of *krasis* and the quintessence, he sought to provide a convincing alternative to the Paracelsian conception of chymical properties by showing their connection with Galenic medicine and the Scriptures. Encompassing the narrative of divine creation in Genesis and Fernel's Platonic emphasis on the celestial form of living beings, Libavius put forward an interpretation of temperament that was faithful to the ancient knowledge (*prisca sapientia*) of alchemists. Strikingly, this account of temperament also emphasized the notion of elemental bodies as "particles" that functioned as homeomerous body parts, chymical principles, and first compounds. As will be shown in our next section, Daniel Sennert was to continue this medical discussion of elemental bodies and their properties within a chymical philosophy that was imbued with atomism.

5 Minima, Atoms and the Superior Form

Daniel Sennert has been the subject of numerous studies in the history of early modern medicine and natural philosophy. Centered on the notions of matter-form, atoms and seeds, his theory of material change was rooted in chymical practice and in Aristotelian natural philosophy.[93] In this section, I will consider two treatises in which Sennert expounded his views on elements, mixture and temperament: the second edition of the *Institutionum medicinae libri V* [*The Five Books of Institutions of Medicine*] (1620) and the second edition of *De chymicorum cum Aristotelicis et Galenicis consensu ac dissensu liber* [*Book on the Chymists' Agreement and Disagreement with the Aristotelians and Galenists*] (1629). The five books of the *Institutiones* proposed a systematic and didactic account of medicine. Dedicated to physiology, the first book explained the body's structure and functioning according to a Galenic framework. As for *De chymicorum consensu ac dissensu*, it proposed a critical appraisal of Paracelsian philosophy and sought to reconcile its most contentious aspects with the medical tradition. On this point, the arguments invoked by Paracelsus's followers such as Petrus Severinus, as well as by his detractors such as Andreas Libavius, were important sources for Sennert's exposition of chymistry and medicine.

accommodarunt. Ab illis acceperunt postea Lullius, Ulstadius, Thomas, Arnoldus, et reliqui recentiores."

93 Michael, "Sennert's Sea Change"; eadem, "Daniel Sennert on Matter and Form."

Like Libavius, Sennert was representative of late-Renaissance physicians who wished to introduce chymistry into the medical curriculum. In his view, understanding chymical properties and the extraction of chymical principles by "separation" were fundamental for making powerful remedies. To integrate this chymical art in learned medicine, Sennert believed that its theoretical foundations needed to be carefully examined in the light of the canonical theories of matter-form, mixture and temperament. I aim to show that in this endeavor, Sennert's medical viewpoint followed Fernel's and Libavius's interpretations, which he combined with his own philosophical inclinations.[94] The Aristotelian influence in his work is often expressed through the doctrines of Julius Caesar Scaliger, expounded primarily in *Exercitationes* (1557) and *De plantis* (1556).

5.1 *Natural and Essential Constitutions*

In the *Institutiones*, Sennert endorsed Galen's definition of health as the ability to correctly carry out body functions at the level of the organic and homeomerous parts. He defined the latter as the simplest "elements" of the body that were accessible to the senses. These compounds (*mista*) formed particles (*particulae*), which were the "proximate principles" (*proxima principia*) of bodies.[95] Although they had an elemental constitution, the homeomerous particles were endowed with specific properties that were related to their substantial form and the innate heat.

Composed of organic and homeomerous parts, the living body was animated by the soul through the innate heat. In keeping with this view, Sennert endorsed the explanation that Fernel had given of temperament in *Physiologia*. The temperament of living beings comprised the elemental constitution of the body along with innate heat as the vital principle and instrument of the soul. As Sennert explained, the elemental nature of the body corresponded to the "natural" constitution of the body, while its living nature constituted its "essential" constitution.[96] Because it depended upon the body's superior form and

94 On Sennert's atomistic conception of the temperament, see Moreau, "Atoms, Mixture, and Temperament."

95 Sennert, *Institutionum*, 1.3, 10b: "Verum cum, [...] corpora nostra non simpliciter et proxime, ut mista, ex elementis constent, sed sua proxima principia habeant, et calore quodam vitali, spiritu vivifico [...] praedita sint, et formam suam atque animam obtineant, a quibus multae nobiles qualitates fluunt." Cf. Scaliger, *Exotericarum*, 218.8, 291r.

96 Sennert, *Institutionum*, 1.3, 8b: "Est autem partis cuiusque constitutio duplex; una essentialis, altera accidentalis. Essentialis est, quae pendet ex materia et forma substantiali, a quibus compositum suam naturam et essentiam habet. Accidentalis vero est, quae illam priorem insequitur, et qualitatum accidentiumque aliorum in singulis partibus est dispositio, et proximum organum, per quod forma essentialis est efficax."

soul, the essential constitution was unalterable so long as the body was alive. By contrast, the natural constitution was due to the body's qualitative balance, and hence it was variable and responsible for alterations in health.[97]

Sennert's distinction between essential and formal constitutions had important consequences for his approach to the temperament. In his view, physicians should follow the principles of natural philosophy, which distinguished substantial form from elemental qualities.[98] Following this distinction, the temperament may vary according to the elemental qualities, which have an accidental status. Its balance results from the battle between qualities, which takes place through a mutual interaction of action and passion, and from the union of elemental substances.[99] Sennert grounded this interpretation in Avicenna's *Canon*, from which he took the notion that the temperament was a quality stemming from the mutual action and passion of the initial qualities, whereby the elements were broken down into minimal parts (*partes minimas*).[100]

When it came to the status of the elements in the formation of the temperament, Sennert, at the time of the publication of the *Institutiones* (1611 and 1620), advocated an Averroistic interpretation of mixture. This stance not only involved the fragmentation of the elemental qualities but also of their forms. In the Renaissance, such a "pluralist" Aristotelian model had been popular among philosophers at the University of Padua, who endorsed the breaking (*refractio*) of the form into a hierarchy of subaltern forms. On this point, Sennert relied specifically on Zabarella's explanation of mixture in *De rebus naturalibus* [*On Natural Things*, 1590]. This treatise described the reduction of elements into small parts of various degrees, which then joined into a homogeneous and uniform body.[101] Their reduction into small parts was synonymous

97 Ibid., 9a: "Itaque in accidentali potius constitutione sanitas collocanda est. Haec enim mutabilis est, et, essentiali integra manente, varias mutationes subit."

98 Ibid., 1.4, 14b: "Temperamentum ad solas qualitates; mistio vero ad ipsorum elementorum essentiam extendenda est: cum formae elementorum substantiales, et qualitates, ut ex philosophia naturali patet, revera genera dissideant."

99 Ibid., 16b: "Sed tum solum fit vera mixtio, ubi mutua qualitatum primarum fit actio et passio, qua pugna et elementorum substantiae uniuntur et qualitates ad quandam concordiam deveniunt." See Scaliger, *Exotericarum exercitationum*, ex. 101, 143v; see also Sakamoto, *Julius Caesar Scaliger*, 145–163; Lüthy, "An Aristotelian Watchdog," 542–561.

100 Sennert, *Institutionum*, 1.4, 15a: "Aliorum autem opinionibus missis, cum Avicenna statuimus, [...] temperamentum est qualitas, quae ex mutua actione et passione contrariarum qualitatum elementorum, quorum partes ita ad minimas sunt redactae, ut cuiusque earum plurimum contingat plurium alterius, provenit"; see Avicenna, *Canon*, 1.1.3.1, vol. 1, 11ab.

101 Michael, "Daniel Sennert on Matter and Form."

with the tearing apart of forms, which united into a plurality of subordinate forms so as to constitute the median form of the *mistum*. Whereas Zabarella viewed these elemental fragments as infinitely divisible, Sennert suggested that these elemental particles were merely united as contiguous, not continuous, *minima*. This reasoning followed Scaliger's definition of mixture as the motion of *minima* towards a mutual contact.[102] Later on, in *De chymicorum consensu ac dissensu*, Sennert honed his account of the elements in the context of his conciliation of the medical tradition with Paracelsian philosophy.

5.2 *The Tria Prima as Perfect Mixtures*

Although Sennert insisted on the primacy of the elements as the first components of beings, he yet acknowledged the special status of the *tria prima*. He reported how the Paracelsian physicians Severinus and Joseph Du Chesne had established that the *tria prima* were endowed with active powers, mostly sensory properties.[103] In this regard, Salt was associated with solidity, colors and flavors, Sulfur with oiliness, colors and smells, and Mercury with fluidity. Moreover, the three principles displayed "occult" properties in physiological phenomena like temperament and digestion. What remained to be clarified, for Sennert, was the status of the *tria prima* in the composition of natural things.

Sennert's interpretation of the *tria prima* was much indebted to Libavius's conception of elemental bodies and seminal powers. The three principles were described as the first compounds (*prima mista*) of animals, plants and metals, into which God had infused seminal powers at creation.[104] Endowed with a superior form, they therefore participated in the constitution of natural beings that were subject to generation and corruption.[105] For Sennert the supra-elemental nature of their substantial form was corroborated by their status as perfect mixtures. He buttressed this view by invoking Scaliger's assertion that the form of any achieved mixture was a "fifth nature," different from the

102 Scaliger, *Exotericarum exercitationum*, ex.101, 143v; idem: *In libros duos de plantis*, 67.

103 Sennert, *De chymicorum consensu ac dissensu*, 11, 127a–135a; see Severinus, *Idea*, 67–68; Quercetanus, *De priscorum philosophorum verae medicinae materia*, 89–130.

104 Sennert, *De chymicorum consensu ac dissensu*, 11, 137a: "Esse illa [principia] non corpora quaedam ante elementa simplicia, ut modo dictum, sed vel prima mista, vel peculiares naturas in prima creatione, cum aliis rerum naturalium speciebus conditas, seu principia sui generis, quae cum inter se, tum etiam cum elementis, vario modo mista, aliis rebus naturalibus, generationi et corruptioni obnoxiis, metallis pura, mineralibus, gemmis, lapidibus, plantis et animalibus, dant convenientem materiam."

105 Ibid., 137b: "Sed peculiares formas habere; et suam cuique formam a Deo Omnipotente in prima creatione, ut aliis rerum naturalium speciebus, datam esse, existimo."

four elements.[106] Accordingly, Sennert viewed the *tria prima* as essential principles, namely first compounds possessing a higher nature.[107] For, indeed, they had a higher degree of perfection than the elements.[108] Their "nobler" matter differed from the immanent (*immanens*) or materialized (*materiata*) matter of the elements, because it was transient (*transiens*). During transmutation, it acquired a new substantial form, in the same way as blood during digestion.[109]

For Sennert, it was the ancient philosophers' ignorance of divine creation that explained the absence of the three principles from the nomenclature of their *primordia*, which was uniquely based on elements and primary qualities. Still, Hippocrates had attempted to relate particular properties such as taste to the notion of power (*dynamis*). Sennert further pointed out that Paracelsian philosophers like Severinus associated the Hippocratic notion of *dynamis* with the active powers of bodies, which they opposed to the sterile qualities enhanced by the Galenic tradition. Following Scaliger, Sennert believed that such powers, including taste, color and smell, resulted from the superior form of compounds.[110]

While associating the chymical properties of the *tria prima* with the active powers of the superior form, Sennert also contemplated their presence in living bodies. To this purpose, he invoked the topos of burning green wood so as to illustrate the separation of bodies into their ultimate components.[111] He borrowed this example from Fernel's *Physiologia*, which describes four phases in the combustion of green wood: volatile, oily, aqueous, and solid.[112] Where Fernel had still followed the medical tradition by relating these phases to the

106 Scaliger, *Exotericarum exercitationum*, ex. 307.20, 405r: "Omnem formam cuiuscunque perfecti misti, etiamsi non est anima, ut in adamante, naturam esse quintam, longe aliam a quatuor elementis."

107 Sennert, *De chymicorum consensu ac dissensu*, 11, 144ab: "Verum […] appellare videntur principia essentialia, ob nobiliores qualitates et actiones, quae in maiore apud ipsos sunt pretio, et ita ad essentiam praecipue medicamentorum, quae inprimis quaerunt, mahis pertinere videntur, quam elementa."

108 Ibid., 151b–152a: "Licet vero principiae haec mista perfecta dicantur, nihilominus aliorum mistorum principia esse possunt. […] et quod in suo genere perfecta sunt, et quod sunt, per suam formam sunt; non tamen talia sunt, quae sui gratia simpliciter sint, et ut seorsim existentia peculiarem corporum naturalium speciem constituerent, sed ut […] aliis corporibus mistis perfectoribus nobiliorem materiam praebeant."

109 Ibid., 152b: "Transiens [materia] est, quae non manet in re effecta, sed mutatur et formam accipit aliam, ita quidem ut priorem recipere non possit."

110 Sennert, *De chymicorum consensu ac dissensu*, 8, 85a; see Scaliger, *De plantis*, 1, 110b. On the role of taste in early modern alchemy, see Ragland, "Chymistry and Taste."

111 Paracelsus had also used the example of burning wood in his *Opus Paramirum*, 1.2, I, 73; see Weeks, *Paracelsus*, 109.

112 Fernel, *Physiologia*, 4.3, 266–7.

four elements, Sennert now suggested that they corresponded with the *tria prima*. In this latter interpretation, it was impossible to experimentally obtain pure elements, as they were accessible only to reason. By contrast, the three principles could be extracted by chymical practice.[113] As a consequence, the separation of a compound equaled its resolution into principles or first compounds (*prima mista*), that is, the *tria prima*.

According to Sennert, the chymical principles released by separation consisted of minimal parts. In his view, this description conformed to Scaliger's approach to the minimal bodies involved in mixture.[114] Such minimal parts mutually acted upon each other through their contrary qualities. Most remarkably, Sennert considered that their forms united under the supervision of the superior form. Within the compound, they remained intact, albeit to an inferior degree with respect to that of the form of the *mistum*.[115] At first, Sennert acknowledged that this description of mixture went against the interpretation of the Franciscan theologian John Duns Scotus (ca. 1266–1308), Averroes and Zabarella as the leading figures of Latin pluralism.[116] Although Sennert adopted the pluralist framework in the first edition of *De chymicorum consensu ac dissensu* (1619), he abandoned it in the late 1620s by calling the fragmentation of forms "pure fiction."[117] He now pointed out that Fernel, too, had mocked this interpretation of mixture in his *Physiologia*.[118] In fact, as we have previously seen, Fernel dismissed Averroes's interpretation of mixture and tacitly

113 Sennert, *De chymicorum consensu ac dissensu*, 11, 149ab: "Peripatetici etiam ex resolutione res ex elementis constare probant: interim prima pura elementa nullibi monstrare possunt [...]. Itaque nec chymici obstricti sunt, ut pura illa principia exhibeant; satis est, si corpora monstrent, in quibus excellunt."

114 Ibid., 12, 210b: "Vero magis consentaneum existimo, in mistione miscibilia in minimas partes redigi, atque ita sibi per minimas partes unita, per contrarias qualitates mutuo agere et pati: non tamen formas suas plane amittere [...] sed ex omnibus unam conflari, aut potius omnes mistas et in unum quasi redactas sub superioris alicuius formae, a qua fiat unum, specie, dominio manere"; see Scaliger, *Exotericarum exercitationum*, 101, 143v.

115 Ibid., 11, 152b: "Re hinc inde diligenter pensitata cum iis iam facio, qui elementa secundum formas integra in misto remanere, ac dum mistum in elementa resolvitur, elementa redire non specie tantum, sed numero etiam."

116 See Michael, "Daniel Sennert on Matter and Form"; eadem, "Averroes and the Plurality of Forms."

117 Sennert, *De chymicorum consensu ac dissensu*, 11, 153a: "Verum refractio illa formarum merum figmentum est, ut a Latinis contra Averrois sententiam disputantibus satis monstratum est; neque ulla Averrois et Zabarella pro ista sententia rationem saltem probabilem afferre."

118 Ibid., 152b: "Neque enim unius Scoti, qui in suam sententiam, quam non sine causa Fernel [...] puerilem, fere inanem et portentissimam nominat, Latinos plerosque, dum veterum philosophorum disciplinam, permistionisque rationem assequi minime voluerunt, [...] pertraxit, autoritate standum vel cadendum mihi existimo. Etiam docti viri fuere

adopted an Avicennian account by postulating the permanence of intact forms in the compound. For Sennert, too, Avicenna's view on mixture proved to be particularly convenient as it overlapped with Scaliger's natural philosophy and Fernel's physiology.

After presenting his account of mixture, Sennert took the further step of identifying the minimal particles constituting the *tria prima* with the atoms of Democritus. He maintained that these were subject to combination (*synkrisis*) and separation (*diakrisis*) for the formation of beings.[119] As Christoph Lüthy has observed, this interpretation violated several fundamental characteristics of Democritean atomism, such as the importance given to the size, shape and position of the various classes of atoms.[120] Moreover, Sennert insisted that the elements provided only the material substrate for mixture. They continued to be directed by the superior form, which imposed its essence on the compound. In other words, the aggregates of elemental *minima* did not follow the logic of a casual encounter of atoms that Democritus had postulated. For Sennert, it was the form that brought about mixture, using the instrument of heat.[121] In light of Sennert's explanation of the *tria prima*, one may in fact conclude that the superior form was also responsible for the active powers of bodies, that is, their sensory and chymical properties.

Despite these divergences from ancient atomism, Sennert asserted that the philosophical tradition had long postulated an atomistic conception of bodies. This tradition did not wholly exclude Aristotelianism, as the *Meteorology* spoke of the combination and separation of corpuscles.[122] As William Newman has shown, this line of reasoning was already present in Libavius's *Alchymia* (1606).[123] Sennert partially adopted it in the pharmacological part of his *Institutiones* of 1611 by speaking of the *synkrisis* and *diakrisis* of minimal parts as homeomerous bodies. But it was only in the first edition of *De chymicorum consensu ac dissensu* of 1619 that he identified these minimal parts with Democritean atoms.

Philoponus, Albertus, Zimara, Avicennas, Fernelius, qui contrarium sentiunt et acriter defendunt"; see Fernel, *Physiologia*, 2.8, 210–211.

119 Sennert, *De chymicorum consensu ac dissensu*, 12, 212a. On Democritus in early modern science, see Lüthy, "The Fourfold Democritus."

120 See Lüthy, "An Aristotelian Watchdog," 552.

121 Sennert, *De chymicorum consensu ac dissensu*, 12, 213a: "Non tamen […] ex fortuito atomorum concursu res naturales fiunt […] sed a forma superiore dirigente, quae caloris instrumento et ope attrahit, retinet, miscet, et disponit omnia, prout opus."

122 Ibid., 211b–212a; see Aristotle: *Meteorology*, 2.9, 369b35–36.

123 Newman, *Atoms and Alchemy*, 85–125.

Most importantly, Sennert maintained that the medical tradition had for a long time presumed the composition of bodies out of the smallest parts of matter. To be sure, Galen had rejected atomism in *On the Elements According to Hippocrates*, but he nonetheless spoke of the resolution of bodies into minimal particles during mixture.[124] For Sennert, Avicenna had further stimulated this interpretation in his definition of temperament (*complexio*) in the *Canon*. Being the overall quality of mixture, the temperament resulted from the reduction of elements into contiguous minute parts.[125] In the *Institutiones*, Sennert elaborated on this description to explain the temperament as a union of contiguous *minima*. Finally, in *De chymicorum consensu ac dissensu*, he stated more boldly that Galenic medicine paved the way towards an atomistic redefinition of elements as minimal particles.

6 Conclusion

In this chapter, we have seen how, in late-Renaissance medicine, the element came to be redefined as a discontinuous particle and a physically indivisible *minimum*, which at times merged with the concept of atom. I hope to have shown that this interpretation was stimulated by the theory of mixture and temperament expounded by Galen and Avicenna. This means that we must add these two medical authorities to the sources of the atomist revival – namely, Plato, Democritus, Aristotle, Epicurus, and Lucretius.

In the perception of late-Renaissance physicians, the atoms entailed, above all, the smallest units of matter. With respect to this definition, Galen's and Avicenna's conceptions of the elements could justify an atomistic interpretation of the body, even though they neither postulated an accidental concourse of elements or the existence of a void, nor indeed reduced the substantial form to the size, shape and spatial arrangement of the elements. This reasoning worked on the assumption that the body's first components, the elements, were contiguous particles. These elemental particles aggregated so as to constitute body parts possessing active powers of a physiological, chymical and sensory nature.

124 Sennert, *De chymicorum consensu ac dissensu*, 12, 212a: "Neque minimas particulas ipse Galenus reiicit. [...] tamen atomos simpliciter reiicere non potest, dum libro 1 de Elementis capite ultimo [...] concludit: quod miscibilia ad minimas particulas confringantur, unde illis mutua actio et passio contingit, et qualitatem invicem participatio, quae tanto promptior et facilior, quanto ad minores particulas confringuntur."

125 Ibid.

It is important to recognize that physicians relied on a variety of sources to support their respective physiologies, and in this they followed the conceptual categories of their time. As we have seen, their accounts went far beyond the simplistic opposition between vitalism and materialism that the historiography of the "scientific revolution" has imposed on them. What was really at stake for late-Renaissance physicians was the intricate relationship between corporeal and incorporeal components, matter and form, body and soul. In this regard, Jean Fernel's account of the temperament was particularly successful in that it explained the twofold constitution of the living body. Whereas its material core resulted from the harmonious mixture of contiguous elemental portions, its substantial form was related to innate heat as the vital principle. Rooted in Avicenna's account of mixture and complexion, this scheme reflected the Platonic side of Fernel's medical philosophy by insisting on the supra-elemental nature of life. Although this living part was tangled up with the body's matter, it was akin to the celestial and divine substance.

The conception of discontinuous elements applied to their first level of organization, that is, to the homeomerous parts, which resulted from the homogeneous mixture of the elements. In Galenic medicine, these first body parts corresponded to flesh, bones, muscles, tendons, nerves, veins, and other body tissues. However, such "perfect mixtures" also characterized natural bodies like metals and, as Libavius and Sennert pointed out, the chymical principles that could be extracted from bodies. As first compounds, these homeomerous parts had to be considered particles endowed with a superior form. This form, in turn, was related to aether, *spiritus* or the quintessence, according to whatever Platonic, Aristotelian and Paracelsian subtexts were being invoked. While Fernel followed a Platonic orientation in exalting the divine origin of the form, Libavius and Sennert cautiously limited this higher provenance to the moment of divine creation. This allowed them to remain close to the Aristotelian analogy between the elemental substance and celestial element.

Remarkably, Fernel, Libavius and Sennert all pointed to the gap between the material status of elemental bodies and the celestial origin of their substantial forms. But by integrating the latter within the teleological horizon of divine providentialism, they could safely postulate a discontinuous or even atomistic structure of the body within a traditional framework, without risking the charge of materialism. It is this philosophical *tour de force* that makes the Galenic physiology of Fernel, Libavius and Sennert so important for our understanding of the atomist revival.

CHAPTER 4

Pores, Parts, and Powers in Sixteenth-Century Commentaries on *Meteorologica* IV

Craig Martin

1 Introduction

Scholarship of the last century has shown that *Meteorologica* IV has held an extraordinary position, both with respect to interpretations of Aristotle's natural philosophy in general and with respect to understandings of the development of early modern matter theory. *Meteorologica* IV is to a large extent thematically disconnected from the first three books of the *Meteorologica*, which are concerned with motions and transformations of the dual exhalations in the sublunary region.[1] These two exhalations – one like water (vaporous), the other like fire (hot and dry) – make up the proximate causes of weather phenomena; fires in the sky, including comets and meteors; optical apparitions, such as the rainbow, double suns, and haloes; formations of bodies of water; and subterranean alterations, namely earthquakes and the formation of fossils and metals.[2] In this last subject there is some continuity between the first three books and the fourth and final book, as the formation of metals is briefly discussed in the last chapter of book III. There, Aristotle explains that metals, by which he refers to homogeneous watery substances, result from the congelation and solidification of the vaporous exhalation that is enclosed within the earth. Elsewhere fossils – that is, minerals dug up from the earth, rather than mined from an ore – form when the hot and dry exhalation acts as an efficient cause upon the underlying matter.[3] In like manner, book IV addresses the role of heat and the characteristics of underlying matter in the transformation and generation of homogeneous substances.

In particular, *Meteorologica* IV addresses the active powers of heat and cold and their roles in separating and uniting matter, specifically in the processes of concoction (*pepsis*) – which includes digestion and ripening – and

1 Wilson, *Structure and Method*, 9.

2 For *Meteorologica* I–III, see Wilson, *Structure and Method*; Taub, *Ancient Meteorology*, 77–115; Solmsen, *Aristotle's System*, 393–439; Gilbert, *Die meteorologischen Theorien*, 176–205.

3 Aristotle, *Meteorologica*, 3.6.378a17–b5; Eichholz, "Aristotle's Theory," 141–146.

 | DOI:10.1163/9789004528925_005

of putrefaction (*sepsis*).[4] These processes are prominent in the generation and dissolution of living bodies, their parts, and their residues. They parallel artificial transformations, being analogous to the cooking heat employed in the culinary arts.[5] Ingemar Düring's description of *Meteorologica* IV as a "Chemical Treatise" was therefore regarded as inadequate by David Furley, who instead saw book IV as more closely connected to the biological works.[6]

Subsequent chapters of *Meteorologica* IV discuss the primary passive qualities of matter, namely dryness and wetness, in addition to the secondary passive qualities of matter. These secondary passive qualities – including malleability, brittleness, combustibility, viscosity, and meltability – are properties, potentially, of the matter of living beings, minerals, and artificial products. The underlying elements, that is, the proportions of earth and water, the two elements most closely correlated to the passive qualities of dryness and wetness, partially explain these secondary qualities, although porousness is invoked as another factor to account for some qualities.[7] For example, materials made up of earth, and possessing pores too small to allow the entry of particles (*onkoi*) of water, become soft when wet.[8] Similarly, wood, wool, bone, and other combustible objects have pores containing little moisture, which fire can readily penetrate.[9] Empty pores also explain why some objects, like sponges, wax, and flesh, reduce in size when pressed.[10] Furthermore, viscosity can be accounted for as the result of interlocking parts that create chain-like structures that inhibit flow.[11]

In 1915, Ingeborg Hammer-Jensen argued that the references to pores and particles in *Meteorologica* IV contradicted Aristotle's attacks on Leucippus, Democritus, and Empedocles in *De generatione et corruption* I.8, in which Aristotle rejected as superfluous and incoherent the argument that pores should be seen as conduits for active powers.[12] Seeing a closer correspondence with the later Peripatos than to Aristotle's own teachings, Hammer-Jensen argued that book IV should thus be considered inauthentic.[13] While notable scholars, including David Ross, Werner Jaeger, and Léon Robin, have agreed with

4 For concoction, see Aristotle, *Meteorologica*, 4.2.379b10–380a9. For putrefaction and natural decay, see ibid., 4.1.379a2–379b9.

5 Lloyd, *Aristotelian Explorations*, 83–103.

6 Furley, "The Mechanics," 73–93; Düring, "Aristotle's Chemical Treatise."

7 For the prominence of earth and water for defining passive properties, see Aristotle, *Meteorologica*, 4.4.382a4–8.

8 Ibid., 4.8.385a28–30; 4.9.385b19–26.

9 Ibid., 4.9.387a17–23.

10 Ibid., 4.9.386b2–11.

11 Ibid., 4.9.387a11–15.

12 Aristotle, *De generatione et corruptione*, 1.8.324b25–326b28.

13 Hammer-Jensen, "Das sogenannte IV. Buch," 118–36.

Hammer-Jensen's determination, after much debate, the scholarly consensus has nevertheless coalesced around the acceptance of the work as genuine.[14] At any rate, book IV's consideration of material properties, its description of corpuscles and pores, its linking of the artificial and the natural, and its appeals to "artificially contrived" experiences, have given the book a distinct legacy in the Middle Ages and the early modern period, particularly in relation to medicine and alchemy, fields that combined theoretical and practical inquiries into matter and its composition.[15]

William Newman has shown that readings of *Meteorologica* IV were crucial to the development of experimental practices and corpuscular theories of matter developed by alchemists beginning with Pseudo-Geber's *Summa perfectionis*, written around the end of the thirteenth century, and continuing into the seventeenth century, when Daniel Sennert and others used the text as a support for atomistic theories of matter.[16] In addition to being widely read in alchemical circles, *Meteorologica* IV was a standard part of the curriculum of medieval and Renaissance universities.[17] Consequently, it was the subject of commentaries by many of the leading university professors. Renaissance commentators frequently addressed the matter theory of *Meteorologica* IV and its significance for alchemy.[18] Niccolò Cabeo's 1646 commentary perhaps best represents this alchemical strain within the commentary tradition; here he used all four books of the *Meteorologica* to serve as a textual foundation for an innovative interpretation of Aristotle. Cabeo believed that his interpretation supported Paracelsian principles, experimental practices, and corpuscular matter theory.[19]

While alchemical theory and practice influenced Renaissance readings of *Meteorologica* IV, during the sixteenth century book IV increasingly came to occupy a transitional place between natural philosophy and medicine.[20] Sixteenth-century Italian universities emphasized medicine, and their instruction in natural philosophy reflected this focus.[21] Many of the topics of *Meteorologica* IV were directly relevant to medicine, including its discussions of

14 Ross, *Aristotle*, 11; Werner, *Aristoteles*, 386; Robin, *Aristote*, 17. For more doubts about its authenticity, see Gottschalk, "Authorship," 67–79. For the acceptance of Aristotle's authorship, see the forceful argument at Furley, "Mechanics," 86; and, most recently, see Popa, "Scientific Method," 317, n. 17; Gill, "The Limits of Teleology," 336, n. 1.

15 For the "artificially contrived situations," see Lloyd, *Magic, Reason and Experience*, 209–210.

16 Newman, "Experimental Corpuscular Theory," 291–329; idem, *Atoms*, 21–153.

17 For example, see Denifle, *Chartularium*, 1:278; Malagola, *Statuti*, 274.

18 Martin, "Alchemy and the Renaissance," 245–262.

19 Idem, "With Aristotelians Like These," 135–161.

20 Idem, "Francisco Vallés," 1–30.

21 Bylebyl, "School of Padua," 338; Lines, "Natural Philosophy," 267–320; Giard, "Histoire," 139–69; Grendler, *Universities*, 268–269.

digestion, ripening (treating also of bodily fluids), and putrefaction. Book IV's matter theory had the potential to help explain the properties and characteristics of flesh, blood, semen, marrow, and other body parts that were subjects for medical instruction, in line with Furley's classification of book IV as being primarily concerned with organic matter. Accordingly, a number of sixteenth-century commentaries, such as those by Pietro Pomponazzi, Lodovico Boccadiferro, Francesco Vimercato, and Francisco Vallés, associated *Meteorologica* IV with medicine and discussed in depth its medical aspects.

In their commentaries on *Meteorologica* IV, Pomponazzi, Boccadiferro, Vimercato, and Vallés cast doubt on the need to use occult qualities to explain material properties. Rather, they emphasized that particles endowed with manifest qualities can account for matter's active and passive powers. In doing so, they undermined the concept of homeomerity, contending that seemingly homogeneous materials are in fact composed of different kinds of corpuscles. They held that the corpuscular structure of these seemingly homogeneous substances played a significant role in defining their passive characteristics. Invisible fibers and pores – and the structures they formed – were crucial to the transformation of material substances, given that they served as conduits for active powers, allowing for the completion of processes such as coagulation and putrefaction. Pomponazzi, and several of his early modern readers, recognized that Aristotle's use of pores and particles was potentially at odds with his dismissal of atomism in *De generatione et corruptione*.

For these early modern commentators, the matter theory of *Meteorologica* IV was directly relevant to medical theory. They held that book IV's references to particles and pores were key to understanding temperaments, complexions, and constitutions. They discussed how blood contains small fibers that play a role in physiological functioning and the determination of temperament. Furthermore, they understood that elemental or vaporous particles enter into the pores of living bodies and alter their composition and the qualities of their parts. Without admitting the existence of indivisible atoms or of void, they contended that small particles, endowed with active and passive powers, were agents and recipients of qualitative change. Thus, the corpuscular interpretation of *Meteorologica* IV was not restricted to alchemical circles but also echoed throughout exegeses of the book that were concerned with, and informed by, medical practice.

2 Pomponazzi's *Dubitationes* and Elemental Parts

In the first years of the 1520s, directly after the controversies over the immortality of the soul and the composition of *De incantationibus* and *De fato*, Pietro

Pomponazzi gave lectures on a number of topics that deviated from the traditional teaching duties of an ordinary professor at Bologna. His contract, which was the result of intensive negotiation and his threatening to depart if his conditions were not met, permitted him to lecture on whatever subject he wished.[22] Taking advantage of this freedom, he lectured on Aristotle's *De generatione et corruptione*, *De partibus animalium*, and the *Meteorologica*. His *Dubitationes in quartum Meteorologicorum* likely stem from lectures given in the years between 1522 and 1524,[23] although they were only printed posthumously in 1563, in the interim between the first and second printings of the *De incantationibus*.

The printer's dedication to Cardinal Ludovico Madruzzo offers few clues as to the motivation for publishing the *Dubitationes*, besides its contentions that Pomponazzi was one of the greatest philosophers of his time and that the work is most erudite.[24] Nevertheless, publishers, purchasers, and readers of this book most likely were well aware of his better-known and more controversial writings, especially after the 1556 release of his *Opera omnia*.[25] To a great degree, these lectures on *Meteorologica* IV followed the materialistic inclinations that were a hallmark both of Pomponazzi's psychological works and the *De incantationibus*.[26]

Avoiding recourse to the supernatural in the *De incantationibus*, Pomponazzi instead gave hypothetical physical explanations for a range of marvelous and seemingly miraculous phenomena and events. In order to explain many of these phenomena, he evoked the occult natural powers contained in herbs, animals, metals, and other substances; he referred to astral influences and to forces transmitted through invisible, but material, spirits and vapors. He thereby replaced a conceptualization of the demonic exertion of active powers on passive subjects with the natural dispensation of these same forces, following the contours of Renaissance theories of natural magic.[27] In the *Dubitationes*, Pomponazzi dedicated himself to explaining material transformations, whether marvelous or not. The causes of these transformations, however, are largely more pedestrian than the accounts of marvelous, preternatural phenomena found in the *De incantationibus*. In the *Dubitationes*, Pomponzzi's causal accounts are mainly in accordance with Aristotle's text and are linked to the elements and sensible qualities. Eschewing occult forces,

22 Podestà, "Di alcuni documenti," 176.

23 Lohr, *Latin Aristotle Commentaries*, 356–357; Nardi, *Studi*, 83–84.

24 Pomponazzi, *Dubitationes*, sig. *2r–v.

25 Doni, "Il 'De incantationibus,'" 183–230.

26 Graiff, "I prodigi," 331–361.

27 Pomponazzi, *De incantationibus*, 7–13. For Pomponazzi and his work's relation to natural magic, see Copenhaver, *Magic*, 273–284.

he circumscribed the active powers to include just heat and cold, emphasizing heat. Furthermore, just as in the text of *Meteorologica* IV, he paid close attention to the passive aspects of matter and the secondary qualities that result from the formations of elemental parts, in particular from the mixing and conglomerations of earth and water.

Despite the differences between the *De incantationibus* and the *Dubitationes*, there are strong parallels between the two works. For example, the *Meteorologica* is among the authoritative texts that Pomponazzi relied upon in the *De incantationibus*. In chapter ten, which contains a lengthy exploration of potential causes of the preternatural, he argued that necromancy could possibly be explained by apparitions similar to the faces of the dead said to be reflected onto the vapors in the night air. In support of this hypothesis, Pomponazzi cited Aristotle's discussion of haloes and rainbows, and Aristotle's statement that such optical reflections do not occur at midday but require dense misty air, composed of small parts of water that have yet to form drops. These were the conditions that Pomponazzi presumed to prevail in the foggy graveyards where necromancers saw what they believed were the faces of dead people, recently buried.[28] Elsewhere in the *De incantationibus*, Pomponazzi referred to pores as key to the transmission of imperceptible agents. In chapter four, he contended that it is more probable that vapors and spirits cause marvelous cures than unguents and plasters because the latter are bulkier and therefore do not so easily enter into the pores and the internal parts of the body as the imperceptible vapors do.[29]

In the *Dubitationes*, Pomponazzi speculated about the natural causes of marvelous phenomena, just as he had done in his early works. For example, in *Dubitatio* LXIIII, in a digression following his consideration of the powers of fire and Aristotle's alleged belief that salamanders live in fire, he speculated that it might also be possible to find a human impervious to combustion, even if placed completely in fire. To support this contention, he cited Plutarch's biography of Pyrrhus, the ancient king of Epirus, whose foot, we are told, would not burn even when engulfed in flames.[30] He went on, recounting seemingly unrelated marvelous phenomena, including his own eyewitness account of a spear

28 Pomponazzi, *De incantationibus*, 91–92; Aristotle, *Meteorologica*, 3.4.373a35–b31; 3.5.377a11–28. Pomponazzi made a similar assertion in *De immortalitate animae* where he contended that visions of the dead often occur in burial grounds where the air is very thick such that it easily receives idols; see Pomponazzi, *De immortalitate animae*, 208; Aristotle, *Meteorologica* 3.4.373b7–10.

29 Pomponazzi, *De incantationibus*, 33.

30 Pomponazzi, *Dubitationes*, 38v; Plutarch, *Life of Pyrrhus*, 3.5. For Aristotle's statement on salamanders living in fire, see Aristotle, *Historia animalium*, 5.19.552b14–16.

being launched and penetrating a knight's armor while itself remaining fully intact despite lacking an iron tip. Having aroused the curiosity of numerous Mantuans, this spear was sold for the lofty sum of 300 ducats. Pomponazzi concluded this section by stating that these marvels are not the result of "spells or evil demons, as many unskilled in natural matters believe."[31]

In other ways, the discourse of the *Dubitationes* appears removed, or at least distanced, from the frequent evocation of the powers of celestial bodies and occult forces found in the *De incantationibus*, where – just as Avicenna had – Pomponazzi linked the products of spontaneous generation to celestial bodies.[32] In the *Dubitationes*, by contrast, he focused on Aristotle's description in *Meteorologica* IV of how spontaneously generated animals were formed by the power of heat lingering in the remnants of putrefaction.[33] By deliberating over the role of heat in generation, Pomponazzi, for the most part, left behind the outsized role of astral powers found in his earlier work.[34] Rather, he theorized about the transformations of earthly matter in terms of mixtures of elements and their manifest qualities.

A recurring *quaestio* among medieval and Renaissance commentators on *Meteorologica* IV was that regarding the *forma mixti* arrived at through the mixture of these elements.[35] In short, the query asks how a new substantial form can arise during the generation of mixture and what happens to the forms and qualities of the ingredients of the mixture. Are the ingredients' forms and qualities destroyed, or do they remain, in a refracted, potential, or blunted manner? Closely related to these questions is the question of whence the new supervening form arises. Does it emerge from the underlying matter, or does it come from outside of the mixture, transmitted by celestial bodies or imposed by God? Although Pomponazzzi did not directly address the *forma mixti* in the *Dubitationes*, his speculations on the accidental qualities of mixtures show that he was to some extent concerned with this question. His glancing treatment of it, like so many of the work's discussions, ends in an indeterminate fashion, even if the propositions of others are dismissed as unsatisfactory. Here, as in many of his lectures dating from this time, Pomponazzi commenced from, and proceeded through, doubt.[36]

31 Pomponazzi, *Dubitationes*, 38v: "Ideo non omnia talia fiunt incantatione, aut malorum daemonum adiuvamine, ut multi imperiti physicarum rerum credunt."

32 Idem, *De incantationibus*, 79; Nardi, *Studi*, 305–19.

33 Aristotle, *Meteorologica*, 4.1.379b6–9; 4.11.389b5–9.

34 Pomponazzi, *Dubitationes*, 35r.

35 Haas, "Mixture," 21–46; Maier, *An der Grenze*, 9–140.

36 Perfetti, "Docebo," 439–466.

Pomponazzi focused on the roles of the prime qualities, and in particular, on the two active qualities – heat and cold – that are necessary for altering, corrupting, and preparing for the introduction of a new *forma mixti*. In *Meteorologica* IV, Aristotle described the power of heat and cold to cause generation by ruling over or mastering the underlying matter.[37] In concoction, internal natural heat acts as the efficient cause. This heat fosters the realization of the end or the perfection of that substance.[38] For Pomponazzi, heat is the key qualitative instrument for bringing about a new *forma mixti*. Yet, in his view, by its essence, heat does not create substantial forms, as it merely prepares the matter through non-essential powers that "divides, separates, rarefies, and unites accidently."[39] The heat, directed by a natural agent that possesses a final cause, arranges the underlying matter "in order that the form of the mixture is introduced." Thus, for Pomponazzi, the other qualities are not part of the active process but are linked to matter, not form:

> I say therefore that heat alone introduces the substantial form into a mixture; moreover, wetness and dryness do not introduce the substantial form, but behave as matter. In the same way, cold does not introduce the substantial form into a mixture but behaves as matter.[40]

Heat's qualities prepare the matter defined by the other prime qualities so they may receive a new substantial form.

In its activity as directive agent in this process of introducing substantial form, nature is overwhelmingly sublunary. Here, Pomponazzi doubted the role of astral influences. In the passage of the *Dubitationes* that most closely resembles an inquiry into the classical problem of the *forma mixti*, Pomponazzi – "for the sake of exercise" – referred to an argument that he "used as a student at Padua," and presented to his teacher, whom he identified as Antonio Trapolino.[41] Pomponazzi supposed, following the Thomistic solution, that only one substantial form actually exists in a new mixture. The forms of the components of the mixture remain only virtually, while the new

37 Aristotle, *Meteorologica*, 4.1.379a1–2.

38 Ibid., 4.2.379b18–32.

39 Pomponazzi, *Dubitationes*, 6v: "Dico, ut prius, quod absolute inquantum est calor, dividit, segregat, & rarefacit, & per accidens congregat."

40 Ibid.: "Dico ergo, quod sola caliditas est, quae inducit formam substantialem in mixto: humiditas autem & siccitas non inducunt formam substantialem, sed habent se ut materia. Sic nec frigiditas inducit formam substantialem in mixto; sed habet se ut materia."

41 Ibid., 7v: "Exercitationis autem causa, adducam ego argumentum quo scolasticus adhuc Patavii usus sum, et adduxi viro percelebri Antonio Trapolino praeceptori meo."

form is introduced to the newly generated substance.[42] Pomponazzi presented Trapolino with the macabre proposition of a man hypothetically thrown into extremely cold water and dying as a result. The question arising from this grim hypothesis was as follows: "therefore, whence is [the new form of the dead man] generated?"[43] It cannot be from heat, because freezing water surrounds the corpse. In Pomponazzi's recounting, Trapolino responded, following Averroes, that through the influence of the heavens, heat from the sun and other stars are impressed on the air, which in turn produces the new *forma mixti*.[44]

Pomponazzi, however, rejected Trapolino's solution, noting that many mixtures are generated deep in water or below ground where there is no air that could convey celestial heat, and, evoking a renal complaint with which he himself was afflicted, that "stones and sand are generated in the kidneys, and [that] worms are generated in man and in other animals, where there is no air."[45] Ending the *Dubitatio* in a state of suspended judgment, he seemingly rejected both the role of celestial influences in the generation of intestinal worms, and the Thomistic view that a new form replaces the qualities of the components of a mixture. This is consistent with his earlier definition, where he stated that "mixture comes to be from heat, for heat mixes the four elements with each other and collects them so that they remain at the same time in the mixture."[46] Thus, in Pomponazzi's definition, the elements remain, seemingly intact. For the dead man submerged, as for Pomponazzi's kidney stones, the heat that introduced the new forms must therefore derive from the accidental qualities of the components that remain. Pomponazzi later clarified this view in a discussion of the spontaneous generation from putrefied matter, where,

42 Maier, *An der Grenze*, 36–40. While it is well known that Pomponazzi studied with the Thomistically inclined Pietro Trapolino at Padua (see Nardi, *Studi*, 104–21), the identity of Antonio Trapolino is mysterious. Eugenio Garin maintained that Antonio Trapolino did not exist (Garin, *La filosofia*, 2:57). Yet, an Antonius Trapolinus is listed as becoming extraordinary professor of jurisprudence at Padua in 1525, the year of Pomponazzi's death and thus well after his student days; see Facciolati, *Fasti gymnasii patavini*, 130.

43 Pomponazzi, *Dubitationes*, 7v: "Tunc accipio hominem, qui exponatur aeri frigidissimo; vel proijciatur in aquam frigidissimam, & moriatur illo frigore; tunc ... ibi generatur nova forma mixti: quaero ergo unde generetur?"

44 Ibid.: "Respondebat praeceptor meus tenendo opinionem Comment. in 12. Met. 18. quod virtute influentiarum coeli influentis calorem solis & aliorum astrorum in aerem fit hoc, & est quaedam caliditas impressa in aer, quae producit hunc calorem, quo demum generatur forma mixti." Averroes, *Metaphysicorum*, 305r.

45 Pomponazzi, *Dubitationes*, 7v: "Item in renibus generantur lapides & arenae, & vermes generantur in homine, & in aliis animalibus, ubi non est aer." On Pomponazzi's own kidney stones, see Nardi, *Studi*, 205–206.

46 Pomponazzi, *Dubitationes*, 2v: "mixtio fit a calido, calor enim miscet quatuor elementa inter se, & ea congregat, ut simul maneant in mixto."

comparing the views of Albertus Magnus and Averroes, he wrote that these animals "sometimes come through the heat in the separated parts of matter, and sometimes through an extrinsic and ambient heat," that is, the "celestial heat [that] is spread throughout the orb."[47]

The discussion in the *Dubitationes* is characterized by the same hesitancy found in the *De incantationibus* to attribute causes to the supernatural. Nature, in the *Dubitationes*, remains largely sublunary; the causes of generation and corruption are the operations of the prime qualities; and their effects are consistent in both artificial and natural entities. His explanation of the powers, characteristics, and qualities of diverse mixtures depends on the existence of parts within them that retain these powers, most often powers similar, if not identical, to the prime qualities of the elements. For example, in his explanation of why strong wine desiccates even though it is itself wet, he maintains that Cretan and Falernian wines have a large amount of spirit and fiery parts within them. Similar accounts apply to the properties of coriander and vinegar. As for the "specific occult property," which played such a strong role in natural magic, Pomponazzi dismissively characterized it as the "universal refuge of all physicians and physicists."[48] Rather, it is in these fiery parts (*partes*) than we find a mundane explanation of the seemingly miraculous. Animals can be nourished and have sensation buried in the snow, "since in the midst there are hot, fiery parts."[49] Animals can be spontaneously generated in stones, because stones are generated from the double exhalation, which contains wet parts susceptible to putrefaction, as is confirmed by both Aristotle's authority and alchemists' extractions of oil and water. Additionally, these parts have corresponding pores. For example, salamanders' supposed ability to live within fire is the result of "such narrow pores within them that fire is not capable of penetrating them and consuming their humidity."[50] Presumably, Pyrrhus' foot could also withstand burning owing to a similar physical structure.

Pomponazzi concluded *Dubitatio* LXII by linking the artificial and the natural. He wrote that "those who are good philosophers can make some things through art that do not dissolve by fire," by using the lessons of the

47 Ibid., 35r: "quoniam aliquando fit per calorem in separatis partibus materiae; aliquando etiam per calorem extrinsecum, & ambientem"; "nam calor coelestis disseminatus est per totum orbem; calor autem est instrumentum productivum genitorum ex putredine."

48 Ibid., 36r: "Tertia responsio quam ponunt omnes medici, est, quod non quia calidum neque quia frigidum & humidum aut siccum; sed quia habet talem proprietatem specificam occultam, & est fuga communis omnium medicorum & physicorum."

49 Ibid., 38r: "quoniam in medio sunt partes calidae, igneae."

50 Ibid., 38v: "quoniam pori in ea sunt tam stricti, quod ignis non potest penetrare illos, & consumere humiditatem illorum."

imperishable salamander's pores.[51] Accordingly, references to the parts, with their qualities, and to pores, which block or convey them, apply to both the artificial and the natural, the living and the dead, as he referred specifically to "artificial mixtures, such as medicinal compounds," in addition to the work of goldsmiths.[52] Yet, most of the examples cited by Pomponazzi are products of nature. Oil becomes white because its water separates, leaving behind "airy parts" that coagulate and thicken.[53] Just as snow contains hot and fiery parts, hail contains air, which makes it foamy and white, and earthy parts, which are evident as residue left behind once the hail has melted.[54] Semen contains airy and potent parts that render it white and foamy immediately upon emission, while their absence renders the remnant fluid and cold. According to Pomponazzi, physicians believed that those who possess a long "rod" (*virga*) are thereby impeded from fatherhood, as the foamy parts of the semen fly off before the complete, potent seed can arrive in the womb.[55]

Pomponazzi's discussion of blood focuses similarly on its decomposition into various kinds of parts. He wrote that blood, taken from the vein, thickens not from the cold but because its hot and foamy parts evaporate, leaving behind its earthy parts.[56] Additional evidence that blood is composed of various parts derives from Aristotle's statement in *Meteorologica* 4.10 that "blood that contains fibers is mostly composed of earth."[57] This sentence had created confusion among medieval commentators. Some manuscripts of Henry Aristippus's twelfth-century *translatio vetus* interpreted Aristotle's *is* (ἴς) as meaning *via*, or path, instead of the more accurate *fibra*.[58] Consequently, in his commentary on the *Meteorologica*, Albertus Magnus, understanding the *viae* to be pores, made the obscure pronouncement that blood that "has many paths of pores, is more earthy than other kinds."[59] Although William of Moerbeke's thirteenth-century

51 Ibid.: "Unde qui boni essent philosophi, possent facere arte aliquas res, quae non dissolverentur igne."

52 Ibid., 2v–3r: "Item in corporibus sive commixtionibus artificialibus ut medicamentis compositis."

53 Ibid., 43r.

54 Ibid., 45v.

55 Ibid., 3r. For similar discussions of the various parts of semen contained in a jar or exposed to the cold, see ibid., 44v; 49r.

56 Ibid., 43v: "dico quod si sanguis sit extractus ex vena, tunc evanescunt partes calidae & spumosae."

57 Aristotle, *Meteorologica*, 4.10.389a20–21. For a general history of fibers in blood, see Haak, "Blood," 295–305.

58 Aristoteles Latinus, *Meteorologica liber quartus*, 10/1, 23.

59 Albertus Magnus, *Meteora*, 320: "Ille *vero sanguis*, qui *habet vias* pororum multas, *magis* est *terreus* quam alius." The italicized words, following Hossfeld's editing, mark Aristotle's words found in the text on which Albertus was expounding.

translatio nova left the word transliterated as *inas*, the continuator of Thomas Aquinas's commentary, perhaps informed by Albertus's reading, nevertheless described blood that has an earthy character as having "paths, that is, pores."[60] Following the translations of Pietro Alcionio (1487–1527) and Theodore Gaza (ca. 1398–1475/76), in the *Dubitationes*, Pomponazzi contended that these fibers cause a thickening of blood, as the foamy parts are led off.[61]

In his lectures on *De partibus animalium*, given in 1523, the same year in which he lectured on the *Meteorologica*, Pomponazzi discussed in greater depth the fibers within the blood, defining them "as hairs dispersed throughout the blood" or "corpuscles of blood."[62] Blood that contains these fibers will thicken and congeal, "since the fibers are of an earthy nature."[63] By contrast, watery blood contains no fiber. The resulting temperament of animals, such as humans and bulls, which have fibrous corpuscles in the blood, is ferocious compared to that of the weak, timid deer and gazelles, whose blood lacks these fibers. In Pomponazzi's account, diverse particles can explain different characteristics in blood. The qualities of venous blood arise "from the mixing with bodies and subtle parts that are invisible and spirituous and that enter and mix with the substance of blood" and thereby provide heat and fluidity.[64] In his view, these particles are analogous to "certain corpuscles" in warm water that render it hot even as it remains water, which is, by nature, cold.[65] Consequently, venous blood, while it appears homogeneous, is really made up of a variety of parts, some subtle and fiery, others larger and earthy.[66] According to

60 The continuator's comments are found at [Ps.-] Thomas Aquinas, *Expositio*, 680–681: "quae habent vias, idest poros." For the continuator and the authorship of the commentary, see Dondaine, "Le commentaire," 81–152. For William of Moerbeke's translation, see: Aristoteles Latinus, *Meteorologica*, 10/2.2, 131: "Sanguis quidem habens inas, magis terre."

61 Pomponazzi, *Dubitationes*, 49r: "Sed ad illud de sanguine occurrit dicendum, quod textus Aristotelis de sanguine habente fibras, non dicit quod congelatur; Theodorus enim legit, inspissatur, vel ingrossatur, sic etiam Alcyonius."

62 Pomponazzi, *Expositio*, 215: "Sunt enim sicut capilli dispersi per sanguinem"; "Nam istae fibrae sunt corpuscula sanguinis." For the fibers of blood, see Perfetti, *Aristotle's Zoology*, 55–58.

63 Pomponazzi, *Expositio*, 216: "Ratio autem istius est quoniam fibrae sunt naturae terreae, ideo ingrossant et congelantur."

64 Ibid., 178: "Nam quando sanguis in venis ex admixtione corporum et partium subtilium quae sunt invisibiles et spirituosae, quae intrant et miscentur substantiae sanguinis et subiectum et substantiam sanguinis calefaciunt et frigiditatem expellunt et eum efficiunt liquefactibilem."

65 Ibid.: "Nam aqua, quantum ad id quod est aqua, est frigida; tamen quia calefit, seu quaedam corpuscula, ideo redditur calida, et recedente caliditate, remanet ipsa aqua."

66 Ibid., 204: "Talis enim massa est diversarum partium. Unde, licet talis massa videatur homogenea, tamen secundum naturam non est ita: sunt enim partes subtiles et igneae et partes grossae et terrestres."

Pomponazzi, this model can explain the qualities of many other substances, including *materia medica*. Rose, rhubarb, and coriander contain diverse particles that are indistinguishable through the senses but generate different effects on the body. Both rose and rhubarb, for example, are composed of both subtle parts that have a laxative effect, and thicker parts that constipate.[67]

Just as the motions, associations, separations, and unions of these various parts explain the qualities and transformations of mixtures, the structures of substances are crucial to their susceptibility to change. Famously and controversially, in *Meteorologica* IV, Aristotle employed pores to explain some secondary passive qualities. Pomponazzi both unreservedly adopted these pores while also questioning the degree to which the promotion of these pores is consistent with the arguments made in *De generatione et corruptione* and in *De caelo*, where Aristotle denounced Democritus' reliance on pores.[68] In *Dubitatio* XCII, Pomponazzi discussed the roles of heat and cold in solidifying and melting substances. Here, he attempted to explain how that which is solidified by heat cannot be melted by heat, and that which is solidified by cold cannot be melted by cold. Anything, therefore, that is solidified by both must be completely incapable of melting. He contended that while heat solidifies *per se*, cold can do so *per accidens*, describing the process as such. First, heat extracts moisture from the body and enlarges it, but leaves some moisture in the body's pores. Then, cold constricts the pores, rendering the body more unified and stronger. Finally, the heat can no longer dissolve the small amount of remaining moisture because the closed pores do not allow it to enter within the body. Anticipating an objection to this explanation, Pomponazzi wrote:

> But then you will say: So, Aristotle Democritizes, although in *De caelo* III and *De generatione* [*et corruptione*] I, he condemns that opinion of Democritus that action occurs through entering through pores. Indeed, Democritus posits that indivisible bodies (*corpora atoma*) enter and exit through pores. I would say that the Philosopher agrees with Democritus, since he concedes there are pores, but he disagrees, since Democritus wants it to happen through indivisibles. Aristotle, however, wants a divisible body that is cold to enter through the pores and in this manner new coldness is generated and acts more powerfully, since it can be better applied in acting on its recipient.[69]

---

67 Ibid.

68 Aristotle, *De generatione et corruptione*, 1.8.324b25–326b28; *De caelo*, 3.8.307b12–18.

69 Pomponazzi, *Dubitationes*, 43v–44r: "Sed tunc dicetis. Ergo Aristoteles Democrizat quum tamen tertio coeli & primo de generatione damnet hanc opinionem Democriti, quod actio fiat per introitum per poros. Posuit enim Democritus, quod corpora atoma intrent & exeant per poros. Ego dicerem, quod philosophus assentitur Democrito in hoc,

Thus, Pomponazzi deemed that Aristotle used pores to explain how active powers transform substances, while rejecting the atomistic premise of indivisibility.

Pomponazzi later expanded upon this assessment of the closeness of the doctrine of *Meteorologica* IV with Democritean views. In his explanation of what makes substances capable of being burnt, Aristotle had written that those that have long pores that run lengthwise are more susceptible to fire.[70] Here again, Pomponazzi presented the potential objection that "Aristotle Democritizes," because he describes action as taking place through pores. Pomponazzi conceded that Aristotle understands the pores as fomenting combustion, "since the fiery corpuscles enter better and are better attached."[71] Yet, Aristotle's views differ from those of Democritus only in the fact that these corpuscles are not indivisible, and that since they are of a small size they are not the principal agents of combustion but only assist in the application of the agent on the passive body.

While noting the differences between Aristotle and Democritus, Pomponazzi conceded that pores play a significant role in determining which living bodies are affected by the powers of the hot and cold. For example, human bodies that have a cold complexion withstand the cold better and do not become sick as often as those with hot complexion, because their narrow pores do not allow the cold to enter.[72] Indeed, he posited that "all animals are full of pores," and "breath" (*halitus*) enters not only through the mouth and nose, but "through pores throughout the entire body."[73] Small bits of air, entering through these pores, are the cause of bodily corruption and putrefaction. Nor should we be surprised, he contended, that even a small amount of air causes corruption, citing numerous examples from experience: even a small bit of air that enters into a bottle of wine turns it bad; spice dealers place fruit under oil to prevent air from affecting their wares; and, chestnuts and the bark of cassia putrefy

quoniam concedit poros; dissentitur vero, quoniam Democritus vult ut fiat per indivisibilia; Aristoteles autem vult ut corpus divisibile quod est frigidum, intret per poros, & ita generetur ibi nova frigiditas & plus agit, quoniam melius applicari potest in actione suo passo." On the significance of this passage, see Lüthy, "Aristotelian Watchdog," 546.

70 Aristotle, *Meteorologica*, 4.9.387a19–21.

71 Pomponazzi, *Dubitationes*, 47v: "Videtur ibidem Aristoteles Democrizare, nam incidit in questionem Democriti. Qui voluit quod actio fiat per poros. Respondeo nolle Aristotelem actionem fieri principaliter per poros; sed quod per poros fiat melior applicatio rei combustibilis; quoniam corpuscula ignea melius intrant, & melius applicantur."

72 Ibid., 46r.

73 Ibid., 15r: "Omnia animalia sunt porosa. In illis enim quae respirant, ingreditur halitus per os & nares, & etiam per poros totius corporis."

when heated, as air penetrates their pores.[74] Pores help explain the process of the aging of fruit, as mature fruit possess many pores, while the pores of unripe fruit are constricted and block air from entering.[75] Although pores may not be the principal conduit for action on passive matter, they nevertheless remained, for Pomponazzi, a useful *explanans* for a variety of transformations of matter. Instead of invoking occult powers and astral forces, in the *Dubitationes*, many marvels of the natural world like amphibians impervious to fire, as well as banalities like rotting nuts, are best explained through the ingredients or parts of mixtures, and structures, such as pores, that allow hot and cold parts to act on passive matter. The parts, powers, and pores of *Meteorologica* IV offered an alternative to the natural magic that characterized much of the *De incantationibus*.

3 Commentaries after Pomponazzi: Boccadiferro, Vimercato, and Vallés

Even though the *Dubitationes* went unpublished until well after Pomponazzi's death, it is more than possible that they had an influence through the diffusion of manuscripts or through the lectures themselves. In the following years, three commentators on *Meteorologica* IV in particular, Lodovico Boccadiferro (1482–1545), Francesco Vimercato (1512–1571), and Francisco Vallés (1524–1580), interpreted book IV in the light of medical theory, just as Pomponazzi had.[76] Furthermore, along with Nifo and Pomponazzi, these three commentators were frequently cited in later commentaries of the sixteenth and seventeenth centuries. These successors to Pomponazzi maintained his views about the persistence of elemental parts and the role of pores in facilitating corporeal alterations.

Lodovico Boccadiferro likely followed Pomponazzi's lectures in the 1510s, and later became extraordinary professor and then ordinary professor, at Bologna when Pomponazzi gave lectures there on the *Meteorologica*.[77] His own lectures, probably given in 1538, followed and expanded on Pomponazzi's views. While Pomponazzi's discussion merely suggested, without stating outright, that the substances to which Aristotle and others referred as homeomerous are in fact composed of various parts that differ from each other, Boccadiferro explicitly

74 Ibid., 15r.

75 Ibid., 31v.

76 Martin, "Francisco Vallés," 1–30.

77 Lohr, *Latin Aristotle Commentaries*, 57–65.

defended this position. In his discussion on putrefaction, he questioned how truly homeomerous mixtures could putrefy, because putrefaction requires difference. The solution, he contended, is to be found in the fact that, "there is no truly homogeneous mixture, because although they appear to sight to be homogeneous, nevertheless in truth they are not homogeneous, because they always have some difference in their parts."[78] Wood, for example, resolves into parts of earth and smoke; gold, lead, rocks, and other substances that appear to be homogeneous are not only made up of different parts but are also susceptible to putrefaction, although only over a "long interval of time."[79] Boccadiferro did not deny the existence of substantial forms. Yet, his conception of them for these seemingly homogeneous bodies is closely linked to prime qualities. For example, in his discussion of why wine turns into vinegar, he maintained that the substantial form of wine has a hot temperament, while its matter has a cold temperament. When the substantial form is corrupted, only the cold temperament of matter dominates, and the resulting vinegar is cold, corresponding to its composition that consists of more cold than hot parts.[80]

Boccadiferro's judgment about the impossibility of a truly homeomerous mixture applied to substances within living bodies, in particular to milk and blood, both liquids prone to coagulate. Their capacity to solidify derives from their earthy parts. Whey, milk's watery part, combines with a more solid part associated with cheese that provides the nutriments. Blood has earthy parts that Boccadiferro associated with the fibers that Aristotle referred to at *Meteorologica* 4.10, and that Pomponazzi had also discussed. Boccadiferro too associated these fibers with earthy parts, describing them as hairs, threads, or gristles, and he took them to account both for blood's tendency to coagulate and for the diverse thicknesses of blood among different species of animals. These fibers create blood that is hot, not in the same way as a flame ignites, but rather in its ability to affect the senses. According to Boccadiferro, heated iron affects the senses more greatly than a lit torch. Analogously, the fibers in the blood convey the heat generated by the heart. For this reason, animals that lack fibers in the blood, like the easily frightened deer referred to in *De partibus animalium*, are rendered cold, timid, and weak. For Boccadiferro, these fibers confirm that no mixtures are truly homeomerous as, despite being earthy, they act as vehicles for the heat that defines living animals.[81]

78 Boccadiferro, *Lectiones*, 50: "nullum potest dari mixtum vere homogeneum, quia licet visu videantur esse homogenea, tamen in re non sunt vere homogenea, quia semper habent aliquam differentiam in partibus eorum."

79 Ibid.: "in longo temporis spatio."

80 Ibid., 238.

81 Ibid., 180–182.

In subsequent generations, other scholars modified conceptions of the homeomerous in ways similar to Pomponazzi and Boccadiferro. Francesco Vimercato studied at Bologna, Pavia, and Padua, before moving on to Paris around the year 1540.[82] Based on his teachings at the Collège Royal, his commentary on all four books of the *Meteorologica*, first printed in 1556, distinguished between living and inanimate mixtures. The concept of a "perfect mixture" was for him relative. Metals, stones, and similar inanimate substances are composed of a "more perfect mixture" but are not "completely perfect," being called so only because they have a form distinct from the elements, even though they pale in comparison to the "most perfect" parts of plants and animals.[83] According to Vimercato's reading of *Meteorologica* IV, Aristotle did not concern himself here with the first causes, namely the soul and the heavens, but rather with the actions of the hot and the cold.[84] For Vimercato, as for Pomponazzi, heat was the primary agent that prepares new mixtures, as its power alone prepares passive bodies to be unified after breaking them down into corpuscles. He wrote that heat "divides bodies, which are mixed into the most minute parts (*in minutissimas partes*) by concocting, so that they are rendered most fit for mixture."[85]

Vimercato's distinction between animate and inanimate homeomerous parts, however, is not consistently applied. Just like his predecessors, he held that milk and blood have constitutions composed of differing parts. Blood serum has more watery parts; the caseous, or cheesy, in milk constitute the more earthy parts, in addition to there being a fatty part that corresponds to butter. He concluded, therefore, that "milk is not unified and homogeneous (*similare*)."[86] Blood is composed of both earth and watery parts, the

---

82 Gilbert, "Francesco Vimercato," 188–189; Del Soldato, "Francesco Vimercato's *De placitis*," 117–120.

83 Vimercato, *In quartum librum Meteorologicorum*, 2: "Perfectiori autem admixtione constant lapides & metalla, quanquam ea omnino perfecta non constent, cum metalla ex aqua una, & lapides ex terra videantur esse: quoniam tamen formam quamdam habent ab elementis diversam, perfecte mixta nuncupantur. Omnium perfectissimae animalium & stirpium sunt partes."

84 Ibid., 11: "Princeps autem causa natura est, seu anima, aut etiam coelum, quae his facultatibus, ut instrumentis, utuntur. Aristotelis hoc in loco institutum non est, causas omnes ortus atque interitus explicare, sed quae sint caloris & frigoris, humoris & siccitatis facultates & opera."

85 Ibid., 12: "Atque hoc quidem caloris solius vi fieri, qui & quae sunt eiusdem generis, alienis secretis, in unum cogit, & corpora, quae miscentur, in minutissimas partes concoquendo dividit, ita ut mixtioni aptissima reddantur."

86 Ibid., 79: "Ac primam quidem crassam esse, e qua caseus conficitur, alteram pinguem, e qua butyrum, tertiam aquam videlicet, serum. Praeterea lac non esse unum & similare."

earthy part being fibers that allow the blood to take on "an earthy, thick, and hot constitution."[87] Noting that Galen said that milk, oil, and wine are anhomeomerous – since they have some parts that are earthy and others that are watery – Vimercato suggested that all "homogeneous substances are in fact completely heterogeneous" because of the elements that remain in them.[88] The idea of a complete mixture or corruption of the elements in a new mixture is therefore untenable.

Just as Vimercato's understanding of the constitution of blood differs little from those of Pomponazzi and Boccadiferro, his discussion of pores allows for them a role in action and passion. In his discussion of why salt and niter are "breakable" (*fictile*), he rejected Olympiodorus' solution to the apparent contradiction between *Meteorologica* IV and *De generatione et corruptione* II.8 that the word *poros* merely refers to parts that are "more suitable for undergoing change."[89] Rather, Vimercato contended, the pores are conduits, and even if action does not result directly from them, "nevertheless they contribute much to the dissolution of bodies and the reception of action."[90] Instead of seeing these pores as necessary for the reception of active powers, they merely facilitate this reception. In discussing Aristotle's statement that substances with pores suitable to the acceptance of fire are capable of being burnt, Vimercato dismissed the parallel suggestions made by Democritus and Plato that the shape of fire particles – proposed respectively as round and pyramidal – was in any way a determinant of which substances will burn.[91] Yet the pores, and the diversity of parts, help explain why some substances burn. In Vimercato's view, when air or a spirit is closed within a body, the earthy parts, having a fiery potency, can change into fire when fire is present. The pores must extend straight in order to provide sufficient respiration for ignition. If the pore is

87 Ibid., 80: "Haec sanguis est constitutio, & ideo lacti persimilis … in sanguine altera pars est veluti liquor sanguinis, sero lactis proportione respondens, altera veluti lutum & faex, quae caseo respondet. Cum autem hac duplici parte constet sanguis, terreus magis est, qui fibras habet … ab his fibris terream accipit, & crassam, calidamque constitutionem."

88 Ibid., 111: "Nam quod Galenus ait, lac, oleum, vinum, dissimilaria esse, quia partem aliam terrestrem habeant, aliam aqueam…. Ac si partes terreas & aqueas dici deberent, similares profecto omnes dissimilares essent, quandoquidem ex elementis, quae diversa sunt, constant." Galen, *De elementis*, 2.2 (1:495–7K).

89 Vimercato, *In quartum librum Meteorologicorum*, 82: "quod Olympiodorus assert, meatus hoc in loco proprie non accipi, sed partes significare, quae ad patiendum magis sunt aptae." For Olympiodorus' reading of the pores, see Viano, *La matière*, 159–163.

90 Vimercato, *In quartum librum Meteorologicorum*, 82: "quanquam meatuum merito actio non fiat, ut alii fuerant opinati, multum tamen rei solutionem, & actionem accipiendam conducunt."

91 Plato, *Timaeus*, 56b; Aristotle, *De caelo*, 3.4.303a13–14; ibid., 3.7.306a26–307b5.

bent, the fire will not ignite; or, if it does, it will be quickly snuffed out.[92] For Vimercato, the constitution (*constitutio*) of substances is in part elemental, depending on the relative proportions of air or moisture that resides within the pores. Yet it is also structural, as shape, despite his protestations, helps ensure that the fire can ignite and remain lit. Moreover, these porous bodies have parts that are distinct from the rest of the body and appear to have not been fully mixed, as only the earthy parts burn.

In his translation and commentary, Vimercato used the word *constitutio* for *dioresis*, whereas William of Moerbeke rendered it as *determinatio*.[93] Renaissance medical theorists used the word *constitutio* in a manner similar to *complexio* or *temperamentum*, applying it to human bodies, the seasons, and medicaments. While Vimercato's terminology resonated with medical theory, the commentary on *Meteorologica* IV by Francisco Vallés explicitly makes the connection between the contents of that book and the temperament of natural bodies.

Even though Vallés lived his entire life in Spain, his fame was widespread, and his writings were well known among Italians. A late edition of his commentary on *Meteorologica* IV was printed in Padua in 1591, decades after its first publication in 1558. Vallés interpreted *Meteorologica* IV as a medical work, importing Galenic vocabulary to clarify Aristotle.[94] The concept of temperament was key to his understanding of the instrumental powers of the hot and cold and their action on passive matter. He wrote that both Galen and Aristotle accepted that active and passive faculties "depend on temperament."[95] Heat and cold, in turn, are not just the causes of the elements but are the faculties responsible for generation and corruption. All substances and their passive qualities arise "by the motions of heat or cold" acting on the wet and dry matter.[96]

Significantly, Vallés considered the main topic of *Meteorologica* IV to be the passive qualities of mixtures and the ways they obtain their capacity to change. Like Pomponazzi, Boccadiferro, and Vimercato, Vallés undermined the notion of homeomerity. He contended that "a given thing (*res*), however

92 Vimercato, *In quartum librum Meteorologicorum*, 102.

93 Aristoteles Latinus, *Meteorologica*, 10/2.2, 105.

94 Martin, "Francisco Vallés," 1–30. For Vallés and the Spanish context of Renaissance debates about matter and form, see Navarro-Brotons, "Matter and Form," 99–116.

95 Vallés, *In quartum Meteorologicorum*, 6r: "Illud unum evidenter se scire Gal. profitetur, quod ex Arist. accipiendum est, facultates tam agendi quam patiendi ex temperamento pendere."

96 Ibid.: "Itaque ortus, aut corruptiones rerum naturales (ut aliae omnes mixtorum passiones, de quibus loquuturi sumus) fiunt calore & frigore moventibus, aut illorum altero; humido & sicco motis; […]."

much it appears simple to the senses, consists of diverse particles, some fiery, some watery, some airy, and some earthy."[97] Using this framework, Vallés contended that putrefaction occurs when the parts that are naturally hot become cold, or those that are naturally cold become hot, leading to the dissolution of the substance. Vallés maintained that the secondary passive qualities, such as friability, fusibility, density, and color, arose out of what he called a "mode of substance" (*modus substantiae*), a concept he traced to Galen.

What Vallés meant by the term "mode of substance" is not readily evident as there was something circular in his definition: "Galen, by the name modes of substance, was accustomed to understanding and accustomed to calling accidents that derive from the mode of substance."[98] The term, however, has a long tradition of usage, in both logical and medical writings. Pietro d'Abano (ca. 1250–1317) appropriated the term from scholastic dialectical discourse, and applied it to complexion, so that it meant a disposition that was inseparable from a substance, without being its essence or substantial form.[99] Vallés' contemporary, Giambattista Da Monte (1498–1551), defined "modes of substance" as natural passive powers, listing the tactile qualities of subtleness, thickness, softness, hardness, smoothness, and roughness.[100] Vallés appeared to adopt both Pietro's definition and that of Da Monte, meaning that these tactile qualities (*modi substantiae*) derive from a singular mode that is similar to complexion or temperament. The mode of substance derives from the proportion of earth and water in the body, and their respective dryness and wetness, that provides the corporeal nature or structure (*corporatura*).[101] The pores as described in *Meteorologica* IV, and the part they play in the corporeal structure, help explain Vallés' modes of substance. Density and rarity respectively result from large or narrow pores.[102] Moreover, bodies made of earth and water "do not become completely compact *per se* but are full of invisible pores."[103] Vallés' pores are not voids, but rather contain subtle matter that allows particles to enter into substances, affecting and transforming them.[104] Vallés was by no

97 Ibid., 14r: "Scire licet … rem quamcunque, quantumvis sensui appareat simplex, constare ex particulis diversis, aliis igneis, aliis aqueis, aeris aliis, ac aliis terreis."

98 Ibid., 34v: "Galenus nomine modi substantiae solet intelligere, soletque appellare accidentia, quae consequuntur modum substantiae."

99 Chandelier, "Nature humaine," 483–484; Klemm, "Medical Perspective," 292–295.

100 Da Monte, *In primam fen*, 174r–v.

101 Vallés, *In quartum Meteorologicorum*, 34v–35r; see also his discussion of modes of substance in relation to temperaments in idem, *Controversiae*, 371–375. For Vallés and modes of substance, see Blank, "Reductionism," 166–168.

102 Vallés, *In quartum Meteorologicorum*, 42r.

103 Ibid., 55r: "Verum haec non fiunt per se tota compacta, sed plena ocultis [sic] poris."

104 Ibid., 57r–v; 58v; 62v.

means an atomist, but his account of *Meteorologica* IV indicated a corporeal structure resulting from the mode of substance and attributed to the presence of hidden particles and pores a significant role in explaining the characteristics of bodies, which he viewed to be made up of diverse, invisible particles.

4 The Medical Tradition

Vallés' commentary was explicitly concerned with medicine, but even those of his Italian contemporaries who taught philosophy, were well aware of *Meteorologica* IV's relation to medicine. During the same years, a number of medical theorists showed themselves to be similarly concerned with passive qualities and the transformations of substances. Avicenna's *Canon*, due to its prominence in the medical curriculum, played an important role in shaping conceptions of elements, mixtures, and complexions. Avicenna's view, as transmitted in the twelfth-century Latin translation by Gerard of Cremona, emphasized that the elements are "the parts of the human body and of others, which can in no way be divided into bodies of different forms."[105] Similarly, his definition of complexion maintained that it arises when "parts [of the elements] are reduced into such small size" that there can be a great degree of contact between them.[106] Late-medieval commentators on the *Canon*, well known to sixteenth-century commentators on *Meteorologica* IV, considered these Avicennian definitions in reference to *minima*. For example, Jacobo da Forlì (ca. 1360–1414) defined mixture as "substantial transformation entailing such a division to *minima*."[107] Like Jacobo, the Sienese physician Ugo Benzi (ca. 1360–1439) adopted the Avicennian position that elements remain intact in mixture. He argued that, if this were not the case, there was no possibility of the occurrence of burning – defined by him as the separation of the thick parts from the subtle ones by the power of heat.[108] A century later, Giambattista Da Monte, a former student of Pomponazzi and professor of medicine at

105 Avicenna, *Canon*, 1.1.2.1, 7v: "Elementa sunt corpora et sunt partes primae corporis humani et aliorum quae in corpora diversarum formarum dividi minime possunt"; Siraisi, *Avicenna*, 242. For the *fortuna* of Gerard's translation in the Renaissance, see Hasse, *Success*, 96–133.

106 Avicenna, *Canon*, 1.1.3.1, 8r: "Complexio est qualitas, quae ex actione ad invicem & passione contrariarum qualitatum in elementis inventarum, quorum partes ad tantam parvitatem redactae sunt, ut cuiusque earum plurimum contingat plurimum alterius provenit."

107 Jacobo da Forlì, *In primum Canonem*, 9v: "Mixtio enim importare videtur substantialem transmutationem connotando divisionem talem ad minima."

108 Ugo Benzi, *In primam fen*, 5r: "cum adustio fit vi caloris subtilis a grosso separatio ita ut subtile sublimetur et grossus descendat."

Padua who frequently cited *Meteorologica* IV, felt it necessary to distinguish Avicenna's definition of elements – as first parts of the body – from the atoms of Democritus and Leucippus, by arguing that indivisible atoms do not contribute to a true alteration of substance.[109] While Da Monte sided against Democritus, he nevertheless considered Avicenna's *minima* to be bodies, describing the formation of mixtures through the division of the ingredients into very small parts that allow for the maximum degree of contact.[110]

The views of Santorio Santorio (1561–1636) emerged from the context of medical and philosophical readings of Aristotle's *Meteorologica* IV, reconciling it with medical conceptions of temperament and elements as found in the *Canon*, while undermining the distinctions between mixture and juxtaposition, and the concept of homeomerity. Santorio's acceptance that corporeal structure underlies all other categories of passive properties, as presented in the 1603 *Methodus vitandorum errorum*, is based on his interpretation of Aristotle's *Physics*. In reference to a passage in *Physics* VIII.7, which he admits that many interpreters have seen as a presentation of the views of others, Santorio contended that Aristotle endorsed the idea that "density and rarity are the principles of all passive qualities," such as heaviness, hardness, heat, and their opposites.[111] Santorio wrote that "if Aristotle is to be believed," rarity and density are prior to the other qualities, including heat and cold, because "just as homeomerous parts in varied ways are arranged because of the alteration of the position and motion of the parts, varied potentials for being cooled or heated arise."[112]

Santorio's evocations of pores that enable the possibility of matter undergoing alteration have no foundation in the text of *Physics* but resonate with sixteenth-century interpretations of *Meteorologica* IV. He knew this tradition, as is evident from his citations of Pomponazzi's *Dubitationes* in his commentaries on Galen and Avicenna.[113] Santorio's use of corpuscular theory emerged

109 Da Monte, *In primam fen*, 24v.

110 Ibid., 56r–57r.

111 Aristotle, *Physics*, 8.7.260b7–15. For a list of scholars who, unlike Santorio, interpreted this passage, not as Aristotle's own view, but as Aristotle's presentation of his opponent's view that density and rarity explain all passive qualities, see Blyth, *Aristotle's Ever-Turning World*, 205–206. Blyth notes that, in contrast to most interpreters, Simplicius – like Santorio – understood the passage to be Aristotle's view.

112 Santorio, *Methodus*, 157v–r.: "licet iis omnibus antecellant raritates, & densitates, quae si Aristoteli credendum est, sunt principia omnium passionum; nam pro ut similares partes vario modo figurântur ob mutationem situs, & meatus partium, variae consurgunt potentiae refrigerandi, vel calefaciendi." For a corpuscularian reading of this passage, see Bigotti, "A Previously Unknown," 29–42.

113 Santorio, *In artem medicinalem*, 109; idem, *In primam fen*, col. 171–172.

not just out of a medical context accustomed to investigating the role of small particles in contagion, as Fracastoro did, but also in a setting in which textual philosophical investigations mixed with theoretical analyses of temperament. In this context, Aristotelians were willing to confront apparent discrepancies in Aristotle's presentations of matter theory by contemplating, as had Pomponazzi, that at times, perhaps "Aristotle Democritizes."

5 Conclusion

By the middle of the seventeenth century, *Meteorologica* IV was regarded by commentators as an authoritative text for a corpuscular theory pertinent to medicine. Niccolò Cabeo, for example, a Jesuit scholar who used his commentary on the *Meteorologica* to reconcile Aristotelianism with corpuscular matter theory and alchemy, also considered the relevance of the work to medicine. He maintained that the pores that Aristotle used to explain secondary passive qualities should not be thought of as small rooms (*camerae*) of air, as some think, but rather as variations within the corporeal substance due to its being composed of qualitatively different particles. Echoing sixteenth-century commentators, from Boccadiferro to Vallés, he wrote that "bodies, however homogeneous they seem to the senses, are not truly homogeneous, and they are not of the same nature (*ratio*) with respect to the whole substance but consist of harder particles and some softer ones."[114] Thus, so-called pores are particles that are affected more easily or differently from the harder particles. Cabeo employed these pores, as well as his conviction that bodily fluids and airs are not truly homogeneous, to explain diseases. He maintained that these fluids and airs are composed of different kinds of ingredients, some of which he defined as spirits, which have active (*activae*) and animated (*vividae*) parts. Putrefaction occurs when a confluence of active spirits enclosed within the body are unable to escape. Their motion agitates body parts, creating an internal war. He applied this explanation to the etiology of putrid fevers, holding that such fevers result in the depths of winter when extreme cold blocks the skin's pores and the active spirits are unable to flow out of the body.[115]

While Cabeo used the *Meteorologica* to reframe Aristotelianism as an experimental and alchemical natural philosophy and to give insight into issues related

114 Cabeo, *Commentaria*, 4:372: "Sed solum dico corpora, quantumvis ad sensum sint homogenea, non esse vere homogenea, nec in tota substantia eiusdem rationis, sed constare ex particulis durioribus, & ex aliis tenerioribus."

115 Ibid., 4:93.

to medicine, others saw Aristotle's use of particles and pores to explain the characteristics and transformations of organic tissues in *Meteorologica* IV as a key to understanding the development of, and changes within, the Aristotelian corpus. The Italian polymath and polemicist Scipione Chiaramonti (1565–1652) – author on a variety of topics including an investigation into melancholy, a local history of Cesena, a treatise on forms of government, and attacks on Tycho Brahe – offered in his commentary on *Meteorologica* IV, printed posthumously in 1654, a potential solution to the objection that Aristotle's employment of pores and particles in that book contradicted his dismissals of similar explanations in *De generatione et corruptione*. He cited, without endorsement, the argument of unnamed scholars who believed that since Aristotle's views had changed or had even progressed over time, *Meteorologica* IV must reflect the historical development of his thought. Furthermore, some of the solutions presented in Aristotle's works should not be understood as absolute; nor should there be any expectation of finding universal consistency in the Aristotelian corpus.[116]

Chiaramonti's comments resonate with modern debates over the possible spuriousness of *Meteorologica* IV. But his attempts to remove any doubts about the work's origins reveal also the thoughts of some of his contemporaries and immediate predecessors. Indeed, Joachim Jungius (1587–1657) in 1642 endorsed the first of Chiaramonti's proposed solutions, maintaining that *Meteorologica* IV promoted a syndiacritical natural philosophy, which could be reconciled with Paracelsian alchemy and opposed Aristotle's earlier polemics against Democritus and Empedocles. For Jungius, a champion of induction who promoted a natural philosophy based on experimentation, the pores and particles of *Meteorologica* IV represent the real, mature thought of Aristotle. The attacks in the *De caelo* and *De generatione et corruptione* were to be considered either as dialectical exercises set down for the sake of his students, or as arguments that were outright corrected in his later works.[117]

The views of both Chiaramonti and Jungius were informed not just by their investigations into nature but also by their deep knowledge of the sixteenth-century commentary tradition on *Meteorologica* IV. Indeed, Chiaramonti's comment on the possibility of explaining the pores through

116 Chiaramonti, *In quartum metheorum*, 315: "Praeterea objicit rationi ex pororum angustia adductae, repugnat enim illis, quae Democrito objicit in primo de Generatione. Aliqui forsan sese extricabunt respondendo progressum Aristotelis hunc, & alios etiam nonnullos esse probabiles unde non mirum si ab integra universitate excedant."

117 Jungius, *Disputationes*, 389; Newman, "Experimental Corpuscular Theory," 327–328; for Jungius's methodology, see Clucas, "*Scientia*," 53–70.

speculations on the development over time of Aristotle's thought, is in response to Pietro Pomponazzi's depiction of passages of *Meteorologica* IV as inconsistent with other books of Aristotle's natural philosophy. In addition to Pomponazzi's commentary, Chiaramonti frequently cited the commentaries of Vallés and Vimercato. Jungius, in turn, had studied in Padua with the philosopher Cesare Cremonini (1550–1631) and the physician Santorio Santorio; he engaged deeply with the writings of Jacopo Zabarella (1533–1589); and his library contained commentaries on *Meteorologica* IV by Vimercato, Zabarella, and the Coimbrans.[118] The *natio germanica* at Padua added Pomponazzi's commentary to its library in 1598 and again in 1611; Vallés' commentary was donated to them in 1603.[119] Jungius gave examples from book IV to show that Aristotle's *onkoi* and pores could, in his view, be considered as evidence of a profession, if not of atomism, then at least of corpuscularism.[120] Moreover, he maintained that the syndiacritical philosophy undermines Aristotle's earlier account of truly homogeneous substances, because simple bodies are said in *Meteorologica* IV to be composed of a variety of particulate components. Specifically, Aristotle's account of milk and blood demonstrates that he believed simple natural substances made in the human body were not truly homeomerous – a view shared by Jungius's teacher Cremonini.[121] For Cremonini, milk is composed of watery, earthy, and fatty particles, which are visible respectively in whey, cheese, and butter. According to Jungius, blood contains watery serum in addition to fibers, revealing that in Aristotle's teaching, blood contains what Jungius called "hypostatical parts."[122]

Jungius's view of the syndiacritical character of *Meteorologica* IV ties in with the traditions of alchemical atomism of the late Renaissance, exemplified

---

118 Di Liscia, "*Operosum negotium*," 215–255; Meinel, *Die Bibliothek*, 117, 192, 195.

119 Favaro, ed., *Atti della nazione germanica*, 2:131, 2:206, 2:332.

120 Jungius, *Disputationes*, 393: "Quid multa? Cap. 8 adeo aperte in castra syndiacritorum transit, ut etiam vocabulo *ipsis* proprio utatur, dum 'ea, quae humoris absentia concrevere, ab humore liquefieri' scribit, 'nisi ita coierint ... ut meatus molibus,' hoc est particulis sive corpusculis, 'aquae minores sint relicti.'" His reference is to Aristotle, *Meteorologica*, 4.8.385a29–30.

121 Cremonini, *Lectiones*, 762: "lac est quasi corpus etherogeneum habens partes crassiores, et terrestres, ex quibus fit caseus, et alias fluividas, ex quibus est serum"; Cremonini, *Lectiones*, 768: "Respondo similiter, quod lac est corpus quasi eterogeneum; habet enim diversas partes, caseum qui habet multum terrenae portionis, et serum quod est propemodum acqua [sic]."

122 Jungius, *Disputationes*, 391: "Porro lib. IV Meteor. Cap. 7 'in sanguine tum ichora sive serum tum fibras' ut hypostaticas scilicet partes 'contineri' docet."

by Andreas Libavius (ca. 1555–1616) and Daniel Sennert.[123] Yet, his examples point to a second arena for the emergence of Aristotelian corpuscularism, namely the intersection of natural philosophy and medicine. From the time of Pomponazzi, sixteenth-century commentaries on *Meteorologica* IV raised doubts about the homogeneity of organic fluids, understood pores to be conduits for small parts that act upon passive matter, endorsed corporeal structure as an explanation for passive qualities, and even saw Aristotle as embracing Democritean frameworks.[124]

123 Newman, *Atoms*, 66–83.
124 Lüthy, "Aristotelian Watchdog," 542–561.

CHAPTER 5

Atoms, Corpuscles, and Minima in the Renaissance: The Case of Nicolaus Biesius (1516–1573)

Christoph Lüthy

Was there such a thing as a 'Renaissance revival of atomism'? The assumption that there was requires that there was such a thing as 'atomism' in the first place, and hence a school of thought that could be, and was, revived. But did the Renaissance – here canonically defined as the period AD 1400–1600 – possess such a category? As was explained in the introduction to the present volume, it did not. To the question of whether there were perhaps Renaissance authors who thought of themselves as 'atomists,' as we have seen, the answer again is no. What we can say with certainty is that we do find Renaissance philosophers and physicians who relied on small bodies – particles, minima, atoms, corpuscles – in their explanations of physical phenomena. Would it be anachronistic to describe their endeavors in terms of a revival of atomism? Yes, it would. But how else, then, should we describe what happened? How should we, as historians, frame the shift in explanatory patterns that these authors seem to illustrate?[1]

By means of a detailed case study, this chapter will try to document that – as historians – instead of pitting 'atomism' against 'hylemorphism' (another term unknown in the Renaissance), we should take note of the multifarious reasons that led to an increased explanatory reliance on small particles, without having to assume that the authors thereby necessarily committed themselves to the revival of an ancient school of thought.[2] We should be alert to the fact that natural philosophers and physicians, under the influence of the most diverse ancient or more recent sources or on the basis of empirical evidence, occasionally and sometimes even routinely invoked small particles in their explanations, sometimes calling them 'atoms' and sometimes something else, without thereby making any ideological commitment to an atomistic ontology or a philosophical sect. By calling these authors 'atomists' by virtue of their use

1 I would like to thank Frederik Bakker, Dilwyn Knox, Elena Nicoli, Carla Rita Palmerino, and Leen Spruit for their precious comments on a draft version of this chapter.

2 On the invention of the term 'hylemorphism' in the nineteenth century, see Manning, "The History of 'Hylomorphism'."

 | DOI:10.1163/9789004528925_006

of this term 'atoms,' however occasionally or frequently the word was used, our view of their natural philosophy is narrowed to the point of distortion. After all, the epithet 'atomist' suggests that atoms were the central explanatory unit of their theories. But there were no Renaissance philosophers before 1610 – the late Giordano Bruno and possibly Nicholas Hill being the only exceptions – for whom atoms served as a central explanatory tool.[3] In fact, in texts where we do encounter the term 'atom,' we are likely to encounter also the four elements, substantial forms, astrological influences, pneuma, sympathies, a world soul, plus an omnipotent God with providential intentions.

In this chapter, I wish to illustrate this point on the basis of the examination of three works by an intriguing author, Nicolaus Biesius, who in the 1560s and 1570s came to invoke atoms, minima, corpuscles, and other small particles with increasing frequency. Biesius – who has so far eluded the attention of historians and plays no role whatsoever in the available 150-year historiography of atomism – documents beautifully how undramatic and gradual could be the shift from an Aristotelian framework to corpuscular explanations. Biesius's works allow us to watch how atoms could make their entry into natural philosophy quite casually, without thereby challenging substantial forms and without causing any public outcry – two aspects in which atoms were engulfed a few decades later. At the latest by 1650, the atom had become a fetish to one party and a scandal to the other, and hence a shibboleth that defined one's adherence to the so-called new philosophy.

Let us first introduce our historical figure and then turn to an examination of the matter theory as it unfolds in three of his works. Surprisingly little is known about Nicolaus Biesius (or Nicolas Bièse), despite his impressive career. Born in the Flemish city of Ghent in 1516, the encyclopedia entries report that he studied philosophy and subsequently medicine at Louvain. In the Preface to *De varietate opinionum*, Biesius tells us that he wished to explore the world. He left Flanders with the intention to continue his studies in Lyon but ended up in Spain for no fewer than ten years.[4] With respect to Spain, the encyclopedia entries mention that he followed courses in philosophy and eloquence at the University of Valencia. In fact, his earliest publication, annotations on a Ciceronian oration, was published in Valencia in 1646.[5] From there, Biesius

---

3 For Bruno, see notably *De triplici minimo et mensura* of 1591. For Hill, see *De philosophia epicurea, democritica, theophrastica* of 1601, as well as Sandra Plastina's chapter on Hill in the present book. After 1610, one begins to see elaborate atomistic ontologies. While present in the private notes of Isaac Beeckman or Thomas Harriot, they appear in print in David Gorlaeus's *Exercitationes philosophicae* (published in 1620, eight years after the author's death in 1612), and in Sébastien Basson's *Philosophia naturalis contra Aristotelem* (1621).

4 Biesius, *De varietate opinionum*, "Praefatio," A4v.

5 Biesius, ed. *Cicero: Pro T. Annio Milone oratio*.

moved on to Italy, where – again according to his own testimony – he stayed for three years, and where, according to the encyclopedia entries, he obtained a doctorate in medicine from the University of Siena. The latter sources further tell us that in 1558, King Philip set up a new chair in Galenic medicine at Louvain and that the appointment went to Biesius. One source also attributes to Biesius the function of academic orator to the Duke of Alva. While I have found nothing to corroborate the foregoing claim, all sources concur that in 1570, Biesius was appointed personal physician to Emperor Maximilian II. Merely two years later, he died in Vienna, either from apoplexy, as most sources report, or "after his clandestine enemies had poisoned him," as Zedler maintains.[6]

His intellectual development was attested by numerous publications. After his Spanish edition of Cicero, he moved on to political philosophy.[7] Thereafter, his publications were mostly survey works in the fields of medicine and philosophy.[8] It is not always evident to today's reader why he felt compelled to publish his manuscripts. It is also doubtful whether his works had the impact for which he must have hoped. One does not find Biesius's name at all frequently mentioned in the natural philosophical literature of the sixteenth and seventeenth centuries, although a few of his works did enjoy second editions. Today, his intellectual achievements are quite forgotten.

There are nevertheless some reasons why we might wish to rescue him from the dungeon of oblivion. Maybe he is the figure responsible for the presence of atomistic thoughts in Louvain in the early seventeenth century that so vexed Libertus Fromondus in his *Labyrinthus, sive de compositione continui liber unus* of 1631 – a period in which the concept of atoms had begun to provoke scandal once again. As we will see below, it was exactly the labyrinthine problem of "the composition of the continuum" that made Biesius edge ever more closely to a discontinuist or atomistic view of magnitudes.[9] There is at least one other aspect that would merit examination, although it cannot be addressed in any detail here – namely, his early acquaintance with, and reaction to, Copernicus. The *De universitate libri tres* of 1556 contains a Copernican diagram (see Fig. 5.1) and discusses the heliocentric model. Although here and later – for example, in

6 See, notably, Saint-Genois, "Bièse, Nicolas," but also Foppens, *Bibliotheca Belgica*, I, 900; Eloy, *Dictionnaire historique de la médecine*, I, 345; Adam, *Vitae Germanorum medicorum*, 194. The details about King Philip setting up the chair in Galenic medicine are from Andrea, ed., *Fasti academici studii generalis Lovaniensis*, 131. The suggestion that Biesius was poisoned is from Zedler, *Grosses vollständiges Universal-Lexikon*, s.v., 3:1809.

7 Biesius, *De republica libri IV* (1556; expanded ed., ibid., 1564). On Biesius's political writings, see Tilmans, "From *institutio* to *educatio*."

8 Biesius, *Theoreticae medicinae libri VI* (1558); idem, *In artem medicam Galeni commentarii* (1560); idem, *De methodo medicinae* (1564); idem, *De varietate opinionum liber I* (1567); idem, *De natura libri V* (1573, 2nd ed. 1613); idem, *De arte dicendi libri II* (1573).

9 On Fromondus, see the observations in our Conclusion, below.

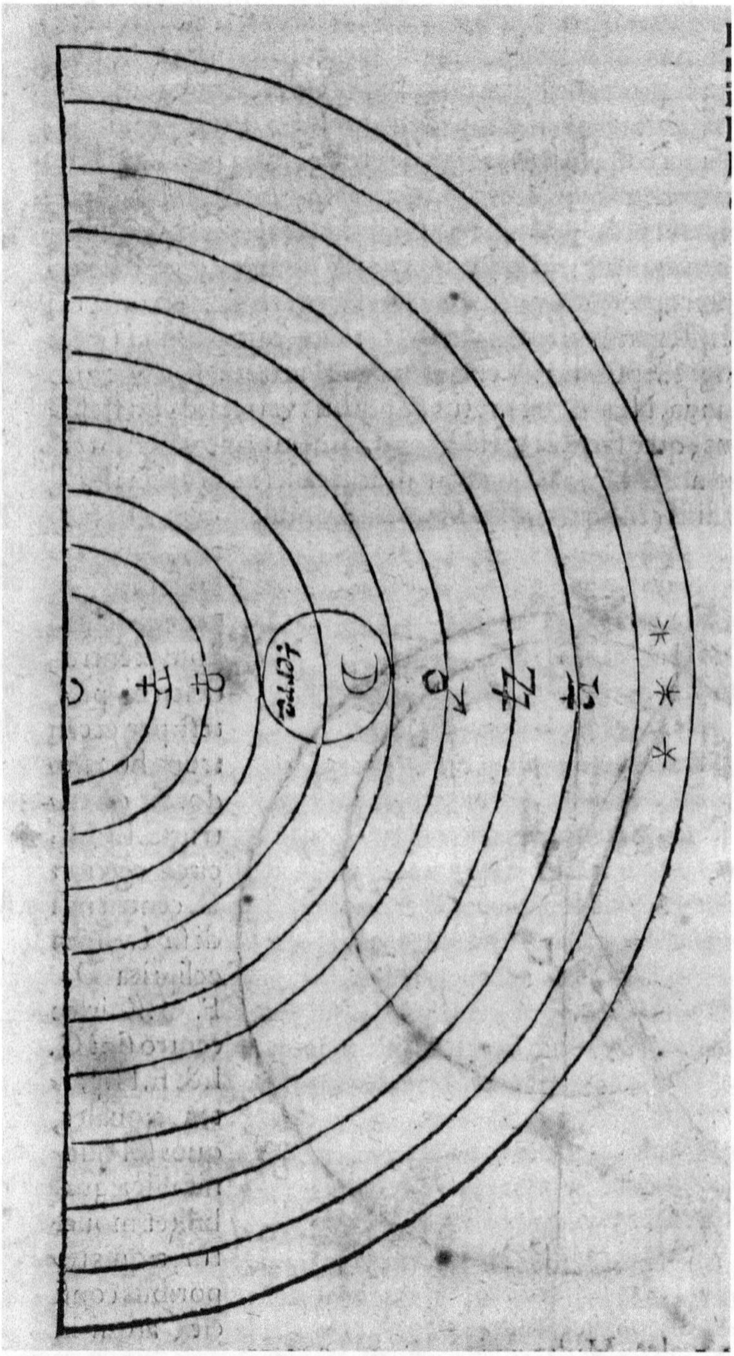

FIGURE 5.1 Copernican diagram. Nicolaus Biesius, *De universitate libri tres* (1556), 131

De natura of 1573 – Biesius does not espouse heliocentrism, his early acquaintance with and interest in Copernicus's recent proposal is conspicuous.[10] One would have to investigate his links to Renierus Gemma Frisius (1508–1555), the physician, mathematician, cartographer, and instrument-maker who was also a member of the Louvain medical faculty, and also his links to Frisius's son, the physician, astronomer and astrologer Cornelius Gemma (1535–1578), who in 1569 was to succeed Biesius in the regius chair of medicine when the latter left for Vienna.

But neither Biesius's interest in cosmology, nor the question of whether he was or wasn't an influential author, is of any consequence for our present argument. What interests us here are his reasons for nurturing ever more positive sentiments about minima, corpuscles, and atoms. By following the development of his argument across three of his works – an early manuscript *De varietate opinionum* published belatedly in 1567, the *De universitate* of 1556, and the *De natura* of 1573 – we hope to be in a position to say something meaningful about why a sixteenth-century author should have ended up invoking atoms without thereby being 'an atomist,' and why he should have done do so in a period when to embrace atoms was quite unusual.

1 De varietate opinionum (1567)

We start our survey with *De varietate opinionum liber unus* ("About the variety of opinions, in one book") of 1567, a very odd work that Biesius penned down, as he explains in his "Preface," during his itinerant years in Spain and Italy. The book is written with a sense of urgency, but unfortunately for us, Biesius is very vague about what it was that so agitated his youthful mind. As we have just mentioned, before becoming professor of Galenic medicine, he had published on political issues, and his official reason for composing *De varietate* is political, as the title of chapter 1 explains: "The variety of opinions is very dangerous to the state."[11] The reader intuits, notably towards the end of the book, that when Biesius writes about "the variety of opinions," about intellectual currents and sects and about the dangers they pose to the state, he is concretely thinking of the confessional divisions lacerating Europe at the time, and even more

10 Biesius, *De universitate*, 129ff., with the diagram at p. 131; see idem, *De natura*, 55b, for an oblique reference to Copernicus, whose proposal however Biesius rejects.

11 Biesius, *De varietate*, title of ch. 1: "Opinionum varietatem admodum esse Reipublicae perniciosam." I have standardized Biesius's spelling throughout, writing out abbreviations and standardizing the use of 'v' and 'u'. I have not changed his punctuation.

specifically about the fast-spreading Protestant confessions. At one point in his treatise, he conjures an image of people who had died at least fifty years earlier, allowed to return to the world and being shocked by the spectacle of division and strife – clearly an allusion to the Reformation.[12] There are, moreover, two chapters in which he specifically defends the Catholic understanding of the Eucharist against "evil and barbaric men."[13] And in the last chapters of his book, he indicates that in the most complex and lofty questions, which are obviously of a theological nature, one has to rely on the eternal consensus of the learned, not on private opinion, which is a standard anti-Protestant argument. For these reasons, Lynn Thorndike has decreed that "the *De varietate opnionum* of Biesius does not present him in an attitude favorable to intellectual development."[14] But with respect to Biesius, as in numerous other cases, Thorndike is a selective and impatient reader.

Still, one would have liked to know how Biesius's concrete experiences in Flanders, France, Spain, and Italy shaped his ideas, and to which authors he thought he was responding. But his considerations remain extremely general. His point of departure is that a society, or a state, needs to have a "common consensus" in order to survive, a point that provides not only his premise but also the conclusion of the book.[15] He presumably fancied that his treatise made a contribution to creating such a "common consensus."

What renders this work interesting for our current purposes is that Biesius, in his search for a consensual view of things, locates the source of error and confusion not in socio-economic circumstances, political developments, or theological disagreements, but in epistemology and ontology. This, in fact, is the context into which matter theory enters, as it provides the glue that holds the fragmented universe together. Note that Biesius is not the only author from that time period who felt that the knowledge crisis that resulted from the clash of the confessional ideologies needed to be addressed from a commonly accepted ontological foundation upwards, and on the basis of clear and generally acceptable ideas about the ultimate units that made up this world

12 Biesius's *De varietate* has no pagination. I will therefore refer to chapters and page number of the chapter quoted; here 21, iv.

13 The defense of transubstantiation is found in ibid., 13, title: "Verbis divinis individuam hominum naturam comprobari, et unam esse verbi substantivi significationem," from where the quote about the "mali et barbarici homines" is taken; the other is chapter 16, whose title reads: "In Eucharistia Christi corpus esse, Christi verba probare."

14 Thorndike, *History of Magic*, 6: 395.

15 Here, 1, ii: "At constat, consistere non posse Rempublicam in qua nullus est de rebus necessariis communis consensus, in qua quisque sibi vel intelligendi vel vivendi legem ponit [...]."

of ours. We find, on the other side of the confessional divide, for example, Nicolaus Taurellus, a philosopher-physician like Biesius, who composed his *Philosophiae triumphus* (1573) for similar reasons. It might be argued that the epistemological turn, which tried to erect an ontology on the basis of an analysis of human knowledge, was a reaction to the intellectual crisis provoked by the strife between the confessions, with "each side trying to sap the foundations of the other," in Richard Popkin's formulation.[16]

Whether Biesius would have recognized himself in this description must, of course, remain uncertain. At any rate, he opens his treatise with epistemological issues and specifically with the problem of obtaining certain knowledge, so that he can later move on to ontological issues. His epistemology begins with the senses. We learn that each of the senses furnishes us with specific information about outside things, and together they produce ideas about the "magnitude, number, position, motion, rest and figure" of things.[17] But not everything we know comes from outside. Our soul, which is "a form capable of all forms," also possesses innate "species and forms," and when the senses furnish their "species" from the outside, our soul merges them in the *phantasia* with our innate "species and forms," enabling understanding.[18]

But regrettably, Biesius continues, we do not often arrive at a proper understanding of things. This failure is due to the obscurity of the things themselves, on the one hand, and to the obtuseness of our intellect, on the other.[19] That is why we have to investigate carefully how we arrive at our judgements. The first step is to see whether we actually use our words correctly, and whether they correlate with nature. In this, we must proceed like Socrates in the Platonic dialogue *Meno*, and lead the intellect from what is well understood to the more recondite truths.[20] The *Meno* is, of course, about mathematical understanding and mathematical proof, and in fact, one often notices how much Biesius would have loved to reduce the world to a construction possessing mathematical certainty and to strip away the bodily imperfections of the material world.

16 Popkin, *History of Scepticism*, 14. On the origin of Protestant ontology, see Leinsle's excellent *Das Ding und die Methode*; see also Taurellus, *Philosophiae triumphus*. In my "Ontology and the Place of Metaphysics," I have tried to show that, in the wake of Taurellus's *Philosophiae triumphus*, a series of thinkers genealogically leading up to Descartes might have sought doctrinal stability in a stable ontology.

17 Biesius, *De varietate*, 2, ii: "Nec tantum proprias differentias suas quisque sensus defert, sed etiam rerum magnitudinem, numerum, positionem, motum, quietem, figuram."

18 Ibid., 2, iv–vi.

19 Ibid., 4, i.

20 Ibid., 6, ii.

In *De varietate*, however, his attempts to find such certainty are hesitant and inconsistent.

It is precisely here, at this stage in his argument, that we find Biesius give voice for the first time to a conviction that he would repeat again and again, also in his later writings. As the title of his seventh chapter explains, "divisible nature is the great reason for the variety of opinions."[21] Given that this is not at all a self-evident claim, the relevant passage deserves to be quoted in full:

> When they take the dimensions to be infinitely divisible, they are found to ruin unity, by which all things are held together and on which principally the cognition of truth is constituted, above all the cognition of the truth of incorporeal and divine things. And thereby they also allow for lots of occasions for highly aberrant opinions. After all, in an infinite division, there is nothing in which the mind can come to rest.[22]

Biesius's Platonic hankering for mathematical certainty seeks salvation in indivisibility and unity. In fact, a sense of horror of infinite divisibility runs through all of the three works by Biesius examined in the present chapter – a sentiment that would eventually drive him into the arms of an ontology of indivisible minima, atoms, and corpuscles. In *De varietate*, however, Biesius still scouts for unity and stable units in the middle of a bodily world rife with infinitely divisible magnitudes of Aristotelian provenance. Biesius's escape route out of this conundrum is, as it were, psychological: he wants us to see in all things their essences; essences, of course, being non-dimensional and extensionless.[23] For, as he never tires of telling us, "all that is bad and false comes from division, just as all that is good and true can be seen to be placed in the one, from which they all degenerate as a result of their divisible nature."[24] And again, with unceasing urgency: "We must free ourselves from these heavy chains of the dimensions," which prevent us from arriving at the divine reasons of things.[25]

---

21 Ibid., ch. 7, title: "Dividuam naturam esse magnam varietatis opinionum causam."

22 Ibid., 7, i–ii: "Cum autem dimensiones in infinitum dividuae statuantur, unitatem, qua cuncta constant et in qua praecipue cognitio veritatis est constituta, maxime cognitio veritatis rerum incorporearum et divinarum, evertere, et plurimas falsissimarum opinionum occasiones praebere videntur: siquidem in infinita divisione nihil est, in quo mens conquiescat."

23 Ibid., 7, ii.

24 Ibid., 7, iii–iv: "Imo vero, ut aliquid etiam in universum dicamus, omne malum atque falsum ex divisione provenire, quemadmodum et omne bonum atque verum in uno positum esse videtur: a quo cuncta degenerant per dividuam naturam."

25 Ibid, 7, iv: "[...] ut nos crassis istis dimensionum vinculis eximamus [...]."

The essential non-dimensional units that Biesius seeks to find in his search for certainty are thus not yet material atoms or minima, as they were to become in his later *De natura*, but spiritual entities – a strange conceptual turn, as essences and spiritual entities are not the same, at least for an Aristotelian philosopher. At any rate, Biesius turns his attention to "matter and form: from which both the remaining accidents and the dimensions are produced."[26] In his philosophy, God created forms, and these forms possess powers (*vires*). These forms are non-dimensional essences, which carry, as it were, pictures of the divine mind into the natural world. As they enter nature, they assume their three-dimensionality and become material bodies. But this very dimensionality is also their weakness, as it makes them corruptible: by entering matter, the non-dimensional essences or forms remove themselves miserably from their divine origin.[27]

Just when one believes oneself to have grasped the Platonic (or possibly even gnostic) leitmotif in Biesius's argument, it takes an unexpected turn: we are told that one can "separate the dimensions from the physical bodies."[28] The proof begins with a dizzying argument about the duplicity of us, human beings, as we are both divisible (because bodily and dimensional) and also indivisible and divine (because of the soul), an argument that echoes Plato's *Phaedo*. Despite our participation in the realm of material divisibility, we can cognize the divine.[29] It is here, in this anthropological context, that the word 'atom' makes its unexpected entry in a monstrous sentence that we have here rendered into three:

> Hence [humans] are here somehow forced to undergo that condition of those lowest forms that are attached to them. But these forms, when they are extended out of their indivisible principle through matter into the triple dimension, produce some minima of nature, which are called "atoms" by the philosophers. And out of these bricks and mortar, this vast house, the structure of this entire bodily universe and its individual parts is constructed to the best of use.[30]

26 Ibid., ch. 8, title: "De materia et forma: a quibus tum accidentia reliqua, tum dimensiones producuntur."

27 Ibid., 8, i–v.

28 Ibid., ch. 9, title: "A corporibus physicis dimensiones separari."

29 Ibid., 9, i–iii.

30 Ibid., 9, iv: "Quamvis hic conditionem infimarum formarum, quibus annexae sunt, quodammodo subire cogantur: quae quidem dum ex individuo suo principio per hanc materiam in dimensionem triplicem extenduntur minima quaedam naturae, quae philosophis atomi dicuntur, producunt: ex quibus, quemadmodum, et ex lateribus, intrito,

Although this is certainly an odd context into which to introduce atoms and natural minima, Biesius seems to nurture the short-lived conviction that he has cleared the way for a notion that we find much more successfully worked out in Giordano Bruno – namely, the idea that the unity of God in all things is somehow mirrored even at the most debased levels of material reality. According to Biesius's explanation:

> Through the potency of division, everything approaches privation and nothingness: for infinite division will be pure privation; which nature cannot tolerate, because of the monad, which holds everything together.[31]

In order to avoid this abyss of the annihilation of "nature, and notably human nature," it is essential to stop the process of division. Biesius, announcing proudly that it is his own invention, wants to call this halt to division by the name of "privation of privation." This negation of a negation also carries the positive names of "monad and one" or "habit and monad," respectively.[32]

At this point, one expects Biesius to unlock for us a Brunian world of atoms, monads, and points, but he doesn't. In fact, the word 'atom' is not used again in *De varietate*, and his idiosyncratic solution to the problem of infinite divisibility is not worked out any further. Instead, Biesius's treatise continues to sail through a syncretistic sea of seemingly incompatible ideas. He remains Aristotelian when speaking of the composition of matter and form, is a Platonist when assuming that the eternal geometry of the heavens gets sullied in the dirt of bodily transitoriness, and a Neo-Platonist when thinking that the material levels of nature are distant emanations of the more spiritual realms. But time and again, minima and monads – but no atoms – come floating along, and they usually fulfill the role of stabilizing an otherwise endangered world of infinite divisibility, privation, and chaos.

---

reliquisque partibus amplissima domus, fabrica totius huius universitatis corporatae singulariumque partium, in maximos usus est extructa."

31 Ibid.: "Sed cuncta per divisionis potentiam ad privationem et nihilum accedunt: nam infinita divisio, pura privatio fuerit: quam propter monadem, qua constant omnia, natura ferre nequit." This use of "privation" is idiosyncratic, as Aristotelian philosophy understood it as one of the principles of change, together with matter and form, and not as the end of division.

32 Ibid., 9, iv–5: "[…] ab ea tanquam abiectissima natura, praesertim humana, debebat in habitum revocari: quem habitum nos monadem et unum appellamus, quod cum individuum et cum aliis non commune definimus, per privationem privationis intelligimus. Dividuum enim privationem significat, quemadmodum iam dicebamus: adverbium vero negandi praefixum privationem eius privationis, hoc est habitum et monadem, ostendit."

However, these units play uneasy roles. For example, Biesius takes recourse to natural minima so as to anchor his four elements. But if natural minima are not just a spatial limit to the divisibility of the quantity of matter for a given form, as they had originally been, but instead actual minimal units – as they became in some Renaissance authors – then they contradict Aristotle's notions of undivided and hence ever divisible continua.[33] Aware of the problem, Biesius tries to offer, rather unsuccessfully, the following solution: natural minima "are numbered as one, when continuous, and as many, when divided, and are yet indivisibles when liberated from the dimensions [...]."[34] But this is a typical case of having one's cake and eating it too: either a continuous magnitude is made up of actual *minima naturalia*, in which case they will stay "many"; or else the term "minimum naturale" designates the lower limit of division, but they are then not the constituents of the elements.

Later in the text, Biesius runs into another problem, when he embraces the Neo-Platonist idea of a God who multiplies himself emanatively into ever lower regions. At the end of the range, where you encounter "prime matter and privation," the specter of nothingness looms large once more.[35] Biesius hopes once again that some kind of "unity" will save him from vanishing into that black hole, but this time it is not *minima* nor atoms to which he turns, but something resembling mathematical numbers: "Unity must be conserved, in all numbers, just as in all other things, so that it doesn't all seep away into nothingness."[36] This is once again a monadic suggestion of sorts: just as in all (natural) numbers, the basic number is conserved, so there must also be an irreducible unit in natural things. But that idea, which was fully worked out by Giordano Bruno with his "triple minimum" of point, atom, and monad, looks in Biesius merely like a desire, a wish, a necessity that was perceived, but which does not yield any concrete ontology, at least not yet in this specific book.[37]

33 Van Melsen, *From Atomos to Atom*, has cast the *minimum* into the role of the bridge between Aristotelian natural philosophy and modern atomism. For a conceptually somewhat more reliable account, see Murdoch, "The Medieval and Renaissance Tradition of *Minima Naturalia*."

34 Biesius, *De varietate*, x, iii: "[...] minima naturae; quae continauae pro unis, divisae pro pluribus numerentur, & tamen a dimensionibus liberae individuae sint, ac in diversis esse possint."

35 Ibid., 14, v.

36 Ibid., 14, vi: "Nihilominus oportet, ne diffluat in nihilum, ut in omnibus numeris, quemadmodum et in reliquis omnibus rebus, unitas conservetur."

37 Cf. Bruno, *De triplici minimo et mensura*. Monti, "Introduzione," is still perhaps the most competent analysis of Bruno's atomism.

2 De universitate libri tres (1556)

We have, a few times already, compared Biesius with Giordano Bruno. In the case of the second work to which we now turn our attention, such a comparison forces itself upon us not only for thematic, but also for stylistic reasons. For *De universitate libri tres* of 1556 is written in the form of a *prosimetrum*, like Bruno's Frankfurt Trilogy of 1591, that is to say, it combines prose and verse parts. We have a few ancient and medieval examples of didactic poems with interspersed prose explanations, including Martianus Capella's fifth-century *De nuptiis Philologiae et Mercurii*, Boethius's sixth-century *Consolatio philosophiae*, or Bernardus Silvestris's twelfth-century *Cosmographia*.[38]

As the full title of *De varietate* announces, it "contains an entire philosophy of nature." Let us begin by stating, first of all, that our impression of *De universitate* is more favorable than that of Louis Brakelants, who concluded that this work is "extremely tedious and seems to have elicited no interest," notably because the rhetorical devices employed "make reading it tiring and at times unbearable"; it is also more favorable than Lynn Thorndike's judgement, for whom the only interesting passages in the work were those on astrology and alchemy.[39] Our own more positive assessment is based on the fact that, although *De universitate* was published in 1556, and thus more than a decade before *De varietate*, it is doctrinally more mature than the latter, which Biesius described as belonging to his *juvenilia*. Above all, *De universitate* documents interesting developments in Biesius's matter theory.

Our author opens his hexametric declamation with an invocation of the divine monad that shines forth in all created forms.[40] In his prose explanation, he argues that Aristotle's contention that the monad (as the basic numerical

38 Granada and Tessicini view Biesius's *De universitate* as the only recent model of a prosimetric natural philosophy, prior to Giordano Bruno's attempts in the Frankfurt Trilogy; see Granada and Tessicini, "Introduzione," 16: "In ambito scientifico e filosofico, ci è possibile riscontrare (almeno per il momento) soltanto un precedente. Si tratta dei tre libri *De universitate* (1556) di Nicolaus Biesius [...]." However, Granada and Tessicini haven't found any evidence that Bruno knew Biesius's work. For a comparison of Bruno's and Lucretius' didactic poetry, see Monti, "Incidenza e significato," and, more recently, Knox, "Wonder and the Philosopher's Perfection," esp. 31ff.

39 Brakelants, "Le *De Universitate* de Nicolas Biesius," 138: "[...] elle est très rébarbative et ne semble avoir éveillé aucun intérêt; c'est à peine si nous la trouvons citée par les anciens bibliographes." Ibid., 142: "les figures de rhétorique [...] rendent la lecture fatigante, insupportable parfois"; see also Thorndike, *History of Magic*, 6: 396–397.

40 Biesius, *De universitate libri tres, quibus universa de natura philosophia continetur*, 25: "Et tamen his imis etiam caelestis origo est / Formis: in cunctis divina relucet imago / Perfectae monadis, quae cuncta amplectitur [...]."

unit) possesses neither body nor weight nor any other natural property and is therefore a mental abstraction, has been contradicted by many philosophers. Biesius in fact prefers to follow Proclus, Plato and other Platonists, when postulating that the unity has to be present in all things, including the "mathematical atom": "For if we take away the monad from the nature of things, everything will at once fall apart into empty numbers [or units] of matter."[41]

This seems a good moment to pause and discuss the three ways in which Biesius uses the term "monad." He uses it, in a Platonizing fashion, to designate the unity that preceded everything else – as the divine fountain of creation, material objects, and worldly multiplicity. He also uses it in the mathematical sense, as the numerical unit that underlies all natural numbers. While the Renaissance sometimes saw these two meanings combined in the Pythagorean and Boethian tradition to which Giordano Bruno was also an heir, it is obvious that the first meaning is essentially metaphysical, the second mathematical. What Biesius is constantly striving towards is a third, physical meaning, that is to say, a monad that can refer to a physical unity that will not fragment further. Note that neither the metaphysical nor the mathematical monad lend themselves easily to this purpose, as they are both immaterial. It is precisely his desire to postulate an unbreakable extended material unit that leads Biesius to thrash about in search of a physical monad – which, in the end, he would feel able to designate with the term 'atom.'

However, at this stage in his intellectual development, the problem is that the concept of a physical monad cannot be extracted from the divine monad. For, if all of nature is ultimately held together by the divine monad, how do we explain the variegated world that surrounds us? In response to this question, Biesius embraces, as also in *De varietate*, a model in which the original, divine, monadic unity is broken and weakened when it enters into matter, becoming divisible and assuming the properties of substance, motion, figure, quantity, and perseverance.[42] Homer captured the situation well, declares Biesius, when he spoke of the golden chain connecting all things to the divine realm. Indeed, that all of creation points to the infinite unity of its original cause can be seen by the fact that the cosmos is round and that its motions are circular.[43] But how, then, did the *machina mundi* come about? In hexametric form, Biesius discusses a variety of cosmogonies, but returns in the end to his own preferred

41 Ibid., 26: "[...] quemadmodum est apud Proclum ἄτομον mathematicum [...]"; ibid., 27: "Nam si tollamus monadem e rerum natura, protinus in vanos materiae numeros omnia distrahentur."

42 Ibid., 29.

43 Ibid.

option, which once more relies on the monad. In his prose exegesis of his verses on cosmogony, Biesius cites Aristotle, Plato, Parmenides, Melissus, Thales, Anaximenes, Heraclitus, and Empedocles, with Democritus conspicuously absent from the list.

There are no indications that Biesius is here purposefully suppressing the obvious source for his speculations. Whenever he invokes atoms in *De universitate,* they don't have Democritean connotations, but they fulfill that third, physical role of the monad, which we have just described and whose unity is somehow due to both its divine origin and to its quasi-mathematical nature. When he tries to argue for it, Biesius depicts the atom as a physical monad which he then connects either with a Platonic notion of indivisibility or with an Aristotelian *minimum naturale*.

In the process, and contrary to the views expressed in *De varietate,* he now begins to believe that matter can be regarded as consisting of indivisible units. It is significant that he introduces that thought via Plato, whose doctrine, in the eyes of Biesius, guarantees the link between individual things and their divine origin:

> Plato puts the forms or ideas together with the magnitude of matter, thanks to whose participation numbers and substances are made from the monad, which in turn are indivisible because of the character of the monad. And he says that the one and the good is only in God, and that it is the principle of all things, but nonetheless that this universe and its individual parts contain the similitude of this monad, so that it is possible to call it by the name of principle.[44]

Trying to summarize what he takes to be a philosophical *communis opinio,* Biesius tells us that "matter is the disappearance of the divine form, and the form is the closest image of the perfect [divine] monad."[45] This "disappearance" takes place, just as it had in *De varietate,* when the infinite and yet non-dimensional divine unity enters into the dimensionality of matter. Matter and the divinity are the opposite poles of a spectrum. Still, something of the

44 Ibid., 32: "Plato formas, seu ideas constituit cum magnitudine materiae, cuius participatione ex monade numeri fiant et substantiae, quae rursus propter monadis characterem sint indivisae. Unum quidem et bonum in solo Deo esse dicit, quod etiam omnium principium sit, sed tamen monadis expressam similitudinem hanc universitatem continere, et etiam singulas eius partes, quae possit etiam principij nomine vocari."

45 Ibid., 33: "Materia est divae tantum defectio formae, / Et forma est monadis perfectae proxima imago."

divinity remains, which is why the indivisibility of its unity is also found in the dimensional world of matter:

> For while the form extends itself in matter, it abandons its monadic indivisibility and produces a divisible magnitude; but the parts of things cohere thanks to the form, which, because it completes the essences of the singular things, it follows that there is nothing that is divided infinitely. Because an infinite division is contrary to the monad, for effectively, it would be nothing.[46]

We have already encountered in *De varietate* Biesius's fear of infinite divisibility as the gate to nothingness. But in *De universitate*, this fear is more explicitly addressed, although Biesius remains uncertain about how to deal with it. In the discussion to which we now turn, he repeatedly invokes the atom, but it is important to recognize that, most of the time, he still understands this term as designating an unextended mathematical unit, while what he is really trying to find is a physical minimum as the roadblock on the downward slope towards infinite divisibility.

We encounter a crucial passage when Biesius evaluates the Aristotelian tenet that all motion takes place in a continuum. The problem with all continua is that, as Aristotle himself had stressed, they are by definition indefinitely divisible. Let us look closely at how Biesius tries to allow for motion while shooing away the nihilistic notion of indefinite divisibility that frightens him so:

> Since motion takes place through a continuum, it doesn't therefore take place through atoms, for two atoms don't produce a continuum, as their parts cannot touch, given that they don't have any such parts, nor does the one extend beyond the other. For this reason, nothing becomes bigger through the addition of atoms, but instead through the addition of minimal corpuscles. For, while the mathematicians imagine that they can be divided according to a more accurate demonstration, they in fact are never thus divided. Hence between two mathematical atoms there will necessarily be a continuum; because otherwise, they would continue alone. And what counts for magnitude, also counts for motion and also time, where the atoms of the single unities [in one category] respond to

46 Ibid., 35: "Nam se forma dum in materiam extendit, monadem suam deserit & magnitudinem dividuam facit: sed partes rerum cohaerent per formam, quae quia singularum essentiam perficit, ideo nihil est in infinitum divisum. Nam infinita divisio monadi contraria, adeoque nihil esset."

> those [of the other], so that if the minimum of motion is divided, so also the minimum magnitude [of space] and the time in which the motion takes place. For the speed growing towards infinity cuts the time and the magnitude [of space traversed] similarly.[47]

If this passage looks opaque to us, this is because Biesius cannot quite find the way out of what Libertus Fromondus was to call "the labyrinth of the continuum."[48] To begin with, Biesius recognizes that *continua* cannot be made out of unextended atoms, because unextended magnitudes don't add up to anything. But fearing infinite divisibility for the reasons mentioned earlier, he prefers extended *minima* that are not actually divided. However, once he presupposes such undivided units, he has to tackle the problem of the isomorphism of space, time, matter, and motion. This isomorphism requires that if the speed of a given moving body increases, the space traversed and the time consumed have to maintain their respective proportionalities vis-à-vis motion.[49] Since Biesius does not accept discrete or, as it were, quantized steps in acceleration, he has to admit continuously growing acceleration and with that also continuously shrinking spaces that are traversed, and time intervals in which the motion takes place.

On the next occasion when he chooses to express himself hexametrically, we see from where some of his ideas of a discrete formatting of all magnitudes derive. In verses that remind us of what Bruno was to write a few decades later, Biesius extolls the beauty of the monadic circle, while also praising the triangle out of which, according to Plato's *Timaeus*, all space and matter is formatted. In fact, Biesius takes us through all of Plato's regular bodies, explains their

---

47 Ibid., 38: "Cum autem motus per continuum fiat, non ergo per atomos: nam atomi duo continuum non reddunt, partibus enim se non contingunt, siquidem partibus carent, nec etiam unus excedit alterum, unde fit ut atomorum adiectione nihil maius fiat, sed minimorum corpusculorum, quae si mathematici dividi posse fingant ad exactiorem demonstrationem, nunquam tamen sunt divisa. Ergo inter duos atomos mathematicos necessario continuum erit: alioqui namque soli continuarentur. Qualis autem magnitudo, talis est et motus et etiam tempus, et atomi singulorum singulis respondent, ut si minimus motus dividatur, similiter et magnitudo per quam et tempus in quo fit motus. Nam celeritas crescens in infinitum & tempus similiter & magnitudinem secat."

48 See the title of Fromondus, *Labyrinthus*.

49 Carla Rita Palmerino, in "The Isomorphism of Space, Time and Matter," has documented that the extended magnitudes of space, time and matter were always treated analogously up to the late seventeenth century. For a continuist, all these magnitudes were infinitely divisible; for a discontinuist, indivisibles had to be postulated not only in matter, but also in time, space, and motion.

identification with the elements, and praises the wise who recognize that the "temple of nature" is constructed out of these bodies.[50]

When he turns to the prosaic exegesis of his own verse, now with some confidence he declares that "the most simple elements are bodies into which all things are finally divided, since they are constructed out of them."[51] Against the Aristotelians who argue that Plato's mathematical shapes have little to do with natural bodies, Biesius attempts a curious defense, by redefining Plato's shapes as forces and expressions of elemental characteristics. With reference to the pointed shape of flames, he argues that "there must be some principle, for which the dimensions of the figures are not divided but retain their force, which extend themselves always into this or that figure."[52] A couple of pages later, he defends Plato's geometrical elements more strenuously against Aristotelian misinterpretations, and attributes to them weights and speeds and other properties.[53] And as if his rather confused utterances about the elements had settled the matter, Biesius continues his treatise as if he had established their indivisibility and figuration: "From these regular, prior figures, the others of the animals, plants and of other things are made."[54] With increasing self-confidence, Biesius now follows a vaguely Platonic matter theory in which the sanctity of the trinity and the quaternity are invoked to explain the composition of the elements, whereby the trinity is more divine and the quaternity more terrestrial. Of course, our eyes are not sharp enough to discern these particles, Biesius tells us, so that we have to view them with the eyes of reason. But he then adds a classic atomistic argument to back up the plausibility of his mental imagery:

> Whoever neglects the subtlety of nature in the way she combines the minimal corpuscles, remains unmoved by her daily and evident actions: for if we consider the simple increase of things, which is carried out through the continuous addition of corpuscles, we will easily recognize that there are minimal parts of nature that are imperceptible for us.[55]

50 Biesius, *De universitate*, 39–40.

51 Ibid., 43: "Elementa simplicissima quaedam corpora sunt, in quae, quoniam ex his constant, omnia postremo dividuntur."

52 Ibid., 44: "Certe si terra semper stabilem ignis expetit figuram acutam, principium aliquod erit, quod quemadmodum dimensiones figurarum non dividatur [for: dividantur] sed vim suam retineat [for: retineant], quae vel in hanc vel in illam se figuram semper extendat."

53 Ibid., 46–47.

54 Ibid., 47: "Ex istis regularibus, adeoque prioribus figuris, aliae animantium, stirpium, et aliarum rerum fiunt."

55 Ibid., 48: "Qui vero naturae subtilitatem in coniungendis minimis corpusculis negligit, is quotidianis et evidentibus eius actionibus non commovetur: nam si vel sola rerum

By way of evidence, Biesius points to the growth of plants and crystals as examples of perceptible growth by imperceptible corpuscular additions. He then turns to that age-old issue of what happens to elements when they mix. In a perfect mixture, according to Aristotle's teaching in *De generatione et corruptione*, the elements should disappear under the new homogeneous *forma mixti*. But then, asks Biesius, why can we extract the former ingredients from alloys? In the Middle Ages, as Anneliese Maier has documented, this question had produced a number of competing solutions, with none winning the day.[56] In the seventeenth century, growing opposition to the Aristotelian notion of the substantial form was to lead to the notion that elements were particles and that these particles would remain distinct entities in the mixt. But this new theory had its own problems, notably that of explaining the emergent properties of the composite structures, as we would nowadays call them. Why – to use an example from modern chemistry – would you get the crystalline structure, taste, and color of table salt out of sodium and chloride if the sodium and chloride particles remained merely juxtaposed, without somehow becoming subjected to a new "form of the mixt?"[57]

How does Biesius situate himself in this uneasy battle between continuist and discontinuist solutions? A rude but not undeserved answer would be: he makes a complete mess of it. He first mentions the Aristotelian standpoint and then turns to more recent authors, whose names unfortunately he does not give us, but who suggest that Aristotle was misled by the limits of sense perception. To be sure, Biesius argues, mixtures do look homogeneous to us, but in reality, they are not: "When simple bodies are mixed, such as water, honey, and vinegar in oxymel, all [the ingredients] are divided into their minimal particles, but the simple natural atoms do not perish."[58] It is unclear whether the terms "minimal particles" and "simple natural atoms" refer to the same entities, and their relation to the four elements is similarly vague. The reason why Biesius then once more gets lost in the "labyrinth of the continuum" is at least threefold. First, as we have seen before, he takes atoms to refer to mathematical points, and out of them, obviously, no three-dimensional body can be constructed,

---

incrementa consideremus, cum adiectione continua corpusculorum perficiatur, facile cognoscemus esse nobis insensiles partes minimas naturae."

56 See Maier, *An der Grenze von Scholastik und Naturwissenschaft*, ch. 1, which examines the various answers available in the fourteenth century to the question "utrum elementa remaneant in mixto."

57 On the history of the argument of the "reductio ad pristinum statum," i.e., the return to the original ingredients in a chemical operation, see notably Newman, *Atoms and Alchemy*.

58 Biesius, *De universitate*, 50–51: "Sed existimant hi posteriores dum simplicia corpora miscentur, ut aqua, mel, acetum in oximel, minimis quidem particulis omnia dividi, simplices tamen atomos naturales non interire." Oxymel is a medicine produced primarily from honey and vinegar.

whether mixed or otherwise. Secondly, he cannot find a solution to the uneasy Aristotelian argument that – in the process of transformation of the separate forms of the elements into the new form of the mixt – there must occur a moment in which the old elements have gone and the new form hasn't yet arrived, so that in that intervening moment, you have only pure, prime matter without any form. And thirdly, he is confronted once again with his constant horror of infinite divisibility, the dissolution of the elements into their ultimate components in the process of mixture, which, because elements are continua and therefore infinitely divisible, would result in their disappearance.

In his bewilderment, Biesius casts about, left and right, looking for answers. At one point, he gestures to physicians who assume that in medicines, there are various substances that act at different times. At another, he gives a nod to the mathematicians who argue that since atoms can't touch, one would need something in between to connect them. Might it be that the atoms really don't touch, but are merely, as it were, united by the form? Biesius likes that idea, but doesn't quite know how to defend it. So it is once again the divine monad that must clean up the mess: "But what joins them [*sc.* the elements] into one, is that form which they call οὐσιώδης, character of the highest monad: in this way, the diverse bodies are not merged, but only the forces of the diverse bodies."[59]

To whom do the theories to which Biesius is here responding belong? This remains his secret. However – and this is relevant for our discussion – he is, *inter alia*, responding to the assumption shared by physicians that medicines contain a series of different substances. It is also interesting that he invokes the evidence of "the chymists," who document that "in quicksilver, terrestrial and water parts are strongly mixed, while in sulphur, there are more fire and air parts."[60] In the end, however, he links the behavior of medicines and metals to celestial analogies, unwilling to rely on the properties of particles, corpuscles, or atoms – entities, moreover, that for the rest of his treatise, in fact disappear entirely.

3 De natura libri V (1573)

Let us now turn to the third and last book of our examination, which was also Biesius's last. In fact, he died in the same year 1573 in which *De natura libri V* was

59 Ibid., 52: "Id autem quod ea coniungit in unum, est ea forma quam *οὐσιώδης* vocant, summae monadis character: ita diversa corpora non confundentur, sed tantum diversorum corporum vires."

60 Ibid., 53: "Aiunt etiam Chymici [...]. In hydrargiro sunt partes terrestres cum aqueis vehementissime commixtae, in sulphure vero partes igneae et aëreae sunt plures."

published. Having in the meantime become regius professor and imperial physician, Biesius dedicated his "Five Books on Nature" to Emperor Maximilian. In his dedication letter, he explains the premise of the book: a good grasp of nature is the foundation for all philosophical and medical studies.

This work is much more clearly about natural philosophy than either of the earlier works, and it shows us Biesius's matter theory at an even more developed stage. In fact, it follows the template of university textbooks, in that it opens with the question of what "nature" means, and by defining the principles of natural philosophy. But in comparison with most sixteenth-century textbooks of natural philosophy, it is once again conspicuous how much Biesius is willing to contaminate the basic Peripatetic structure with Platonic, Neoplatonic, medical, mathematical, and religious arguments. He says that he follows "mostly the Platonists and Aristotelians" – note the order of the two schools! – but by no means do these two schools exhaust the list.[61] In his few but precious text references in the margins, we encounter the writings of Aristotle, Plato, Plotinus, Simplicius, Hippocrates, Galen, Dionysius Areopagita, Averroes, and once even Hermes Trismegistus. In the last two books, which deal with physiological and medical issues, he also cites three recent authors: Gabriele Falloppio, Andreas Vesalius, and Albrecht Dürer.

In keeping with the doctrines of his earlier writings, Biesius once again speaks of the cosmos as created by, and in the likeness of, the original monad. All beings in the created universe are "unities," which are "images" of the primordial unit.[62] Biesius claims that he is following Simplicius's commentary on Aristotle's *Physics*, Book 1, when contending that "one and being are the same" and that all numerical multitude is made "of unities."[63]

A similar train of thought would lead David Gorlaeus, forty years later, to an atomistic ontology, in which all that existed, God included, was conceived of in terms of actually existing, extended units.[64] But Biesius's ontology is far less economical. Unlike Gorlaeus, he does not reject prime matter and substantial forms, and he couches his natural philosophy in what might be called an emanationist theology, in which the divine unity is fractured into ever more remote domains. While the divine is without matter and dimensions, the created world is material. Matter guarantees corporeality, but – and here the horror of infinite divisibility encroaches once again – it also poses an existential

61 Biesius, *De natura*, 3b: "[…] Platonicos ac Aristotelicos maximè secuti."

62 Ibid.

63 Ibid., 3b–4a; see von Oeing-Hanhoff, *Ens et unum convertuntur*.

64 Gorlaeus, *Exercitationes philosophicae*; on the doctrines of this philosopher, see Lüthy, *David Gorlaeus*. See also Kuni Sakamoto's chapter in this book.

menace, as its infinite divisibility tends to push nature towards the abyss of potential disappearance. We have encountered this basic fear repeatedly in both Biesius's *De varietate opinionum* and his *De universitate*, and we recall that there, it has been the form (sometimes in the guise of the monad) that vouchsafed the survival of things. Being mirror images of the divine unity, the forms turn out to prevent, as also in *De natura*, nature's collapse into nothingness.[65]

One might take the following passage to express Biesius's worldview in a nutshell:

> It is clear that the divine mind expressed itself in various characters, which we call forms. Among those, some approach their cause more closely, others recede from it farther. But since they are all produced by the same cause, the closer they are to it, the more they are one and the same, the more remote they are, the more forces they retain and the less they are exposed to influences by which they bend towards their ruin. But the farther they are [from the divine origin], the smaller are their forces and the more short-lived they will be and more exposed to mutual influences. This is why incorporeal forms display a greater unity and are more divine [...].[66]

In such a framework – or so one is inclined to conclude – in which unity is found primarily in the immaterial realm and in which the material realm is only saved from disappearance by the presence of form, there cannot be any space, let alone any role, for material particles. And yet, Biesius warrants a chapter in the present book precisely because there were contrasting forces that, in the course of his life's work, drew him ever more clearly towards such particles. Soon after the passage just quoted, he elaborates on the origin of the various substances encountered in our world. And, apparently out of nowhere, the atom makes its reappearance. The origin of the differentiation found in nature, Biesius now explains,

65 Cf. e.g. Biesius, *De natura*, 6a: "[...] quae tamen, quoniam nec formae corporeae degenerantur in infinitam, hoc est, in nihilum, enadis necessarium characterem semper adiunctum habent, quemadmodum prius de numeris dicebamus."

66 Ibid., 8a: "Ex ijs autem quae prius diximus constat, e divina mente varios esse characteres expressos, quos formas appellabamus: inter quos alij propius ad causam accedant, alij vero longius ab ea recedant. At cum ab una cuncta producta sint, quanto sunt illi propriora, tanto magis una quaeque sunt, latius patent, plusque virium obtinent, et minus affectionibus obnoxia sunt, quibus ad interitum vergant. Quanto vero longius distant, tanto minoris virtutis arctiorque sunt, et affectionibus mutuis magis subiecta. Quapropter incorporeae formae magis unae divinaeque sunt [...]."

> lies in this, that the incorporeal forms extend more widely, and that those most remote, obviously material forms are very narrow, and that nothing is narrower than the natural atom: but still, it is the principle of all corporeality.[67]

The marginal title accompanying this passages states: "The generation of the dimension. The atom degenerates most," by which Biesius means that it deviates most from the original unity of forms.[68] This is a very interesting turn of events: however meek and miserable the atom might be, it now constitutes for the first time a kind of end-point of degradation, a stop to division, and a "principle of corporeality."

Let us compare this assertion to the one, quoted above, from his juvenile *De diversitate*: "all that is bad and false comes from division, just as all that is good and true can be seen to be placed in the one, from which they all degenerate as a result of their divisible nature."[69] This comparison may provide the clue to the story that is told in the present chapter: while in *De diversitate*, there was no way to prevent the degeneration of the unity once it was placed in divisible three-dimensionality, in *De natura*, the atom – however degenerate it might be in itself – is itself a unity and thereby calls a halt to all further degeneration.

Clearly content with this result, Biesius elaborates further:

> Every atom has its own proper species, and any particle, even the smallest, of fire, air, water and earth is – and is also called – "fire," "air," "water" and "earth." And this must also be so understood in the other similar [i.e., homogeneous] bodies, in metals or bone, with the difference that their *minima* are bigger, for they are made up of elements.[70]

In contrast to earlier works, where the term 'atom' referred to an extensionless mathematical point, Biesius now treats it as a physical *minimum*. This perhaps is surprising enough, but the second surprise is that he now also allows for atoms of various orders: there are not just elementary atoms, but also

67 Ibid., 91: "Eius autem origo haec esse videtur, quod incorporeae formae latius pateant, quodque remotissimae, nimirum corporatae formae sint angustissimae nihilque sit angustius atomo naturali: quod quidem iam est principium omnis corporaturae."

68 Ibid.: "Dimensionis generatio. Atomum maxime degenerat."

69 Biesius, *De diversitate*, 7, iii–iv. The Latin was quoted above, in n. 24.

70 Ibid.: "Iam, quodlibet atomum propriam suam speciem habet, & quaelibet ignis, aëris, aquae, terrae vel minima particula, ignis, aër, aqua, terra tum est, tum nominatur: quod etiam in reliquis similaribus, in metallis, in ossibus intelligi debet: nisi quod horum minima maiora, iam constant ex elementis."

upper-level atoms of homogeneous mixtures – an idea that is often ascribed to the early-seventeenth century thinkers Isaac Beeckman and Sébastien Basson.[71]

In the course of *De natura*, Biesius becomes increasingly affirmative about his *minima* or *atomi*:

> From all these minima of this bodily nature, the elements and the celestial bodies in their totality are constituted, as if from all their parts, which under the moon are variously combined and separated. But since in all of nature there cannot be any vacuum, as we will later demonstrate, it is necessary that all these minima touch one another. And it often happens that the forms, which touch each other through the minima, are of the same type, as in their [natural] places those [*minima*] of fire, air, water, and earth. And often they are of a different type, as when different elements are mixed. But the forms of composites such as animals and plants, while they are indivisible, are each poured into several continuous atoms.[72]

Admittedly, this string of assertions leaves open more questions than it answers. Still, the atoms or minima have now become the ultimate constituents of things. It is thanks to them that we have three-dimensionality. Recall that this characteristic had previously been ascribed to general matter, not to atoms. The forms – in some way that remains unclear – supervene upon them and "touch each other through the *minima*." In the respective spheres of earth, water, air, and fire – that is, in Aristotle's so-called "natural places" – these atoms are all of the same type. When they are mixed, as in metals or bones, to result in homogeneous bodies, we seem to have elementary particles clustered into upper-level atoms. And in heterogenous animated bodies such as plants or animals, it seems that the overarching form – the form of the lion or of the edelweiss – is "poured" into the atoms. One wonders whether Biesius

71 Kubbinga, "Les premières théories 'moléculaires'."

72 Biesius, *De natura*, 9a–9b: "Ex omnibus autem huius corporatae naturae minimis elementa corporaque caelestia tota, tanquam ex omnibus suis partibus, constituuntur: quae sub luna varijs modis tum coniunguntur, tum separantur. Quia vero nihil in uniuersa natura vacuum esse potest, quemadmodum postea sumus demonstraturi, necessarium est, ut haec omnia minima sese contingunt. At hic saepe fit, ut formae, quae se per minima contingunt, eiusdem generis sint: ut in suis locis ignis, aëris, aquae, terrae. Saepe quoque diversi generis sunt, ut cum elementa diversa commiscentur. Formae vero compositorum, ut animantium et stirpium, cum individuae quoque sint, eadem in pluribus atomis continuatis versantur."

assumes that in the process, the atoms still maintain their primary, elementary and their secondary, mixed natures, and how the added-on lion-ness or the edelweiss-ness affects their natures. Whatever the definition of their nature, Biesius's conclusion is that they are no longer just mathematical points, as he had previously taken them to be, but the very units, however humble and abject, that "through conjunction and continuation constitute and fill the entire space from the center of the Earth to the highest sphere of the heavens."[73]

As for Biesius's fear of endless division and fragmentation, we recall how in his earlier books, he tried to employ the forms to keep under control the nihilistic tendencies of matter. But it is only in *De natura*, thanks to his new theory of physical minima, that his angst for the first time becomes conceptually manageable. He is now positive that the unitary character of the divinity is impressed on all things, including material *minima* or atoms.[74] For this reason, he adds that

> in the nature of things, there are some minima, which though they can be imagined as divisible, are in reality not divided further, and they are maybe different in different bodies, which is why Plato said that earth is constituted by larger and fire by smaller parts. And out of the heap of all these minimal particles of nature the mass of all bodies and of the whole universe is composed.[75]

What a move! Biesius here abuses the Aristotelian concept of *minima naturalia* – which, in truth, was not about constant entities but about the limits of material divisibility for a given form – and joins it to an atomistic conception of matter, buttressing this construction with a Platonic account of the elementary regular solids.[76] To the mathematician who insists that out of extensionless atoms, you cannot construct a world, Biesius now offers the confident rejoinder that "natural *minima* do have parts, which are not natural, but infinite mathematical ones. And these natural *minima* have not just parts,

---

73 Ibid., 9b: "[…] ex minimorum physicorum […] tum coniunctione, tum continuatione, quibus totum spatium a centro terrae ad summum caeli tum constituitur, tum impletur."

74 Ibid., 35-a–b.

75 Ibid., 38b–39a: "Sed in natura rerum sunt aliqua minima, quae quamvis ut dividua intelligi possint, ulterius vere non dividantur, eaque fortassis in diversis corporibus sunt etiam diversa: propterea qua Plato terram maioribus, ignem minoribus partibus constare dicit. Ex illis autem minimis naturae particulis coagmentatis constat omnium corporum et totius universitatis moles."

76 On *minima naturalia*, see again Murdoch, "The Medieval and Renaissance Tradition of *Minima Naturalia*."

but also shapes," which is why they can fill space.[77] And if the mathematician continues to insist, arguing that motion requires infinite divisibility, because it can become indefinitely faster or slower, Biesius follows his earlier defensive logic but now also adds an answer that we would have expected from a seventeenth-century reader of Pierre Gassendi: "while the fastest body covers one natural minimum [of space], the slowest will hardly traverse a thousandth of a thousandth of that particle, which shows that the natural minimum is divisible into parts."[78] Biesius thus accepts the mathematical divisibility of his *minima* while insisting on their physical indivisibility, and he successfully moves towards a theory that relates degrees of speed to his basic physical units with their mathematical sub-units.

As mentioned before, up to the late seventeenth century, matter, space, time, and motion were always treated isomorphically: all of them were conceived either continuously or atomistically, but before Newton it was impossible to believe in an atomic structure of matter while being a continuist for space.[79] For this reason, we must not be surprised to find that Biesius, too, has now become a minimist in all respects. His marginal titles leave no doubt in this respect: "Time is not divided into infinity"; "motion doesn't take place through a continuum"; and "there is no space without corporeal bodies," which, in turn, are constructed out of *minima*.[80] In one sentence: we measure the motion in terms of the number of *minima* traversed in a given minimum of time.

But when, in Book II of *De natura*, Biesius must address the nature of the four elements, his new faith waivers. How must he deal with the Aristotelian idea that elements are defined by their qualities and can therefore transmute into one another, such that when water is warmed it turns into air and when air is cooled down it turns back into water in the form of rain? This is difficult to reconcile with natural *minima* as previously defined, notably when clad in the garments of Plato's regular solids. Biesius attempts to defend Plato's geometrical notions for a page or two, but recognizing that they can't really be made

77 Ibid., 39a: "[…] respondebo, minima naturalia partes habere, non quidem naturales, sed mathematicas etiam infinitas. Nec tantum haec minima naturalia partes, sed et figuras habent […]."

78 Ibid.: "Cum enim corpus celerrimum minimum naturale conficit, interim tardissimum vix millesimae millesimam eius particulam percurrit: unde videri potest naturale minimum in partes esse divisum."

79 See again Palmerino, "The Isomorphism of Space, Time and Matter."

80 Biesius, *De natura*, 40b–41a: "Tempus non est dividuum in infinitum"; "Motus non fit per continuum"; "Spatium nullum sine substantia corporea."

to fit into his new minimist ontology, he abruptly ends the discussion with an impatient last word: "enough!"[81]

Vexed by this impasse, however, he feels a compulsive need to tend to his Achilles' heel, and so, he unexpectedly returns to the question of how points, atoms, and *minima* relate to one another. Under the marginal title "the point is the principle of dimensions," and with an explicit reference to Proclus' commentary on Plato's *Timaeus*, Biesius tells us that the word "point" has a double meaning, whereby "point" and "atom" are once more used analogously, at least initially. We are told that for a body to have extension, it needs dimensions, and those require as their foundation "the atom,"

> which is a sort of unity, which has a position in matter and place, not through itself, but through the bodies, to which it is connected thanks to the dimensions. It is however double, partly a natural minimum, partly mathematical, as what remains in thought after an infinite division: the latter is in the mind, while the former is in the bodies and in the very nature of things.[82]

Whether this clarifies things is not in the least evident, but we now find Biesius, more clearly than before, invoking the term 'atom' both as a mathematical unit, which is a merely mental construct, and as the name of the ultimate physical magnitude that makes up all bodies. His next marginal title therefore proclaims: "The atom is similar to unity."[83]

Let us stress the importance of this assertion. In his earlier works, Biesius had recognized the substantial form as the remote image of the divine unity. In *De natura*, its place is now taken up by the physical minimum. Biesius has discovered a way of avoiding the infinite fragmentation of matter, by positing an atom that is no longer merely an unextended point, but one where if you "join two natural *minima*, they form a very short physical line, which can be divided not so much by nature as by thought," and this is how the physical world is constructed.[84]

---

81 Ibid., 49a–51a; 51a: "Sed hactenus de figuris elementorum satis."

82 Ibid., 56a: "[...], in qua principium quoddam etiam constituendum est, quod atomum appellamus: idque tanquam unitas quaedam est, habens positionem in materia atque loco: non quidem per se, sed cum corporibus, quibus per dimensionem est connexum. Est autem hoc duplex, partim naturale, minimum, partim mathematicum, post infinitam divisionem cogitatione relictum: quod tantum est in mente: cum illud, propter dimensionem, sit in corporibus, atque adeo in ipsa rerum natura."

83 Ibid.: "Atomum unitate simile."

84 Ibid: "Sed coniuncta duo minima naturalia breuissimam lineam physicam constituunt, quae, non quidem natura, sed tantum cogitatione diuidi potest. productiores vel lineae,

After that, corpuscles, *minima*, points, and atoms disappear for a great many pages, in which they give way to celestial circles and terrestrial triangles, to souls and to influences. They return only in the treatment of mixture. We find Biesius here clearer and more determined than in his earlier *De universitate*. Mixture, he now tells us, requires that the ingredients act upon one another, and they do so through their primary qualities, when the *minima* of the elements touch each other, jointly producing a new set of properties.[85] It is interesting to see that Biesius still accepts the Aristotelian notion of a homogeneous *forma mixti*, but that he now feels that Aristotle erred in assuming that this meant that the elementary ingredients had to perish in the mixt. If this were the case, Biesius argues, we would not be able to recover the original ingredients.[86]

Emily Michael has pointed out that one of Daniel Sennert's arguments for a plurality of forms within a given material body relied on an analogy with the medieval Scotist assumption that different souls could co-exist in one animate being. According to this so-called pluralist view, a human being doesn't just possess a rational soul that comprises all other vital functions, but also possesses separate vegetative and sentient souls.[87] Interestingly enough, Biesius invoked the same analogy, albeit a few decades earlier than Sennert, but his reference is not to Duns Scotus, but to Plato.[88] Here is his conclusion to the argument:

> It is therefore not absurd that mixture is brought about by physical atoms, and that in one body several forms are present, namely those of the elements and of the entire composition, of which one has the principal place and the other take the place of matter. They might say: hence several substances come together in one. And I will answer: this is quite true, since detachable forms are also substances, which endure because

---

quoad plura vel pauciora minima naturalia continent, magis aut minus diuiduntur, et ex his coniunctis corpora physica constituuntur."

85 Ibid., 78b: "Ad physicam autem mistionem necessarium est, ut quae miscentur in sese vicissim agant. Agunt autem et afficiuntur qualitates tactiles: praesertim, cum elementa, quae materiam communem habent, sese per minima tangunt tanquam ad aequabilitatem actionis et affectionis tandem redacta."

86 Ibid., 79a: "Nova vero forma compositionis per minima mistis elementis accedens, cum unitatis character sit, mistique partes in unum continuum plane comprehendit: quae forma compositi cum formis elementorum esse potest, quemadmodum clarissimi philosophi iudicarunt: quamvis diversum Aristoteles dicat." William R. Newman has repeatedly stressed the importance of the argument from the *reductio ad pristinum statum* for atomism; the argument is made most extensively in his *Atoms and Alchemy*.

87 Michael, "Daniel Sennert on Matter and Form."

88 Biesius, *De natura*, 78b.

> of their own nature, just like the human souls, which together with the substances of the elements bring about human nature.[89]

It seems as if this double doctrinal development – his recognition of atoms as extended physical *minima*, and his espousal of the notion that a plurality of forms can co-exist in one body – gave Biesius new confidence. He now feels that he can explain a whole range of things by means of his atoms, whereby he continues to use the terms "atom" and "natural minimum" interchangeably. Under the heading "the metals have their atoms," he tells us that "just as elements are entirely composed out of certain atoms, so we encounter also certain *minima* in metals which are constituted out of the *minima* of the elements."[90] And when he turns to animate beings – that is, plants and animals – he continues in the same vein. He tells us that the "*minima* of elements" bring about homogeneous parts; plants and animals are made up of a range of homogeneous parts such as bone or wood; and this means that the seeds of things need to contain "particles" that bring about these homogeneous parts, and these particles might in fact be called "*minima* of bones and nerves."[91] He now also feels that the growth of metals, plants, and animals may be explained in such terminology, as "a continuous increase, in which to minimal times and minimal bodies some further *minima* are added."[92]

Curiously, in the last two books of his *De natura*, this entire vocabulary of *minima, particulae, corpuscula,* and *atomi* disappears again. There, Biesius is interested in the functioning of the body, of our senses and our mental faculties, as well as in our physiology and physiognomy. And for these issues, the ultimate structure of matter seems to have no role to play.

---

89 Ibid., 79b: "Non est igitur absurdum, quod ex atomis physicis mistio fiat, et uni corpori plures, ut elementorum et totius compositionis formae praesint: inter quas una teneat principatum, et aliae sint materiae loco. Dicent, ergo plures substantiae convenient in unam: respondebo id quidem esse verum: cum formae separabiles etiam substantiae sunt, quae nimirum suapte natura consistant: velut animae humanae, quae cum substantiis elementorum constituunt naturam humanam."

90 Ibid., 84b: Marginal title: "Metalla suos atomos habent." Text: "Quemadmodum vero diximus ex atomis quibusdam elementa tota componi, sic etiam in metallis minima quaedam sunt ex minimis elementorum constituta, quae metallorum minima, cum constent ex minimis elementorum in unum conspirantibus."

91 Ibid., 88b–89b.

92 Ibid., 91a: "Continuum vero fit incrementum, quod minimis temporibus, minimis quibusque corporibus, minima quaedam adijciantur [...]."

4 Conclusion

In Book III of *De natura*, the book in which he relies most heavily on *minima* and atoms, Biesius suddenly interrupts himself. It is as if it abruptly occurred to him that a suspicious reader might mistake him for an Epicurean. In very much the same way that René Descartes was to insert into the *Principia philosophiae* (1644) a paragraph in which he clarified that despite his massive reliance on corpuscles and world-producing vortices, he did in no way follow Democritus, so Biesius, seventy years earlier, inserts a disclaimer so as to snuff out the flickers of suspicion before they could turn into flames:

> Although we make everywhere much use of similar *minima*, no one should therefore believe that we agree with the Epicureans or Anaxagoreans. For they posit *minima* of the same kind, accidentally moving down through an empty space, and homogeneous parts that are everywhere the same, and not *minima* of diverse elements which fill this entire space and through certain analogies according to God's providence.[93]

Biesius makes a valid point. In the extant works of Epicurus and Anaxagoras, we encounter no divine providence, and as for Epicurus, his atoms bear no analogical relation to celestial and immaterial monadic unity, and they are indeed chunklets of the same universal matter descending through the vacuum instead of representing the units of discrete elements and other substances. In this sense, we can take Biesius's protestation as a useful admonition also for ourselves, insofar as we must indeed guard against automatically viewing early modern authors who occasionally mention atoms as followers of some ancient thinker.

At the same time, of course, it is significant that Biesius feels that he must issue such a disclaimer in the first place. It appears that, even in his own time, his terminology and his explanatory strategies might have raised eyebrows. The imaginary reader accusing him of being a follower of Epicurus or Anaxagoras might have done so for two reasons. The first was simply terminological: if you speak of 'atoms,' you must needs be a follower of Epicurus. The second reason

93 Ibid., 89b: "Quamvis autem ubique minimorum similarium multam mentionem facimus, propterea nemo credere debet, nos cum Epicureis vel Anaxoagoreis consentire. [N]am illi minima constituunt eiusdem generis, per vacuum spatium fortuito delata, similaresque partes ubique pares, non elementorum diversorum minima, quae compleant hoc spatium universum, et certis quibusdam analogijs, secundum providentiam Dei." Cf. Descartes, *Principia philosophiae*, Book IV, § 202, marginal title: "Democriti Philosophiam non minus differre a nostra, quam a vulgari."

is more complex, because it relies on the logic of rejecting a rejection. In many of his treatises, Aristotle confuted the notion that continua had a particulate structure, a notion that he associated with Leucippus and Democritus. Against them, Aristotle pitted his own continuist theory of extended magnitudes. So, if a Renaissance thinker deviated from this Aristotelian notion of ever-divisible matter, time, space, and motion, he could be seen as rejecting Aristotle in favor of the very theory Aristotle himself had rejected – and hence in favor of a theory of discrete atoms.

That Democritus or Epicurus would not have recognized their own theories in Biesius's *De natura* might, moreover, have been less evident to the Renaissance reader than it is to us. Readers who are used to viewing theories as representing sects – something Biesius himself did when saying at the beginning of *De natura* that he "mainly followed the Platonists and Aristotelians" – will not easily read Biesius as proposing an independent theory, but will try to group him with an ancient sect, hence as an 'Epicurean,' a label that was not easy to wear at the time. The extent to which it was difficult to shake off this type of 'group thinking' became evident in the middle of the seventeenth century, when those who refused to follow any one authority but wished instead to take the best of all authors, were quickly categorized into a sect of their own, namely that of the "the eclectics," those, that is to say, "who selected."

So, what should we say about Biesius's invocations of atoms and other *minima*? To begin with, it may well be that he and, closer in time, his students and the students of his students formed the group of Louvain atomists that Libertus Fromondus was so strongly reacting to in 1631, but whose identities have so far eluded discovery.[94] "Who exactly the Louvain atomists were, remains however unclear"– so concluded Carla Rita Palmerino, having conducted her own research into their identities; "for although Fromondus frequently refers to specific atomist positions held by some of his colleagues," she continues, "he never mentions any of them by name". Palmerino herself points to Biesius as a possible source, but adds that

94 Fromondus, *Labyrinthus*, "Ad Lectorem": there, Fromondus claims that the notion that the continuum was made up of indivisibles was widespread in Louvain. But given the dominance of Aristotelianism at that university, historians have found that claim baffling. Could this have had anything to do with Gassendi's trip to the Low Countries in 1628/29? Attempts to identify the authors at which Froidmont took aim in 1631 have been undertaken, notably by Vanpaemel, "Libert Froidmont et l'atomisme," and Palmerino, "Libertus Fromondus' Escape from the Labyrinth."

> it is not clear, at least for the time being, whether he represented a tradition that continued until the days of Fromondus. To the best of my knowledge, no trace of atomism is found either in the works of Petrus Castellanus, *professor medicinae* and *linguae graecae*, or those of Thomas Fienus, *professor primarius* in medicine, whose *Dissertationes de cometa anni 1618* were published together with those of Fromondus.[95]

In order to discover whether Fromondus provides indeed the link that historians have been looking for, one would need to look at the medical works published in, and the medical disputations held at, Louvain in the period 1570 to 1630. Alas, the destruction of the Louvain archives in the First World War renders the search for medical disputations from the period futile, at least in Louvain itself.

There are two further things that deserve to be stressed by way of conclusion. The first is that atoms, for Biesius, never played the integral role that they were to play for Giordano Bruno or David Gorlaeus. To be sure, all three authors – Biesius, Bruno, and Gorlaeus – viewed God as an ultimate unity. But, as mentioned above, Gorlaeus compares this divine unity to the material atom, because they have in common that they are numerically one, actually existing, spatially extended, and yet physically indivisible. In other words, he embraces an ontology of extended but indivisible units.[96] Bruno also espouses an ontology in which the atom and the divinity share properties, but in his case, the trick is brought about by means of the Cusanian notion of the *coincidentia oppositorum*: the minimum and the maximum, the smallest units in nature and the all-embracing divinity, coincide in their properties despite being opposites, for example, by being spherical – minimally or maximally so. For Bruno, the small, spherical, ensouled atom is a mirror image of the all-embracing, divine, creative force that animates the universe. Both views, however different they may be, attribute to the atom a fundamental role.[97]

Biesius shares with Gorlaeus and Bruno an un-Aristotelian preoccupation with the relation of the material bodies to the original divinity, but it is not the physical atom as the ultimate material particle to which his attention is initially drawn; this owing to the fact that, in the first place, he regards the form, not matter, as the representative of divine unity. As we have seen in our survey of his doctrines in *De diversitate* and *De universitate*, matter is dangerous;

95 Palmerino, "Libertus Fromondus' Escape from the Labyrinth," 6.

96 Gorlaeus, *Exercitationes philosophicae*, esp. 225ff, exerc. 13: "De atomis."

97 On the visual aspects of this 'figurative' coincidence, see Lüthy, "Centre, Circle, Circumference."

because of its infinite divisibility, it threatens creation and all that is in it with annihilation. In that phase of his thinking, he could have chosen to attribute form to the atom, as Daniel Sennert was to do in his later writings, but for Biesius up to about 1560, the notion of the indivisible atom was occupied by the mathematical unit and could not refer to an extended part. So this way out of the conundrum, at that time, was barred.

We have seen that his almost lifelong disquiet regarding the potential disappearance of unity and identity of things in the face of an ever-divisible matter came to an end in *De natura*. Here he embraced a theory of physical *minima*, which he would call by various names, including "atom." More of his works need to be examined, and notably his medical work, to render our hypothesis more robust. But on the basis of the three works examined in this chapter, it would seem as if Biesius had started to turn to ontological questions partly because of a perceived need to stop confessional fragmentation and to stop what we would nowadays describe as the polarization of public opinion. We recall that he claimed that it was matter's infinite divisibility that was the ultimate cause of the intellectual chaos: "the divisible nature is the great reason for the variety of opinions," and "in an infinite division, there is nothing in which the mind can come to rest."[98] There was originally an ethical weight attached to both unity and disparity: "all that is bad and false comes from division, just as all that is good and true can be seen to be placed in the one, from which they all degenerate as a result of their divisible nature."[99] When examining *De varietate*, *De universitate*, and *De natura* together, one feels that the preoccupation remains the same but that Biesius comes to recognize that with the notion of an extended physical minimum or atom, he has found a logical safety net. Not that these atoms do much work for him. They serve, above all, to put an end to division, they guarantee elemental identity in mixtures, and they help to account – in a very unspecific way – for organic growth. But once Biesius has found a stopping device for the much-dreaded disintegration into nothingness, he is quite willing to neglect his units entirely as he turns to physiology or psychology, where atoms, *minima*, or particles – for him – have simply no role to play.

98 Biesius, *De varietate*, ch. 7, title: "Dividuam naturam esse magnam varietatis opinionum causam." Ibid., 7, i–ii: "Cum autem dimensiones in infinitum dividuae statuantur, unitatem, qua cuncta constant et in qua praecipue cognitio veritatis est constituta, maxime cognitio veritatis rerum incorporearum et divinarum, evertere, et plurimas falsissimarum opinionum occasiones praebere videntur: siquidem in infinita divisione nihil est, in quo mens conquiescat."

99 Ibid., 7, iii–iv; the Latin is in n. 24, above.

This chapter has been about Nicolaus Biesius, an author who deserves more thorough study than he has hitherto enjoyed. But however incomplete this present examination has been, it has nonetheless served the purpose of demonstrating that a Renaissance thinker could use the term 'atoms' without therefore being 'an atomist,' either in the sectarian sense of the word – that is, in the sense of subscribing to the doctrines of Democritus or Epicurus – or in the ontological sense, by subscribing to a world in which all was built up from indivisible chunklets of matter in motion. *Quod erat demonstrandum.*

CHAPTER 6

Mechanical Arts and Biological Development on the Sixteenth-Century World Stage: The Paracelsian Mechanical Philosophy of Petrus Severinus

Jole Shackelford

Historians of science have long associated the introduction of mechanics – both the Archimedean mathematics of statics and the reduction of natural philosophy to sequences of motions and interactions of impenetrable material parts, like the cogs in a clockwork – with a profound transformation in scientific conceptualization and practices that can be subsumed under the term "scientific revolution." While *the* Scientific Revolution of the seventeenth century is now a contested term among historians of science, the transformative role of "the mechanization of the world picture," as E.J. Dijksterhuis termed it, is still cogent, as is the natural-philosophical reductionism captured by the term "mechanical philosophy," introduced in the late seventeenth century. "Mechanization," "mechanical," and "mechanics" are therefore terms and concepts that remain to be engaged critically in our construction of the historical development of natural philosophy between the Renaissance and the modern period. I argue here that the terms and ideas associated with mechanics, mechanism, and mechanical actions were diverse in the sixteenth century, with a range of meanings that incorporated vitalist as well as instrumental notions, as is evident from an analysis of the writings of the Paracelsian physician and philosopher Petrus Severinus.

Severinus likened the world – the cosmic world at large – to a theater. It was not so much a material thing as it was a space or place in which the drama of life unfolds from its preordained script, mechanically, in the same sense that he perceived ritual Christian liturgies to be carried out in the theater of the church. In this worldly theater, the actors take on new roles in the endless processes of generation and corruption. Insofar as dramatic art, like medicine, was a mechanical art, and Severinus envisioned physiology as mechanical process, the mechanization of the world picture was not so much an imposition of a new conceptual framework on an Aristotelian natural philosophy as it was a narrowing of the broad meaning of mechanical to exclude the mechanic – the vital agent – from the machinery of the world.

 | DOI:10.1163/9789004528925_007

1 The "Mechanical" in Petrus Severinus' *Idea medicinæ* – Some Philological Considerations

Historians of science who are familiar with the sixteenth-century iconoclastic Swiss-German physician Paracelsus and the role traditionally assigned to him by post-Enlightenment historiography will wonder at the implication of Paracelsus in anything mechanical and may balk at the idea that Paracelsus' conception of the "mechanical" could have been relevant in the mid-seventeenth century. To be sure, Paracelsus' medical and philosophical treatises were by then widely available in Latin – and some even in English – translations, but his controversial reputation, the obscurity of his prose, and the utterly alien nature of many of his ideas, distant from much seventeenth-century natural philosophical discussion, led Francis Bacon and others to regard his ideas as absurd.[1] But that does not mean that they were wholly without influence when mediated by authors respected for their erudition. To understand how Paracelsus' ideas were transmitted to posterity, we must look at intermediary texts, those that were read and valued as legitimate and scholarly by seventeenth-century readers. For readers of Latin, perhaps no Paracelsian author was generally more influential in the first half of the century, or more widely read, than the Danish royal physician and contemporary of Tycho Brahe, Petrus Severinus (1540/2–1602).

Severinus' central metaphysical doctrine was that the visible world is a continual playing out of invisible, seed-like potencies (*semina*) that were planned into creation by the Creator. This theory wove together Paracelsian conceptions of the chemical operations of nature with Aristotelian natural philosophy, Neoplatonic idealism and emanation theory, and Hippocratic Medicine, and it was broadly utilized and commented on by seventeenth-century medical and philosophical writers. Insofar as Severinus described as mechanical the process by which bodies come into being, grow, and manifest characteristics of the natural world, a process conducted according to what he termed "mechanical knowledge," his book, *Idea medicinæ philosophicæ* (*The Ideal of Philosophical Medicine*), was one vehicle for the transmission of an important sixteenth-century understanding of "mechanical." This book was available to, and read by, Robert Boyle and his contemporaries, who began to construct what would later become known as "mechanical philosophy" in the second

1 Francis Bacon's debt to Paracelsian thought is ambiguous. He generally disdained Paracelsians but made an exception for Severinus. On his relationship to Severinus' ideas, see Shackelford, *A Philosophical Path for Paracelsian Medicine*, 257–264.

half of the seventeenth century. Consequently, examination of what "mechanical" meant for the Dane, in the context of his background and sources, will provide additional and overlooked clues to the rich intellectual heritage that the creators of mechanical philosophy confronted beyond the agentless austerity of classical Greek atomism and the Spartan dualism of Descartes. For Severinus, "mechanical" was embedded in a long tradition of performance, the agency of players on the stage, and the familiar rites of the Christian church.

"Mechanical" could refer to mechanics in the abstract (*mechanica*) or the actions or functions of a human agent, a mechanic (*mechanicus*). As a plural neuter noun, *mechanica*, it often referred to mechanical works, products of the mechanical art, for which the term *mechanisma*, a transliteration from Greek, also applies.[2] Since Severinus was schooled in a period of humanist Neo-Latin revival in Denmark and had a reputation for his erudite Latin composition, these are a fair indication of the Renaissance and early modern meanings he would have encountered in the texts he studied in his youth. Mechanics as referring to mechanical works is found in the title that Severinus' contemporary and one-time friend Tycho Brahe gave to his 1597 book *Astronomiæ instauratæ mechanica* (*The Mechanics of a Restored Astronomy*), which describes his observatories and instruments, the machinery or tools that he used in his restoration of astronomy, and Severinus was surely familiar with this use of the term. But, before we assume that Tycho and his contemporaries understood "mechanical" mainly in this modern sense of an inorganic mechanism or machine, we should also consider traditional medieval uses, and the uses to which mechanical was put in the medical literature of Paracelsus and his followers, in order to recover the full range of cultural meanings of the term available to the seventeenth-century scholar, meanings that were embedded in the ecclesiastical and pedagogical institutions of Severinus' youth, and in the medicine and philosophy he studied as an academic.

"Mechanical" in medieval scholarly discussions emerges clearly in the recurring debates about the distinction between art (*ars*) and science (*scientia*), which itself was derived from Aristotle's distinctions between *technē* and *epistēmē*. The dialectic between these two, in particular, was an enduring feature of the academic medical curriculum, as each generation of students was

2 Lewis and Short, *A Latin Dictionary*, 1122, s.v. *mechanicus*, et al. Latham, *Revised Medieval Latin Word-List*, 293 lists *mechanica* as meaning "workmanship" (1432) and *mechanicum* as "device," or "system" (1115). Howlett, ed., *Dictionary of Medieval Latin from British Sources*, fascicle VI, 1740 lists meanings that correspond closely to the classical meanings, referring to "a skilled manual worker" (masculine noun), "a mechanical art" (feminine noun, *ars mechanica* or *scientia mechanica*, usually in contrast with *ars/scientia liberalis*), and as "a mechanical device" or "mechanical system" (neuter noun).

called upon to consider whether medicine was an art, a science, or perhaps both. As a medical student, Severinus was likely to have been well aware of this perennial question and the stock answers that students were expected to produce and explore, and he would have understood "mechanical" in relation to the Aristotelian and Galenic conception of *technē* as craft, as a discipline in which general knowledge is applied to specific artisanal ends, such as healing the human patient. "Mechanical" in this sense is intimately tied to the essence of the *mechanicus*, the craftsman who applies his mechanical knowledge to bring the raw materials he confronts to purposeful completion as a product. Paracelsus and Severinus thought of the mechanic in this sense. But they also extended the concept to natural philosophy.

Severinus used *mechanicus*, *mechanica*, and related forms in his description of how the things of the natural world – bodies – have come into being from incorporeal forms that are instantiated in seed-like entities that he simply called *semina*. In the *Idea medicinæ*, when he used the terms "mechanical process" (*mechanicus processus*) and "mechanical knowledge" (*scientia mechanica*) to describe seminal development, we can assume that he had in mind an analogy between the human craftsman with artisanal knowledge and his own version of Paracelsus' *archeus*, the seminal efficient agent that is equipped with *scientia mechanica*, the knowledge needed for development.

Paracelsus introduced the *archeus* to explain the spiritual agency at work within matter, carrying out the predestined completions of the transformations in beings that were left in an unfinished state at Creation. Paracelsus referred to this general principle of transformation as *Vulcanus*, but when ascribing it to specific entities and processes, he used the term *archeus*, *archei* being disseminated throughout the material cosmos. These inner efficient agents were equipped with seminal intelligence and knowledge and were intimately embedded in material nature. As Walter Pagel puts it, Paracelsus' *archeus* "is matter that is alive and at the same time it is spirit that is of finest corporeality." In his reading of Paracelsus, *semina* and the *archei* were closely related agents that conferred specific development on things.[3]

Severinus elided any ambiguity from Paracelsus' writings by conferring agency to the *semina* themselves. *Semina*, possessing specific mechanical knowledge (*scientia mechanica*), direct the mechanical processes that bring bodies into actual existence in nature according to preordained plans. This is at one level the Aristotelian theory of generation, that the seed of an oak tree contains within it the form that will unfold in development as the mature oak, but Paracelsus and Severinus meant it more generally. This metaphysics

3 Pagel, *Paracelsus*, 85, 105–106.

explained the degree of regularity and order observed in natural development and the apparent sudden appearances of new species – from new diseases to new stars – but it did not imply a rigid determinism.[4] For Severinus, a preordained course of development might be altered by the contingencies imposed by the nature of the raw materials or by the interference of external, supervening knowledge or agency, which then produced what he called a "transplantation."[5] In fact, Severinus introduced this spiritual, mechanical efficient agent into his metaphysics to ensure a measure of vitalism in nature and to deny what he took to be Galenic materialism, which was incompatible with his understanding of Christianity. In a providential cosmos, diseases, like other organic developments, must result from foreordained plans, or rather corruptions of the development of these plans resulting from divine will.[6] Nevertheless, the basic meaning that Severinus invested in "mechanical" as an adjective modifying the nouns "knowledge," "spirits," and "process" is that of orderly, regular organic development. On occasion, he referred to this process as *lithurgia*, a *lithurgia mechanica*.[7]

2 Mechanical Liturgy

Severinus' use of the term *lithurgia* in connection with mechanical activity is somewhat perplexing, but sheds light on the meanings he wished the term "mechanical" to carry in his natural philosophy. Two semantic variants are educed. If by *lithurgia* Severinus meant the art of working with stones – stone masonry – then the term "mechanical lithurgy," used to describe seminal

---

4 Speculating on why the "new" star of 1572 had not been observed previously, Tycho Brahe considered Paracelsus' idea that it might have existed in a state of potency in the primaeval *Illiadus* until it was newly revealed, an idea that corresponded to Severinus' explanations for apparent novelties in a world in which God had created all things in the beginning. See Shackelford, "Providence, Power, and Cosmic Causality," 55.

5 On *semina* and transplantation, see Shackelford, "La transplantation," 35–40; Idem, "Transplantation and Corpuscular Identity," 229–253.

6 Severinus explicitly stated that his idea that the forms of bodies do not originate externally, coming from stars or elsewhere, but lie hidden providentially within the *semina*, is Paracelsian and consistent with Christian religion (Severinus, *Idea medicinæ*, 81, transl. Shackelford, *A Philosophical Path for Paracelsian Medicine*, 174).

7 Shackelford, "Seeds with a Mechanical Purpose," 24 n. 21: The seventeenth-century English translator of Severinus *Idea medicinæ* as "A Mappe of Medicyne or Philosophicall Path," translated Severinus' *mechanicos processus* as "ordenary processes" and *spiritus mechanicos* as "workeing spirits" (fol. 35r), although adding "mechanicke" above the line in each case, indicating an uncertainty as to how to render the term.

development, conjures up the vision of a craftsman responsible for constructing nature's bodies out of basic building blocks of homogeneous but solid prime matter.[8] The Greek passages that Severinus included in the *Idea medicinæ* imply that he knew enough Greek to figure out the root words of this term. However, by the late-sixteenth century, Danish had already long dispensed with the "th" phoneme denoted by the Anglo-Saxon and Old Norse thorn, and he likely pronounced *lithurgia* as *liturgia*.[9] Those authors who commented on Severinus' text or adopted its *semina* theory in the seventeenth century more often chose the spelling *liturgia*, suggesting that he was understood by his readers to be using a term from ecclesiastical ritual or possibly ancient public works.[10] Severinus used other terms to describe the metaphysical procedures of nature that carry an ecclesiastical connotation, *ministerium* (ministry or administration) and *officium* (duty, office, function), so we should consider the possibility that he meant *lithurgia* to evoke in the reader's mind both the mechanical art of stone masonry and the programmatic, scripted church rituals of *liturgia*.[11] It is clear from the title of the eighth chapter of his book, where he referred to *seminum mechanica lithurgia*, the "mechanical lithurgy of the seeds," that *lithurgia*, whatever its precise meaning, is mechanical in some sense.[12] But further on in this chapter, he wrote that forms or species "carry out the lithurgy of the world comedy" – *mundanæ comoediæ lithurgiam* – "through the ministry of generations, transplantations, and mixtions."[13] Even if we

8 "Stonemason" is the translation of λιθουργός given in Liddell and Scott's *Intermediate Greek-English Lexicon*, 473 and was also the reading given by the twentieth-century historian of science Walter Pagel and by Severinus' seventeenth-century commentator, William Davidson, who wrote that Severinus took *lithurgia* from the mechanical arts and used it as metaphor. For details, see Shackelford, "Seeds with a Mechanical Purpose," 23–24, and *A Philosophical Path for Paracelsian Medicine*, 179–180, 448.

9 Skautrup, *Det Danske Sprogs Historie*, vol. 1, 252.

10 Lewis and Short, *A Latin Dictionary*, 1072, defined *liturgus* as "one who fills a public office, a servant of the state," the offices of whom is evidently compatible with the ecclesiastical use of the term liturgy.

11 William Davidson, in his 1660 commentary on Severinus' book, *Commentariorum in sublimis philosophi & incomparabilis viri Petri Severini Dani ideam medicinae philosophicae ... prodromus*, and Køning, *Dissertatio de rerum principiis et mechanica seminum liturgia* used the spelling *liturgia* when citing Severinus' *lithurgia*, although Davidson vacillated on this, retaining the original spelling in his second (1663) commentary. *Commentaria in Idæam Medicinæ Philosophicæ Petri Severini Dani.*

12 Severinus, *Idea medicinæ*, ch. 8: "De generatione rerum naturalium et seminum mechanica lithurgia." See Shackelford, *A Philosophical Path for Paracelsian Medicine*, 179.

13 Severinus, *Idea medicinæ*, 81: "Unde profectæ, quo pergant Formæ, Species, quæ nobiscum definitis temporum intervallis negotiaturæ, Generationum, Transplantationum & Mixtionum ministerio mundanæ Comoediæ lithurgiam peragunt."

concede that *ministerio* may merely mean "through the service of" or "through the performance of" in this context, Severinus must have been aware of the ecclesiastical resonance of the term when he chose it.

More puzzling in this sentence is the juxtaposition of "lithurgy" and "comedy". Whereas stone masonry would seem to have little in common with comedy, church liturgy and theatrical comedy are both performances, and there is a formidable secondary literature linking the two, again arguing for reading *lithurgia* as "liturgy." An anonymous seventeenth-century English translator of Severinus' *Idea medicinæ* appears to have thought that this was Severinus' meaning when he rendered the phrase *Generationum, Transplantationum & Mixtionum ministerio mundanæ comoediæ lithurgiam peragunt* as "doe by meanes of generations, transplantations, & mixtions *play their parts* in this worldely commedy."[14] The English translator's association of *lithurgia* with an actor playing a part in a comedy is striking, and harks back to Severinus' other references to seminal development as something occurring on a stage, the world stage.

Severinus used the term stage, *scena*, eleven times in the *Idea medicinæ*, beginning in the second chapter and recapitulated in the postscript, which serves as a kind of summary:

> Generation is the development of seeds, in which these protect the continuity of species through the ordained unfolding of bodies from their sources, abysses, and vital principles, with the renewal of individuals stepping forth onto this world stage [*in mundanam scenam*] at determined times: in this process [*in hac lithurgia*] the visible is made from the invisible, the corporeal made from the incorporeal, by the power of the vital and immortal knowledge thriving in all of nature.[15]

As a summary, this portion of the book would have been read even by the most casual reader of the book, who would have readily grasped the connection between the liturgy of Eucharist and the making of the incorporeal corporeal through transubstantiation. Most of the references to *scena* come in

---

14 Shackelford, *A Philosophical Path for Paracelsian Medicine*, 179; "A Mappe of Medicyne or Philosophicall Path," fol. 31r. Emphasis added to stress that "play their parts" is the seventeenth-century English translation of *lithurgiam peragunt* in this instance.

15 Severinus, *Idea medicinæ*, 411: "Generatio, progressio est seminum, in qua, ex fontibus, Abyssis, & vitalibus Principiis, ordinata corporum explicatione, in hanc mundanam scenam, definitis temporibus, progredientia, Individuorum renovatione, Specierum perpetuitatem custodiunt: fiuntque in hac lithurgia, ex invisibilibus visibilia, ex incorporeis corporea, potestate vitalis immortalisque Scientiæ, in universa Natura vigentis."

chapter eight, where Severinus elaborated his doctrine of seminal development as the basis for all apparent changes in nature, healthy as well as pathological. But the reference to the world stage in chapter twelve occurs in the context of explaining pathology in terms of his *semina* metaphysics, namely how diseases arise from changes in *semina* in the body, the idea of which is entangled with his concept of internal *astra* or stars within the body. Severinus discussed the role of these "invisible stars" in governing the timings of diseases through their risings, settings, and conjunctions. He must have had in mind common medieval and Renaissance concerns about the theological implications of astral determinism – limiting God's power and human free will – when he wrote that diseases are not actually caused by the visible stars above, but rather by their invisible correlates that are disseminated in the terrestrial world, which with the passage of time become operative in their material *matrices* (wombs) and produce diseases. Such diseases may be of short duration – such as plague – or may linger for years, producing various symptoms and undergoing further changes or transplantations, such as is the case with syphilis and leprosy: "Thus new anatomies of death and diseases are introduced onto the world stage with each revolution of the ages," he wrote.[16]

In this passage, the "world stage" is the place where the cosmic actions unfold that are foreordained in the movements, aspects, and alignments of the stars. That this theory was integral to his conception of the world and its development in time as providential – a legacy of the influence of Philipp Melanchthon on Lutheran education in Denmark – is evident from a passage in chapter five of the *Idea medicinæ* where Severinus described God, in the beginning of the world, placing the seminal reasons for all created things into the four Aristotelian elemental *matrices*, as recounted in *Genesis*:

> Into these four incorporeal, empty, and vacuous substances, the Creator placed the light and the seminal reasons for all things, by an incomprehensible magic, by the virtue of that Word and spirit that was moved over the waters, imparting the principles of bodies, in which those things about to come forth onto the world stage were clothed. These vestments of the elements have bewitched the eyes of many philosophers: having supposed they had seized upon the kernel of things, they grasped the husks and shells.[17]

16 Severinus, *Idea medicinæ*, 289: "Sic morborum mortisque Anatomiæ novæ, in mundanam scenam, singulis seculorum revolutionibus, introducuntur."

17 Ibid., 41–42: "In his quatuor Naturis incorporeis, inanibus, vacuis, Lucem & seminales rerum omnium Rationes, incomprehensibili Magia imposit Creator, uirtute Verbi &

This particular passage drew the attention of Thomas Erastus, who appropriated it without identifying Severinus as its author and used it somewhat disingenuously as evidence that Paracelsians professed magic. This same passage was cited in various forms by Kort Aslakssøn (1597), Gregor Horst (1612), Andreas Libavius (1615), Ambrosius Rhodius (1643), and Mouritz Køning (1663), and must therefore have been relatively well known to readers of the *Idea medicinæ*.[18]

In keeping with this stage metaphor – which I take to be a *theater* metaphor for reasons adduced below – Severinus referred to the development of seeds into material bodies as a process by which the originally immaterial *semina* clothe themselves in matter, becoming physical bodies with characteristic properties: "Therefore, by separation, the seeds lying peacefully in the elemental wombs are aroused at the appointed times, appear on the world stage, and with remarkable knowledge (*scientia*) they regulate the ebbing and flowing of generation and corruption."[19] As seminal potencies, the seeds contain the knowledge, namely the seminal reasons (*rationes seminales*), to make the clothes and instruments that suit their abilities, and this knowledge finds expression in development as physical, chemical properties. For Severinus, these clothes and instruments are the chemical properties or Paracelsian signatures by which one recognizes and identifies the nature of diseases and the medical uses of substances. Severinus' account of the genesis of diseases in what was once God's perfect creation sheds light on the fundamental place of this idea in his worldview.

The doctrine of seminal reasons is itself derived from Neoplatonic and early Christian theology, given expression by St. Augustine, and elaborated throughout the Middle Ages.[20] But, explains Severinus, although the seeds that God sowed into the world in the original creation were purely good, they were subverted or perverted by "tinctures" introduced by God's wrath at the disobedience of Adam and Eve in the Garden of Eden. As a consequence of the primary pair's fall from grace (the Fall), these tinctures caused transformations

---

Spiritus illius, qui super aquas ferebatur, Principia corporum adiungens, quibus induerentur in mundanam scenam proditura. Hæc Elementorum vestimenta, oculos multorum Philosophorum fascinarunt: nucleum deprehendisse se rati, cortices & siliquas obtinuerunt."

18 See Shackelford, *A Paracelsian Path for Philosophical Medicine*, 218–219, 297, 305–306, 322, 351, and 383 for these references.

19 Severinus, *Idea medicinæ,* 86: "Separatione igitur semina in Matricibus quiete delitescentia, digestis temporibus suscitantur, in mundanam Scenam prodeunt, admirabili Scientia Generationum Corruptionumque fluxus & refluxus moderantur."

20 See Shackelford, *A Philosophical Path for Paracelsian Medicine*, 170–176.

in the seeds, with the result that some of them carry seminal programming for diseases and death. These pathological seeds are joined intimately with the original, good seeds by a kind of affinity that Severinus expressed as being based on the similarities of the vestments in which they are cloaked: the diseases maintain their malevolence hidden from sight by taking on vestments resembling those of the "pure and first roots of things," namely the bodies that sprout from the original seeds. In short, diseases go undetected for so many ages and suddenly appear on the world stage because they pass for a long time incognito, actors dressed up to look like good seminal productions, masquerading as healthy bodies and processes.[21] These disease-seeds eventually reveal themselves by taking on the vestments of fevers, for example, which are the manifest symptoms by which we recognize them, much as a Renaissance audience might recognize the role an actor is playing on stage by the costume he is wearing.[22]

Severinus did not elaborate this theater metaphor, which we can tease out of his words and the verbal images they conjure, but his intentionality is unmistakable: He made an explicit reference to the comedies of Terence, indicating a Renaissance dramatic context for his allusion to the mundane stage as the site where the world comedy is played out. This suggests the importance of a late-Renaissance literary context for Severinus' philosophy that his readers would likely have recognized – a medieval tradition in which the theatrical arts were listed among the *artes mechanicæ* that paralleled the *artes liberales*, and in which actors were perceived as vital and mechanical agents who enacted a scripted comedy. I will elaborate this point later, but first we need a better understanding of the sense in which Severinus conceived of *semina* as operating mechanically and why this is a Paracelsian conception.

3 *Scientia mechanica* as a Paracelsian Concept

The most obvious place to look for sources of Severinus' use of *mechanica* in connection with the operation of seeds is in the writings of Paracelsus, since Severinus consciously identified his medicine as Paracelsian and cited several of Paracelsus' works, which were just then becoming publicly accessible in printed editions. Although Paracelsus seldom used variants of the Latin

21 Severinus, *Idea medicinæ*, 216–218.

22 Ibid., 251. This especially true for the Commedia dell'arte, which flourished in the sixteenth century and used stock characters, recognized by their costumes; see Brockett, *History of the Theatre*, 147–149.

noun and adjective *mechanicus* / *mechanica*, and the secondary literature on Paracelsus has not drawn attention to his use of them, they do appear here and there in his writing, conspicuous as Latin lexemes afloat in the sea of his German prose. They occur in treatises he composed during the years after his hasty and indignant departure from Basel in 1529, in *De caduco matricis* and *Opus paramirum*, where he set down elements of a theory of pathogenesis, and they may therefore reflect his effort to establish his medical ideas on a theoretical foundation as a response to criticisms leveled at him by university-educated physicians.

In *De caduco matricis*, an extension of his ideas about epilepsy applied to the feminine disease hysteria, Paracelsus defined *mechanicus* as a fabricator, human as well as celestial, and described mechanical art – *kunst mechanica* – as "astral"; it is this *kunst mechanica*, moreover – and *not* the dead Galenic humors and complexions – that should be understood as the true agent of change in the body.[23] What this means is that Paracelsus' understanding of an astral-"mechanical" efficient cause, which is analogous to a human workman, is integral to a basic metaphysical assumption of his medical thought, namely that Galenic medicine is wrong-headed and un-Christian in its materialistic and qualitative foundations.

To illustrate the need for such a metaphysical-mechanical agent as efficient cause in biological processes, Paracelsus drew on an agricultural analogy: The production of milk is a cooperative effort, a "conjunction" of the cow and the grass it eats; neither the cow nor the grass makes the milk by itself. It is rather an *astrum mechanicum*, a "star-mechanic" or maybe "mechanical heaven" resident in nature that produces the milk.[24] To make the connection between this astral-mechanical agent and that of a human craftsman, Paracelsus wrote that just as a human can paint something white, or black, so can the heavens; meaning that both the heavens and the human are mechanics (*mechanici*).[25] Paracelsus equated this term *astrum* with *himel* (heaven), and the 1603 Latin translation faithfully retains his term *astrum mechanicum*, while rendering

23 Paracelsus, *De caducis liber secundus, nemlich de caduco* matricis, in Sudhoff, ed., *Theophrast von Hohenheim genannt Paracelsus sämtliche Werke*, Abt. 1, v. 8 (hereafter: Sudhoff 1, 8). For a deeper understanding of contemporary readings of Paracelsus' German, I have compared Sudhoff's edition to the 1603 Latin translation, *De caducis liber secundus, nempe de caduco matricis*, 271–308 in Paracelsus, *Operum medico-chimicorum sive paradoxorum*, tome 4.

24 Sudhoff 1, 8, 333: "Dan zu gleicher weis wie ir sehent, das gras und ein ku beids zusamen milch geben. Nun macht das gras die milch nit, noch auch die ku; do ist ein astrum mechanicum, das in der natur ligt, dasselbig machts wie angezeigt ist."

25 Ibid., "Dan der himel, das ist das astrum und der mensch seind beide mechanici."

himel as *coelum* and flagging the concept in the margin, "*Coelum Mechanicum est*" – "the heaven is mechanical."[26] This suggests that the Latin translator and editor believed that readers would find this idea noteworthy. Paracelsus also seemed to be aware that his use of the term was unusual, for he immediately added: "Moreover, what mechanical is, Philosophy will teach, and for this reason I pass over it here."[27] Nevertheless, he left clues to suggest what he meant by this term in his explanation of physiological processes that constitute healthy and morbid functions, which were taken up and elaborated by Severinus.

The cosmic nature of agency in physiological processes – in the production of milk as well as in the production of epileptic and hysteric paroxysms – sets Paracelsus' medical theory apart from traditional academic humoral medicine, and it is significant that he likened this agency to the human *mechanicus*, the craftsman. Paracelsus tells us that we are to understand the activity of the stars in relation to the medical treatment for epilepsy as being like that of a carpenter's activity with wood and iron. This efficient agent functions as we do when we use iron to cut and build with wood. This astral-mechanical process is no different in the stars than it is in us – a clear statement of the macrocosmic-microcosmic continuity that runs through his medicine. There is no Galenic complexion at work, nor a quality or humor, but rather a mechanical art (*kunst mechanica*), which is an astral mechanics (*mechanica astralis*).[28]

Paracelsus believed that all of God's creations were good, or at least potentially so, but that they might need preparation to bring them to goodness. This preparation was to be understood as an astral operation, as it was carried out

26 Paracelsus, *Operum medico-chimicorum sive paradoxorum*, tome 4, 281: "in natura astrum quoddam Mechanicum latet." Paracelsus' use of the singular noun *astrum* is unusual. Classical use, including in the Vulgate, was in the plural (*astra*), where it meant tropologically "heaven." It may be that Paracelsus is using the singular *astrum* as a Latin equivalent to the singular *himel* – indeed he equates these words – much as we often use the plural "heavens" or "stars" to refer to the singular "himel" or firmament. Although *mechanicum* is likely used as an adjective modifying *astrum* here, elsewhere he uses *mechanicum* and *mechanica* as nouns. The sense of *astrum mechanicum* as an efficient agent might well be rendered "star-mechanic."

27 Sudhoff 1, 8, 333: "Was mechanicum sei, wird die philosophei lernen, darumb ichs hie auslass."

28 Ibid., 334: "Auf solchs in solcher heilung diser hinfallenden krankheiten wissen wie gemelt stehet, so verstehet klerlich ein wirkung im gestirn, zugleicher weis als uns beschaffen ist, das wir mit eisen zimern und hauen das holz. Nun ist das kein complex, kein qualitas, kein humor, sonder es ist aus der kunst mechanica, das <ist> mechanica astralis." The editor of the 1603 Latin translation, Paracelsus, *Operum Medico-Chimicorum sive Paradoxorum*, tome 4, 282, found this idea important enough to flag it in the margin (*Mechanica Astralis*) and index it.

by the *astrum mechanicum*.[29] Just as the carpenter is the efficient cause in building something with nails and boards, so too there are inner astral efficient causes transforming raw materials into finished products within human bodies, and these could be healthy or morbid processes, resulting in good and bad products respectively. The efficiency of these astral mechanics cannot lie in the materials themselves or in their attributes, which is why Galen's humoral pathology, with its emphasis on surfeits, deficiency, and qualitative intemperances, provides a false foundation for medicine.

Paracelsus' conflation of celestial astral agents and their operations in humans and the terrestrial world was completed in Severinus' use of the term *astra* to describe seed-like efficient agents that were disseminated throughout Creation, linking terrestrial operations to celestial causes. We find this idea similarly reflected in the duality of Tycho Brahe's more or less contemporary emblems for *astronomia celestis* (astronomy) and *astronomia terrestris* (alchemy).[30] The efficiency of the physician's preparation lies not in the material ingredients, but in the astral agency or power within them, which brings about the desired effect. The carpenter analogy does not adequately convey the sense of perfecting an object through preparation, and Paracelsus turned to a new analogy to make his meaning clearer, the ironworker who separates iron from slag and then smiths it into a product; "Such is the whole power of the star-mechanic."[31]

Paracelsus understood the cure of diseases like epilepsy and hysteria in this same astral-mechanical sense, not as coming about from a humoral or complexional nature or condition.[32] Just as the carpenter needs nails in order to build, and not heat or cold, but rather his tools, so should the physician

29 Sudhoff I, 8, 335: "Nun ist dasselbig gestirn also zu erkennen. Ir sehent das alle geschöpft etwan zu gut seind, iedoch von inen selbst nicht sonder durch bereitung dohin gebracht! Nun muss dise bereitung in astralisch art verstanden werden durch das astrum mechanicum, also das wir die astralisch operation zu irem effect bringen." Paracelsus, *Operum Medico-Chimicorum sive Paradoxorum*, tome 4, 282–83: "Id vero hoc modo cognoscendum est. Videtis scilicet omnes creaturas quandoque nimis bonas esse, non quidem ex seipsis, sed quod per præparationem eo deducantur & redigantur. Hæc iam præparatio Astrali more intelligenda est per astrum Mechanicum, ita ut Astralem operationem ad suum effectum referamus."

30 On Tycho's use of these complementary concepts, see Shackelford, "Tycho Brahe, Laboratory Design, and the Aim of Science," 211–230.

31 Sudhoff I, 8, 335: "Als so ein eisen gescheiden wird vom seim schlacken und geschmit zu einer art, nun ist es iezt in seiner wirkung. Solchs alles ist die kraft astri mechanici."

32 Ibid, 335: "Nun also auf mechanisch art sollen ir verstehen die cur solcher krankheiten und nicht aus humoralischer complexischer &c natur oder eigenschaft. Dan der himel in solcher operation ist mechanicum und nicht complexionatum." Paracelsus, *Operum Medico-Chimicorum sive Paradoxorum* (1603), tome 4, 283: "Eadem Mechanica ratione

understand the *astrum mechanicum* in medicine. In all things one must take into consideration both the astral conjunction and the mechanics, the first being theory and the second practice, "without which things the physician is to do nothing," Paracelsus wrote.[33] Two things stand out in this description: first, the role of astrology in medicine, since the stars within us correlate with those above; and secondly, the comparison of medicine, which is a perfective art that restores the sick patient to health, to the mechanical arts. The practice of celestial astrology was commonplace in medieval medicine, so Paracelsus' extension of it to the body was a reconceptualization of accepted practice.[34] The second point, that medicine is like mechanics, that the physician's practice is a kind of mechanical art not unlike the art of theater, was also common in medieval theory.

In the winter of 1530–31, during which Paracelsus traveled to St. Gallen and took up residence there, he composed several related texts that lay out his ideas on pathology. These were eventually published together under the title *Opus paramirum*. The first two books, published in 1562 as *Das Buch paramirum*, concerned the chemical foundations of diseases in the three primary principles, salt, sulphur, and mercury, which are often used today to characterize Paracelsus' medicine. Three years later a third book, on the origins of disease from tartars, was published separately and with a treatise on the invisible causes of disease under the title *Liber secundus de causa et origine morborum* (*Second Book on the Cause and Origin of Diseases*), and it is here that Paracelsus again mentioned "mechanical" efficiency in connection with pathology.[35]

---

curam quoque horum morborum intelligere debetis, non ex natura ac conditione humorali aut complexionali, &c."

33 Sudhoff 1, 8, 335: "Und als wenig der zimermann bedarf kalts oder warms im eisen zu seim zimern, sonder alein er bedarf den wezstein und das mezger messer, das stricheisen, also sol der artz in der arznei astrum mechanicum verstehen, das ich hie herein für, aus ursach das ir der constellation dest bass [sic] underricht seind, wie angezeigt ist, das in allen dingen sol betracht werden coniunctio und mechanica, das ein als ein theorik das ander als ein praxim, ausserthalb welchen der arzt nichts sol." Paracelsus, *Operum medico-chimicorum sive paradoxorum* (1603), tome 4, 283: "Sicut enim tignarius ad fabricandum ferri sui nec calore, nec frigore opus habet, sed tantum indiget cote ascia, regula, &c. ita eodem modo Medicus quoque in Medicina Astrum Mechanicum indicare debet. Quod hic explico, ut constellationis certiorem notitiam habeatis, sciatisque quod ut dictum est, in rebus omnibus coniunctio & Mechanica considerari debeant: una quidem, ceu theorica: alterum ceu praxis: sine quibus Medicus nihil est."

34 French, "Astrology in Medical Practice."

35 Paracelsus, *Das Buch Paramirum darinn die ware ursachen der Kranckheyten*; idem, *Liber secundus de causa et origine morborum*. Books 1–3 were supplemented by a fourth, on diseases of the womb, and published in Latin translation in 1575. This collation, titled *Opus paramirum*, was adopted by Huser for his 1589 edition of Paracelsus' works and is

The diseases described in this third book are those that result from a failure of the chemical processes that convert raw materials into their final forms. Failed separation results in a coagulation of tartar, which can accumulate in various parts of the body as granular, stony deposits that cause physiological dysfunction. Paracelsus described the efficient agent in this process as a *mechanicus*, which in German he identified with a blacksmith (*das ist der schmid*).[36] In the Latin translation, the term *mechanicus* is retained, and *schmid* is rendered *faber*, conveying the identification of natural mechanical agency with the art of the skilled human craftsman to a broader, Latinate scholarly audience.

Paracelsus applied this craftsman / alchemist analogy to explain the separation of urine from nutriment, which he likened to the generation of rain in the macrocosm: just as the water of rain does not fall out of the sky all at once, but does so drop by drop, as the drops are individually generated (*als in seinen mechanicis*), so is the matter of the nutriment drawn off bit by bit, attracted to the liver.[37] What does the parenthetical comment "*als in seinen mechanicis*" mean? The earliest printed German text (1565) reads "*als in seiner mechanica*," the notation suggesting a dative singular for the feminine mechanica (*mechanicæ*), but otherwise reads much the same as the standard edition.[38] The 1603 Latin translation rendered Paracelsus' German as "*sicut in Mechanicis traditur*," suggesting that the plural refers to 'mechanics' as a subject of study, much as

---

followed in the modern Sudhoff edition of his collected works. Thus, what was in 1565 *Liber secundus de causa et origine morborum* is now Sudhoff I, 9, 121–175: *De morborum utriusque professionis origine et causa Aureoli Theophrasti ad D. Joachimum Vadianum, medicum.* [*Liber tertius paramiri, de morbis ex Tartaro*]. On the publication history, see Sudhoff, Ibid., 7–14.

36 Sudhoff I, 9, 127: "So wissent auch damit, das der mechanicus im leib ist, der es darzu bringt und sein ultimam materiam im leib eröfnet." "Wie nun obstet, vier sind der genera calculus, arena, bolus viscus … So sie nun im leib komen, so werden sie da geboren, nachdem und der spiritus da ist, der do ist mechanicus an dem ort, das ist der schmid solcher dingen."

37 Sudhoff I, 9, 140: "Zu gleicher weis als ein regen, des tropfen fallen, wenn sie sich generiren und nit das ein ganz wasser sei, sonder es ist ein tropfende generation, dieselbige falt also herab (als in seinen mechanicis), also auch da die materia des nutriments, so zur lebern gehört, dieselbig ist vermischt mit dem urin und aus dem urin wirts an sich zogen, und also bleibt der harn alein, der get durch sein eigen virtutem expulsivam zu blatern hinaus."

38 Paracelsus, *Liber secundus de causa et origine morborum* (1565), Diii, verso: "zu gleicherweiss als ein regen / des tropffen fallen / wenn sie sich generieren / unn nit das ein gantz / Div recto/ wasser sey / sonder es ist ein tropffende generation / dieselbige feldt also hinab (als in seiner Mechanica::) also auch do die materia des nutriments so zum leben gehört / dieselbige ist vermischt mit dem Urin / und auss dem Urin wirts an sich gezogen / und also bleibt der harm allein / der geht durch sein eigne virtutem expulsivam zur blatern hinauss."

we use the term today, or possibly to a treatise on mechanics. If my interpretation is correct, then Paracelsus associated the physics of separation, whether the separation of water in the atmosphere to form droplets through condensation, or the separation of portions of nutriment from portions of urine in the microcosm, with a craftsman-like efficient agency operating at the droplet level, and that the Latin translator more than a half century later construed this as mechanics in the abstract sense.[39]

In sum, Paracelsus regarded mechanical agents, sometimes labeled *archei*, as governors of diverse metabolic processes throughout the body. These processes were characterized as separations of pure nutriment from impure waste product, and failed separations were thought to result in local buildups of harmful chemical residues, usually and typically called "tartars." These were also manifest in the patient's urine, examination of which could yield information about the chemical condition of the entire body, Paracelsus reasoned. His implication of a mechanical agent responsible for chemical separations associated with the production of urine reveals the fundamental role that this concept of mechanical action played in his understanding of human physiology and in the pathology of a broad class of diseases.

This physiological theory linked macrocosm to microcosm and incorporated medieval astrology and uroscopy into a new chemical biology, which was elaborated and put into practice by Paracelsus' followers.[40] The fact that tartar diseases were not only isolated, specific instances of kidney or bladder stones, but also constituted a wide class of effects produced almost anywhere in the body under the guidance of astral mechanics, meant for Paracelsus that no proper physician could be ignorant of the patient's astral, temporal, and essential nature; as he repeatedly said, the true physician must be an astronomer and a philosopher.

Severinus was an ardent student of Paracelsus' ideas and read several of the latter's works, bringing an erudition in Aristotelian natural philosophy, Galenic medicine, and Neoplatonic metaphysics to bear on interpreting the essential core of Paracelsian mechanical agency that is sketched above. He shaped Paracelsus' vision that all development was akin to the sprouting and growth

39 Paracelsus, *Opus paramirum*, book III, tractatus tertius, in Paracelsus, *Operum medico-chimicorum sive paradoxorum* (1603), tome 1, 125: "Æque enim ac pluvia stillatim decidit si generatur, non universa simul aqua, sed guttalis generatio ea est, quæ taliter etiam decidit, sicut in Mechanicis traditur: sic pariter accidit cum materia nutrimenti ad hepar pertinente. Ea cum urina permixta est, & ex urina attrahitur, ita ut urina iam sola remaneat. Hæc tandem per propriam virtutem expulsivam ad vessicam, & inde foras exturbatur."

40 See Shackelford, "Paracelsian Uroscopy," 13–35.

of seeds into a thoroughgoing biological metaphysics, which he described at length. The conception of "mechanical" that Severinus deployed in his book is therefore very probably Paracelsian in the first instance. But we should also consider the cultural background of this idea more broadly, to allow us to approach the wider spectrum of meanings that Severinus may have attached to it, beyond merely "seminalizing" the Paracelsian craftsman, the *archæus*.

A broader range of meanings of "mechanical" is hinted at by one of the rich and colorful characterizations applied to Paracelsus by one of his editors in the 1603–05 Latin edition of his works. We find there that besides being a Philosopher of Paradoxes, a Ruler of Mysteries, a Master of the Arts, and a Professor of (both) Medicines, Paracelsus is labeled "The German Trismegistus of the Mechanical Muses."[41] What did this mean? "German Trismegistus" is relatively straightforward: it identifies Paracelsus as a German version of the famous Hermes Trismegistus, whose name was associated with Hermetic chemical medicine in general by the end of the sixteenth century, but also with early religion, owing to Marsilio Ficino's translation of the *Corpus Hermeticum* and his hermeneutic approach to it as a vehicle for understanding early Christian religion. But "mechanical muses"? Taking muses in the ancient Greek tradition as a metonym for academic studies, this must refer to the mechanical disciplines or arts. With this, we return to the theater and to the stage.

4 Lutheran Education and Providential Play on the World Stage

Plato identified fine arts such as portraiture and poetics, which included drama, as *mimesis* and therefore held them to be of low value in his *Republic*. Basically, he argued that whereas the craftsman creates an object by translating an idea into an artifact by imposing forms on natural materials, the fine artist merely portrays or copies what has been crafted; he is imitating imitations – a further departure from true reality. Consequently, comedy and tragedy, as mere imitative fine arts, should have no place in Plato's ideal state.[42] But by the late Renaissance and Reformation, Plato's assessment of the value of theater

41 In the front matter to Paracelsus, *Operum medico-chimicorum sive paradoxorum*, tome 5, Paracelsus is denoted as "Philippus Theophrastus Bombast Hohenheimensis ... Philosophus Paradoxus: Mysteriarcha: Artium Magister: Medicinarum Professor: Musarum Mechanicarum Trismegistus Germanus."

42 Carlson, *Theories of the Theatre*, 16–17, 27. Plato discussed this in the tenth and last book of *The Republic*. Fortunately, Plotinus, an author much admired by Severinus and contemporary Renaissance Platonists, rehabilitated the arts somewhat by arguing that artists imitated spiritual objects, not material ones.

had largely been displaced by that of Aristotle; theater thereby enjoyed an improved status among the learned, and found a place in Lutheran education.

During Severinus' youth, Renaissance theater arrived in Denmark as part of the humanist educational reforms spread by the followers of Philipp Melanchthon. Denmark had been reformed by the Philippist theologian Johan Bugenhagen, and under its earliest Protestant leaders, drama was brought from Wittenberg into the Latin School curriculum in Denmark's provinces as part of Lutheran classical studies.[43] We can suppose that Severinus was exposed to basic Latin drama – the comedies of Plautus and Terence, and commentary by Donatus – while attending the Latin school attached to the cathedral in Ribe, his home town in southern Jutland.

However, theater was more than a dramatic textual form or a vehicle for teaching eloquence and classical culture; it was a performing art. While the new Renaissance dramas may have been limited mainly to school performances and occasional entertainments at court, medieval religious plays had been staged in European churches and at town festivals since the High Middle Ages, and historians are doubtful that these traditions – so ingrained in community practices – were immediately expunged by the Reformation. Lutheran reformers purged the outward and obviously idolatrous elements of Catholic liturgy, and with them the passion plays and other highlights of Catholic ritual, but theater continued to serve as a vehicle for moral teaching and as an exemplar of fine classical humanist art.[44] Thus, Severinus, a capable student of Latin at one of Denmark's reformed episcopal seats and Jutland's bustling trading hub, was likely exposed to the dramatic arts, both in the classroom and sitting before the stage. To appreciate the metaphysical meanings Severinus attached to theater, especially to comedy, we should begin by asking in what sense theater was an art, and what medieval philosophical discussions about the relationship between art and nature can tell us about Severinus' employment of theater as a metaphor for agency in the natural world.

Much as Plato saw the craftsman as making a material replication of the ideal, Severinus used the terms "comedy" and "world stage" to evoke a theatrical metaphor for the processes of natural development, the work of inner mechanics who transformed seminal ideas into actual bodies. In his vivid description, the seminal *archei* make reifications of the seminal ideas by dressing them up in vestments suitable to the parts they will play. They act in a

43 Albeck and Billeskov Jansen, *Dansk Litteratur Historie*, vol. 1, 178, 184; Billeskov Jansen, *Ludwig Holberg*, 56–57.

44 For discussion of continuity in medieval Catholic theater into the Reformation period, see Harris, *Medieval Theatre in Context* and Tydeman, *The Theatre in the Middle Ages*, 235.

dramatic cycle, from their initial appearance to their exiting the world stage. Severinus also frequently likened this cycle to the agricultural growth cycle, from sprouting to maturity and decay, but the two metaphors of field and stage reinforce each other in a dramatic unity.

As a poetic art, drama – and, by extension, its practice on stage as histrionics – was a natural medium for discussion of the human capacity to imitate nature. But what about theater as a metaphor for the processes at work in the natural world, both natural and artificial processes? The use of the term "theater" to represent the abstract concept of an inclusive world was commonplace in early modern literature.

Ann Blair points out that in the late sixteenth and seventeenth century, "theater" was used both to mean a book that captured the essence of a subject that it presented to its reader-audience in a collection of sorts, as evident from book titles such as Theodor Zwinger's *Theatrum vitæ humanæ* (1565), Heinrich Khunrath's *Amphitheatrum sapientiæ æternæ* (1595), and Lazarus Zetzner's *Theatrum chemicum* (1602), but also to convey the sense that nature itself was a providential theater, presenting divine beauty and order. She claims that "these two metaphors appeared in the sixteenth century" and points to Philo of Alexandria's *On the Creation of the World* as a "precedent for this new metaphor of the theater of the world," also noting that Plotinus used the theater as a metaphor for the world in *Enneads* 3.80–105, a work admired by Severinus and his close friend and fellow Paracelsian student, Johannes Pratensis.[45] She argues that whereas Philo regarded the theater of the world as a place for action, "the emphasis of Renaissance natural philosophical authors is rather on the static qualities of the theater of nature – its vast expanse, intricate order, and elaborate construction."[46] Here she had in mind specifically Philipp Melanchthon and, by extrapolation, the Lutheran natural philosophical tradition that emanated from Wittenberg at his direction in the middle of the sixteenth century.

But Severinus did not use the term *theatrum* or *theatrum mundi*, choosing instead *mundana scena*, the world stage, and he did indeed make it a place for action – *the* place for the seminal development. To this end, he foreshadowed William Shakespeare's statement that "all the world's a stage" in Act II, Scene 7 of *As You Like It*. For Severinus, the stage conveyed more than the variety and inclusiveness of *theatrum* as a book genre; he used it to denote the world as a physical place where nature's story, written by God, was put into action. The roles were scripted by Providence and played out in the generation and

45 Blair, *The Theater of Nature*, 153, 159–166, 176. See Shackelford, *A Philosophical Path for Paracelsian Medicine*, 171 n. 67.

46 Blair, *The Theater of Nature*, 154–155, 283 n. 16.

corruption of the bodies that populate the natural world, appearing on stage from their seminal lodgings, unfolding their stories in seminal development, and exiting to obscurity when their stories concluded. Moreover, he described these acts as *mechanical*, carried out by Paracelsian mechanical agents using Paracelsian mechanical knowledge. How did Severinus come up with such an idea, a juxtaposition of mechanics and theatrics?

5 Theater and Liturgical Ritual as Mechanical Arts

The concept of theater arts as mechanical arts had medieval precedence. Classical pagan theater had been discouraged by the early Church Fathers, who shared the Neoplatonists' view that drama appealed to the passions and thus misled the pious soul, which should seek simplicity, unity, and divine virtue.[47] The suspicion that theater had nothing good to offer to the true Christian began to fade in the ninth century with Amalarius, Bishop of Metz, who drew a comparison between the liturgical ritual of the Latin Mass and ancient theater, justifying allegorical interpretations of ecclesiastical liturgy in general.[48] Amalarius' legitimation of drama as a way of thinking about church rituals, and by implication the use of dramatic gestures and structures in liturgy, apparently struck a chord, since it generated criticism that the clergy inappropriately employed histrionics in their celebration of the Mass.[49]

In the early twelfth century, Honorius of Autun strengthened Amalarius' association of the church with the theater, viewing the Mass as an active dramatic form. Honorius used ancient theater as a model for how human arts could help churchmen make the meaning of the Mass vivid for parishioners.[50]

Honorius' idea was taken up by Hugh of St. Victor a couple of decades later, who not only legitimized the idea that the physical labor of the craftsman assisted his spiritual formation, but recognized that the mechanical arts (*artes mechanicæ*) – an extension of the traditional liberal arts – were becoming important to medieval courts and cities.[51] These he classified into seven

---

47 Carlson, *Theories of the Theatre*, 28. Given the agreement of Tertullian and St. Augustine, Christians into the Middle Ages remained deeply suspicious of the theater.

48 Dox, *The Idea of the Theater*, 50.

49 Hardison, *Christian Rite and Christian Drama*, 78–79.

50 Ibid., 40–41; Dox, *The Idea of the Theater*, 75, 78.

51 Dox, *The Idea of the Theater*, 86: Hugh extended the traditional liberal arts, comprising the *quadrivium* and *trivium*, to include "*physica*, *economica*, and *mechanica*." This argument would push Edgar Zilsel's argument for the importance of the crafts laborer in knowledge-making back to the High Middle Ages.

types, the sixth mechanical art being *medicina* and the seventh *theatrica*.[52] Interestingly, Hugh regarded mechanics as basically corrective or restorative, analogous to the role of medicine in assisting nature to restore the body.[53]

The definition of theatrics as a mechanical art or craft appealed to the thirteenth-century churchmen Bonaventure and Vincent de Beauvais. After a period of oblivion, it surfaced again at the end of the fifteenth century in works by Angelo Poliziano and Gregor Reisch. Reisch's *Margarita phylosophica* (1496), which was published in numerous editions in the first quarter of the sixteenth century and sporadically until 1600, listed *theatrica* as the seventh *ars mechanica* and reproduced Hugh of St. Victor's description of theatrics.[54] Thus, the identification of the "stage" with "mechanical art," and medicine as a mechanical art, were concepts that were available to Severinus during the formation of his medical ideas in the third quarter of the sixteenth century and help us interpret the nexus of theater, liturgy, and medicine that we find in his book on Paracelsian vital philosophy.

The identification of the Mass with what we today might call "the theater of the church" is controversial among church historians, and certainly church drama as it developed in the High and late Middle Ages is not to be confused with classical comedy and tragedy. Nevertheless, the anthropological view of liturgy as a form of theater provides a good framework for understanding how ritual related to dogmatics and was used as a means to celebrate and communicate the fundamental truths of Christian salvation. The use of drama in the Mass was implicit already in early church ritual and was ingrained in late medieval practice until the Reformation, as is evident in artists' depictions of fifteenth- and sixteenth-century Masses.[55]

In what sense is the Mass theater? Clearly the liturgists themselves can be regarded as actors, and the church as the place of performance. But the identification of one with the other runs deeper; even church vestments, replete with symbolic meanings, can be seen as costumes put to sacred uses.[56] As a dramatic structure, the Mass is *comedy*, beginning with a descent toward a crisis, followed by a reversal, and a joyful resolution. In terms of the paradigmatic Christian Mass – the Easter Passion – the comedic cycle is crucifixion, entombment, and resurrection.[57] This cycle of incarnation, worldly life, death,

52 Olson, "The Medieval Fortunes of 'Theatrica'," 267.

53 Ibid., 267. In this regard, medicine was comparable to alchemy; see Newman, *Promethean Ambitions*, 69–72.

54 Tatarkiewicz, "Theatrica, the Science of Entertainment," 268–271.

55 Hardison, *Christian Rite and Christian Drama*, 79.

56 Ibid., 79.

57 Ibid., 83, 284.

and resurrection – parallel to the waxing and waning of the year, the birth and growth of elemental beings from seminal potency to mature creature and then contraction, resolution, and ultimate return to body-less perfection – offers perspective on Severinus' use of the term "mechanical lithurgy" or liturgy to refer to biological processes.

In Severinus' lifetime, the dramatic arts had become mechanical in another, more literal sense. The actual production of theater had become more physically mechanical, in terms of the construction of the material stage in a defined physical space, the mechanics of orchestrating the production, and the moving of any backdrops and theatrical scenery. Scholars agree that there were no dedicated, purpose-built standing theaters in the Middle Ages, just as there were no specific, permanent theaters for anatomical dissection before the late sixteenth century, but that interest in the architectural aspects of theater appeared with the humanist revival of the classical Roman architecture of Vitruvius. Prior to the construction of permanent theaters, whether for the presentation of dramas or public dissections, stages were erected outdoors or in churches or noble halls as the occasion demanded, and the term "theater" applied more to the space, the place for the spectacle, than to a physical structure.[58] However, the *scena* or stage was a physical structure, and this is the term that Severinus used when he wrote of the "world stage" (*mundana scena*). The term had been used by Isidore of Seville to denote a small, house-like structure with an attached platform, called an *orchestra*, on which players or dancers performed. The house-like *scena* provided a booth for the poet or narrator and screened the actors while they waited to emerge onto the orchestra-platform to mime out the poets' words, as needed, and to which they retired when their parts were finished. This method of dramatic presentation was attributed to Terence; it was employed in Europe in medieval productions and saw continued use late into the sixteenth century.[59]

The fifteenth-century Italian revival of classical theatrical forms, with an emphasis on dramatic unity and a central focus for the play's action, was assisted by the publication of Vitruvius' *De architectura*, the fifth book of which was taken up with the design of theaters. Instead of the booth-like *scena* of Isidore and the medieval theater, the Vitruvian stage featured a *scenæ frons* at the rear of the stage, a concealing screen with multiple doors through which

58 Tydeman, *The Theatre in the Middle Ages*, 46.

59 Ibid., 48, 235. In his introduction to *The Comedies of Terence*, xix, Matthew Leigh notes that in Terence's day, before Pompey's theater opened in 55 BC, temporary stages and seating was erected for performances, and that actors came and went from houses at the back of the stage.

actors could enter and exit the play space as demanded by the script. This is a form familiar to audiences of classical Shakespeare productions today. The Vitruvian Renaissance arrangement was designed to focus the audience's attention on a single area of play and to support the appearance of realism.[60]

The production of late-medieval and Renaissance plays was apparently scripted in great detail, and toward the sixteenth century required increasingly sophisticated machinery for effects.[61] One can readily understand how Severinus might liken a fully planned and scripted play – in which no detail was left to chance – to the great providential comedy cycle charted by the Bible, from Genesis to Daniel and Revelations, which was being played out on the world stage. In Severinus' day the operators of the mechanical contrivances were provided with detailed cue sheets that specified the appearance, movements, and disappearances of special effects on stage, and one can well imagine Severinus likening the emergence of rudimentary bodies in nature, their growth and decrease, to mechanical processes being carried out by stage mechanics with the requisite mechanical knowledge.[62]

The Renaissance revival of classical theater on Vitruvian stages entailed the production or adaptation of classical pieces of dramaturgy, the earliest and most common of which seem to have been the comedies of Plautus and Terence, printed in incunabula and early sixteenth-century volumes with woodcuts depicting actors emerging from the *scenæ frons* and standing before an audience in a neoclassical-style amphitheater space. It is this image that gives cogency to the depiction in the frontispiece to his pioneering anatomical textbook, *De humani corporis fabrica* (1543), of Andreas Vesalius as a performer of a great dramatic work. It has been suggested – plausibly – that the physical space in which Vesalius is shown dissecting was a depiction of the Temple of Peace in Rome, where Galen had dissected, but the more immediate iconographic model may have been an illustration accompanying contemporary printed editions of the plays of Terrence (1497) and Plautus (1511) published in Venice.[63] Indeed, the similarity is striking, down to the detail of the figures supporting

60 Tydeman, *The Theatre in the Middle Ages*, 235–236. Brockett, *History of the Theatre*, 132, notes that interest in classical theater had already revived in the fourteenth century, but that it was not staged before 1486, when the Roman Academy employed a Vitruvian design. On the influence of the Renaissance revival of Vitruvius among theater architects and its impact on the design of Tudor and Stuart theaters in particular, see Yates, *Theatre of the World*.

61 Brockett, *History of the Theatre*, 99.

62 Ibid., 108; Brockett notes that the staging of the morality play, *The Castle of Perseverance*, in the early fifteenth century charted human development from birth to death (112).

63 O'Malley, *Andreas Vesalius of Brussels*, 140. See also Ferrari, "Public Anatomy Lessons," 84.

themselves upright by holding onto the columns. The adaptation may have been made for artistic convenience, but given the symbolic and personal representations of Vesalius' frontispiece more generally, and the arguments that have been advanced for Vesalius' intentionality with these elements, Vesalius likely expected his readers – his viewers – to understand the theater metaphor for the public dissection and to identify such dissections as grim comedies unfolding on a stage, with the formally-vested Vesalius and the fully divested female body as principal actors.

Vesalius had the engraver of his frontispiece depict him as the principal actor on the anatomical stage, using his scalpel to reveal the providential design of the human body, much as his master Galen had done. But, for Severinus, Providence was not only evident in the static design of the world body, but also in its temporal unfolding – anatomy, in the larger sense of the term as it was used by Paracelsus.[64] To this end, Severinus' world stage was not an anatomical theater, but rather a liturgical theater, where the action took place mechanically, with the dependability of liturgical ritual.[65] Perhaps this choice of metaphor reflects Paracelsus' perspective that the dissection of cadavers reveals not the nature of the living body, but the anatomy of death, doing little to inform the physician about health and disease, which were dynamic processes.

For Severinus, whose natural philosophy was shaped by his readings of Plotinus and Proclus, and who viewed the material world of bodies as emanating from an immortal world of seed-like ideas and potencies, stone masonry provided a meagre, yet perhaps too concrete metaphor for the generation and corruption of natural bodies. When he described this process as a mechanical lithurgy, he may have had in mind Paracelsus' *werkmeister*, the *architectonicus* who followed the Grand Architect's plan to shape nature's beings into material things, much as the mason builds with stones or bricks. But if so, the emphasis was on the mechanical agent, not on the pre-existing building blocks. His natural philosophy was not a materialist philosophy, but a vital, mechanical philosophy, a Christian philosophy, in which the physician was an artist and medicine was a traditional mechanical art; it provided a model for organic

64 Paracelsus used the term "anatomy" to mean generally the nature of something that might be revealed by analysis, e.g., the anatomy of the world, the anatomy of diseases, the anatomy of remedies.

65 Perhaps this characteristic of liturgy was impressed on him during his travels in France and Italy. Here much of his *Idea medicinæ* took shape, doubtless under the influence of his observations of the Catholic Mass, which was liturgically richer than Protestant ceremonies. The theatrical nature of liturgy may also have occurred to him from observations of carnival or the Feast of Fools or Feast of Asses, or other Bakhtinian inversion of the traditional roles. See Tydeman, *The Theatre in the Middle Ages*, 16–17.

development, interpreted as the acting out of prescribed roles on the comedic stage, much as the liturgist performed scripted rituals on the ecclesiastical stage. Severinus completed the *Idea medicinæ philosophicæ* in Florence in 1571, while he and his friend Johannes Pratensis were concluding their medical studies. One can imagine Severinus as a young Danish medical student attending such performances of the classical comedies in the very heart of the European humanist Renaissance while writing his *Idea medicinæ philosophicæ* and drawing an analogy between the appearances, roles, and exits of actors in comedic cycles on the stage, and the life-cycles of individual animals, vegetables, and minerals in the macrocosm.[66]

6 Seminal Agents as Actors on the World Stage

Severinus invoked this image of actors on stage to help explain his biological theory. For him the fundamentals of development were "elements or wombs, seeds rich in seminal reasons, and the principles of bodies," a clear statement of Paracelsus' basis for physical nature on the four Aristotelian elements, now interpreted as passive receptacles, *matrices*, and the three primary principles (salt, sulphur, and mercury), but with an added emphasis on the *rationes seminales*, which gave Severinus' biological theory its distinctive explanatory power.[67] These seeds, empowered by the spiritual principles and "endowed with many functions," as Severinus wrote, are disseminated in the elemental *matrices* or wombs, a metaphor for the relationship between the passive Aristotelian elements and the three active spiritual principles that Paracelsus founded on the classical understanding of the heterosexual reproduction of sexually dimorphic types. Much as the Aristotelian seminal form directs the epigenetic development of the fetus *in utero* (*in matrice*), Severinus' *semina* use elemental matter to construct bodies as they grow.

66 Yates, *Theatre of the World*, 128–135, 189 noted the obvious association of the Renaissance Vitruvian theater with its painted "heaven," central stage of action, and zodiac geometry with the macrocosm / microcosm relationship between the greater theater and the actors on the stage. Shakespeare alluded to this in *As You Like It*, Act 2, Scene 7, where he identifies the world with the stage on which the microcosmic human drama unfolds from cradle to grave (ibid., 167), a metaphor also expressed in Thomas Heywood's 1612 poem *Apology for Actors* (ibid., 164–165).

67 Severinus, *Idea medicinæ*, 82: "Decet enim Philosophum in Generationum doctrina … primum radicem, materiam, fundamentum (quomodocunque appellare libuerit) adducere…. Multa vero necessaria esse ad Generationis lithurgiam diximus: Elementa vel Matrices, Semina Rationibus foecunda, corporum Principia."

Seeking to explain Paracelsus' use of many terms to describe the potencies of matter, the wombs or *matrices* on which the seminal spirits operate, he wrote:

> It is very difficult, and on the other hand very easy to give a name to this matter that is pregnant with endowments for so many functions. One who will try to explain the treasuries of such a rich nature with one name will meet with difficulty: for it cannot be done. One who has understood that one nature submits to the call of so many and such different names will wonder at the ease. If it is called one thing, he understands a unity; if many, he reveals a multitude. If two, three, four, etc. he extemporaneously sets out so many Syruses and Parmenos.[68]

Although it might elude the modern reader, with his mention of "Syruses and Parmenos" in this passage Severinus explicitly adduced Renaissance theater to explain this difficult concept of how an essentially unified natural potency reveals in its development a multiplicity that can be characterized by diverse names, a point on which Paracelsus drew censure, his critics accusing him of creating unwarranted neologisms.

Syrus and Parmeno are characters in various of Terence's comedies, servants who, despite their inferior status, are crucial to the unfolding of the narratives of these plays. A Parmeno shows up as the slave of Laches in "The Mother-in-Law" and the same name is given to the slave of Phædria, Laches' son, in "The Eunuch," so it is possible that the same character was summoned into the action of two different plays. Syrus likewise appears in both "The Self-Tormentor" as the slave of Chremes, and in "The Brothers" as the slave of Micio. Shakespeare must have had in mind a similar overlapping of parts when he chose the same name for both of the twin servants Dromio in *A Comedy of Errors*, probably not a random association.[69] Thus, Severinus' comparison of ubiquitous but anonymous seminal agents to the Syruses and Parmenos of

---

68 Ibid., 82–83: "Hanc Materiam tot officiorum Donis gravidam, nominare difficillimum est, & rursus facillimum. Difficultatem inveniet, qui uno nomine, thesauros tam divitis Naturæ explicare conabitur: nec enim fieri potest. Facilitatem admirabitur, qui tot, tam diversorum nominum appellationibus, unam Naturam obtemperare conspexerit. Si una dicatur, unitatem intelligit: si multa, multitudinem profert: si duo, si tria, si quatuor, &c. tot Syros, Parmenones, subito emittit."

69 T.S. Dorsch's introduction to Shakespeare's *A Comedy of Errors*, 9 focuses on Plautus as one of Shakespeare's main sources for this play, but reflects scholarly opinion that Shakespeare was also influenced by John Lyly's play *Mother Bombie*, in which there is a character named Dromio. Dorsch notes in passing that in Lyly's play the servants are crucial to the development and that they "recall Terence's witty slaves and old men."

Terence's comedies – characters of relatively undefined social backgrounds serving similar roles in various of his comedies – provides an apt analogy, easily grasped by his humanistically-trained contemporaries. An actor may indeed play several roles in such dramatic performances, taking on different costumes to designate each, much as the *semina* in Severinus' biological metaphysics unfold their scripts by taking on the vestments of bodies, terms that Severinus deployed when explaining his theories.

In a sense, for a Neoplatonist like Severinus, comedy provides an obvious model for the metaphysical passage from birth to death. The cycle from the latency of non-body to the actuality of body and the subsequent return to non-body *was* comedic, inasmuch as comedies were required to introduce a problem, resolve it, and provide closure, ending on an optimistic note. The traditional Christian literalist, who associated the elemental body with sin but required it to be resurrected at the end of the cycle, might have objected to the anti-materialism implicit in the ultimate dissolution of the body, but this would not have been a problem for a follower of Paracelsus' religion, which assumed a distinction between the Christ-like, exalted and glorified immortal body and the Adamic elemental body, which was doomed to end up as food for the worms.[70] Paracelsian anthropology fit well the alchemical cycle, as reflected in the religious imagery of late-sixteenth- and early seventeenth-century alchemical, iatrochemical, and mystical texts.[71]

7 Conclusion: Mechanical Philosophy and the Stage without Actors

The radical nature of the intellectual and cultural transformation of the medieval worldview into modern science and scientific cosmology was characterized as "the mechanization of the world picture" by E.J. Dijksterhuis, who identified mechanics as the route by which mathematics was introduced into natural philosophy, which he regarded as essential to the emergence of

70 On Paracelsus' anthropology and its place in radical Protestant religion, see Daniel, "Paracelsus on Baptism and the Acquiring of the Eternal Body"; Rudolph, "Hohenheim's Anthropology in the Light of His Writings on the Eucharist"; and Gilly, "'Theophrasta Sancta' – Paracelsianism as a Religion, in Conflict with the Established Churches." On Johan Arndt's relationship to Paracelsianism, the Rosicrucians, and the origins of Pietism, see Shackelford, "Western Esotericism."

71 Szulakowska, *The Sacrificial Body*, has argued that the presentation of such images is associated mainly with Lutherans who were dissatisfied with Lutheran orthodoxy and sought to reintroduce mystical participation in the Mass via an alchemical Eucharist.

modern science.[72] To understand the cultural profundity associated with this transformation, just consider the reflections of C.S. Lewis, who recounted how in the expansive cosmic world of Nicholas Copernicus, Giordano Bruno, and Galileo Galilei, the Earth and its inhabitants were no longer central and were perhaps not unique; as a consequence of the discarding of the medieval, geocentric and human-centered cosmic image, the world thus revealed both to seventeenth-century poets and philosophers alike seemed an emptier and colder place.[73]

We find in the work of Carolyn Merchant another sketch of the broad consequences of a scientific and cosmological shift in the early modern period that complements Lewis' portrayal: in the seventeenth-century disenchantment and mechanization of nature, she saw deeply significant changes in attitudes toward the exploitation of natural resources, the subordination of women, and the subjugation of individual workers to capitalism and a mantra of progress. The continued cogency of this perspective for historians of science is evident from the 2006 *Isis* "Focus" section celebrating and explaining the enduring legacy of Carolyn Merchant's *The Death of Nature*.[74]

This chapter has argued that what Dijksterhuis and subsequent historians of science have explained as the imposition of a mechanical worldview or philosophy, which supervened and largely displaced Aristotelian natural philosophy and metaphysics, and which Merchant used as a point of departure for *The Death of Nature*, was not the creation or imposition of mechanism as much as it was a deadening of the meaning of mechanism, depriving it of spirit, soul, or other internal agency and leaving behind a dead, inorganic machinery. That is, notions of the 'mechanical' and 'mechanics' were not introduced into natural philosophy in the seventeenth century, but rather were reinterpreted. Mechanics, mechanical knowledge, and mechanical processes

72 Cohen, *The Scientific Revolution*, 66–71. Modern emphasis on a mechanical view of nature as a crucial component of the emergence of modern science from medieval natural philosophy owes much to Dijksterhuis, *The Mechanization of the World Picture*; see, for example, Hooykaas, *Religion and the Rise of Modern Science*, 13: "The model of the world as an organism was replaced by that of the world as a mechanism; the whole development from Copernicus to Newton has rightly been called the mechanization of the world picture."

73 Lewis, *The Discarded Image*.

74 Merchant, *The Death of Nature*, 193; Cadden, ed., "Focus: Getting Back to *The Death of Nature*." The centrality of mechanization to Merchant's interpretation of 'the death of nature' is underscored in the opening line of the abstract that accompanies her own rereading, as part of this Focus group, of her own seminal book: Merchant, "The Scientific Revolution and The Death of Nature," 513: "*The Death of Nature* ... presented a view of the Scientific Revolution that challenged the hegemony of mechanistic science as a marker of progress."

happily coexisted in what might be usefully called the vital philosophy of the sixteenth-century Paracelsians, exemplified in the work of Paracelsus' most able early commentator, Petrus Severinus, for whom vital agents operated mechanically. What became the devitalized mechanical philosophy of the early Enlightenment required not only a rejection of what the German chemist Andreas Libavius identified as Severinus' "vital philosophy," but a narrowing of the meaning of the term "mechanical" so as to remove its earlier connotation of agency.[75]

The dramatic stage was an apt metaphor for Severinus' world, which was richly complex and restlessly dynamic, with the appearance and disappearance of things, as of actors entering and exiting the natural scene. Moreover, it found resonances in the disquieted world of Counter-Reformation Europe. In 1643 Ambrosius Rhodius, who had come to Denmark from his native Saxony and married Severinus' granddaughter, took up Severinus' metaphor to help make sense of the world he experienced. Rhodius had begun university in Wittenberg, but left to pursue medicine, first at Königsberg and then at Copenhagen, without attaining the M.D. During these years, his homeland was ravaged by the Thirty Years' War and his family fortune was dissipated; his means of livelihood now lay in employment as a German-speaking city physician in Christiania, the capital city of Norway under Danish dominion. In the dedication of his 1643 commentary on Severinus' *Idea medicinæ philosophicæ*, pondering the vicissitudes of fate and the variability of the world, he toyed with Severinus' theater metaphor:

> For just as God turns the fortunes of men by a wondrous change, so too do natural things play under the changeable condition of fate on this world stage. One thing disappears, another comes back.... During the time one compound is dissolved, another living one is born. For mixable things begin to exist by themselves. Soon these once more suffer their fate and are mixed with the earth. From the earth they are again attracted by plants and they also undergo remarkable changes. Worms appear from decaying plants, and from these are born chrysalides and from those are born butterflies, and other little animals. All things undergo this changeable spectacle as if they are acting out a theatrical play.[76]

75 The term "vital philosophy" was an actor's category applied to Paracelsian philosophy, specifically to Severinus' version of Paracelsian philosophy, in the early modern period. See Shackelford, "Rosicrucianism, Lutheran Orthodoxy, and the Rejection of Paracelsianism," esp. 202–204, and idem, *A Philosophical Path for Paracelsian Medicine*, 288–298.

76 Rhodius, *Disputationes supra Ideam medicinæ philosphicæ Petri Severini Dani*, ii–iii: "Sicuti enim Deus mirabili conversione fortunas hominum versat, ita etiam sub varia sortis

Five years earlier, Rhodius had published a dissertation on the Pythagorean transmigration of souls. His text was not primarily concerned with human souls, but rather with souls as forms of living beings.[77] In the uncomfortable transition from Aristotelian explanations of material transformation, which required the termination of the forms of reagent bodies, and the emergence or imposition of new forms in reaction products, to the corpuscularian explanations that were developed in the second half of the seventeenth century – a transition with which his teacher at Wittenberg, Daniel Sennert, had grappled – one problem that demanded particular attention was the question of what happened to forms or souls of entities when they underwent mixtion and transformation into other entities. This problem was perhaps less obviously fraught in the case of what we consider inorganic reactions involving metals, acids, and salts, but what about the metamorphosis of a larva into an adult butterfly, as mentioned specifically by Rhodius? Was there no continuity of the insect's form as it underwent these so obviously radical changes?

Addressing this question in 1638, Rhodius cited Severinus' *Idea medicinæ philosophicæ*. Presumably he was so taken with this book, written by the King of Denmark's one-time royal physician and a rich element of the social legacy of his new in-laws, that he undertook his much lengthier *Disputationes* in 1643 to explore, explain, and defend Severinus' metaphysical biology of seminal development. For this, the theater metaphor was well suited. As agents of activity and change in the mundane world, actors presented a continuity from their existence as potential characters before they enter the scene, through whatever changes their roles required on stage – perhaps even the exchanging of roles, to their returning to an invisible and latent state of existence after they have exited the scene. This provided a much-needed constancy in a world of change, alteration, generation, and decay. These seminal actors were by definition agents and brought agency to material bodies. But Severinus, elaborating Paracelsian concepts, had clearly portrayed them as *mechanical* agents, and there was no scope in his vital philosophy for a mechanized world in the sense defined by corpuscularian philosophers, who would abandon the ancient concepts of form and matter, and, in so doing, set new limits to the meaning of "mechanical."

---

conditione in hac mundana scena res naturales ludunt. Hoc abit, illud revertitur.... Dum compositum dissovitur, aliud vivens nascitur. Nam miscibilia per se incipiunt existere. Mox hæc rursum sua fata patiuntur, terræque miscentur. Ex terra rursum a plantis attrahuntur, adeoque mirabiles subeunt vicissitudines. Ex plantis putrescentibus prodeunt vermes, ex his nascuntur chrysalides ex istis papiliones, & alia animalcula. Omnia tam vario spectaculo peraguntur, ac si lusum scenicum agerent."

77 Rhodius, *Dialogus de transmigratione animarum Pythagorica*.

CHAPTER 7

Democritus in Francesco Patrizi and Giordano Bruno

Leen Spruit

Francesco Patrizi (1529–1597) and Giordano Bruno (1548–1600) are among the best-known sixteenth-century opponents of Aristotle and the Peripatetic tradition. Driven by panpsychic and panvital orientations, both developed non-Aristotelian worldviews, but with significant differences. Patrizi hearkened back to late Neoplatonism (including Proclus and Damascius) and the tradition of Christian Platonism, while Bruno, who based his thought on Neoplatonic notions and an original interpretation of Copernicus's new astronomy developed the idea of an infinite, physically homogeneous universe as the effect of an infinite cause that had only a very remote affinity with the scholastic idea of a first cause.

The two philosophers also became involved in the lively contemporary debate over the legacy of ancient atomism. In his *Discussiones peripateticae* (1581), Patrizi defended Democritus against Peripatetic criticism. Moreover, as has been underlined in recent studies, while it is certain that Patrizi was not favorably disposed towards atomism, he did have recourse to atomistic views in his doctrine of space.[1] Like Patrizi, Bruno also rejected Aristotle's natural philosophy, but unlike Patrizi, he openly advocated a most peculiar form of vitalist atomism. His view of the constitution of physical entities is characterized by the idea that, in every atom, a vital force is contracted, capable of aggregating around itself – as in a center – other atoms to form a composite body.[2] In this chapter, I do not address in any detail, however, the connections

1 Inter alia, Patrizi rejects the atomistic-mechanical approach to the heliothermal theory of the tides; see Patrizi, *Nova de universis philosophia*, fol. 144ra. In the context of a broader range of scientific and cosmological issues discussed in that book, Patrizi rejects Aristotle's denial of the possibility of a vacuum, presenting instead an argument for a position closer to the atomistic view that a material void is possible, albeit couched in terms of his own conception of space; see Henry, "Francesco Patricius da Cherso's Concept," and idem, "Void Space, Mathematical Realism."

2 For a discussion, see, among others, Atanassievitch, *Le doctrine métaphysique*; Michel, "L'atomisme de Giordano Bruno"; Lüthy, "Bruno's *Area Democriti*"; Gatti, "Giordano Bruno's Soul-powered Atoms"; Mucillo, "La concezione dello spazio"; Matteoli, "Materia, minimo e misura"; Gatti, *Essays on Giordano Bruno*, 71–90; Amato, "Atomo."

 | DOI:10.1163/9789004528925_008

Patrizi's and Bruno's thinking had with ancient and contemporary atomism; the focus is instead on their more specific references regarding Democritus. As we shall see, in their works, Democritus is mentioned mostly for his atomistic matter theory, but he is also mentioned for his notion of the plurality of worlds, and finally for his view on the dynamics of perception and knowledge.

Although Patrizi was basically a Platonist, one may recognize numerous sources in the history of thought that may have influenced his philosophical and scientific views. In both his *Discussiones peripateticae* (1581) and in *Nova de universis philosophia* (1593), Patrizi strenuously defends the pre-Aristotelian tradition, including a large and varied group of authors, from ancient theologians and Orpheus to Pythagoras, Democritus, and Plato. Section 1 of this chapter is devoted to an analysis of his 'defense' of Democritus in the third volume of *Discussiones peripateticae*, with particular attention paid to Democritus' view on the principles of natural reality and the psychology of perception.

As for Giordano Bruno, in both his Italian and Latin works, Democritus is frequently referred to for a wide range of issues, including the infinite worlds, atomism, the principles of natural reality, and the relationship between chance and necessity. On several occasions, Bruno also mentions Democritus' views on perception and knowledge. Relevant passages will be analyzed in section 2. Bruno also refers to Democritus' reputation as the "laughing philosopher," and as a prominent ancient alchemist, but these occasional allusions will not be investigated here.[3]

1 Patrizi's 'Defense' of Democritus in the *Discussiones peripateticae*

In Patrizi's philosophical works, the study of sources in the history of philosophy forms an essential part in the development of his own philosophical views. Not content to interpret major sources such as Aristotle and Plato, or to limit himself to the tradition which sought to delineate the major differences between those two major schools of Greek philosophy, Patrizi used his philological and historical skills to recover, study, and make available a great many sources which he believed to have dated to the pre-Platonic and Presocratic period.

Patrizi was among the most vehement and probably also one of the most systematic critics of Aristotle in the second half of the sixteenth century. In

3 On the laughing philosopher, see Bruno, *Candelaio*, in *Opere italiane*, I: 277; *La cena de le ceneri*, in *Opere italiane*, I: 432; *De la causa, principio et uno*, dial. I, in *Opere italiane*, I: 638. On Democritus as alchemist, see *Candelaio*, in *Opere italiane*, I: 299.

his *Nova de universis philosophia*, he replaced the four Aristotelian elements with space, light, heat, and humidity. Space (*spatium*), the first principle of the corporeal world, is viewed as prior to all bodies, even light. The universe consists of three separate worlds: an infinite space filled with light (*Empyreum*); the sphere containing stars and heavenly bodies (*aether*); and the "elementary," sublunar realm. Thus, the Earth lies at the center of an infinite expanse of light-filled space which is positioned beyond the material realm. This worldview has several sources in the history of science and philosophy. The infinity of space and the existence of a vacuum, which have been maintained by the ancient atomists, are cases in point. And Patrizi's idea of the centrality of light as an intermediary level between the corporeal and incorporeal has solid roots in the Platonic tradition.[4]

Patrizi discussed at length Aristotle's relation to earlier Greek philosophers in his *Discussiones peripateticae*, the first part of which appeared in Venice in an abridged version in 1571, while it was issued in a four-volume set by Pietro Perna in 1581 in Basel.[5] Patrizi's *Discussiones* is one of the most significant texts of sixteenth-century philosophical historiography, not only for the critical and philological examination of Aristotle's works, but also for the detailed picture it offers of Greek philosophy prior to Aristotle.[6] By listing a large number of ancient, late-medieval and Renaissance sources, Patrizi sought to replace the dominant Aristotelianism of his era with a new and challenging alternative, that is, a Platonic conception of natural and transcendent reality.

In his historical reconstruction, Patrizi does not set his philosophy against that of Aristotle but rather intends to destroy the myth of Aristotle's originality and, by extension, that of his alleged primacy in the field of philosophical investigation. The idea that Aristotle was an intellectual fraud and that he plagiarized the views of earlier philosophers is a central notion in the *Discussiones*

4 Purnell, "Francesco Patrizi"; cf. Leinkauf, *Il neoplatonismo di Francesco Patrizi*.

5 In vol. I of *Discussiones peripateticae*, Patrizi offers a detailed account of the life and works of Aristotle and of his followers. Vol. II contains an "Aristotelis et veterum philosophorum concordia" and sets out to prove that Aristotle's works also contain true and useful insights, although all are borrowed from earlier philosophers, including Plato, Pythagoras, Orpheus, Hermes, and the Presocratic philosophers. In vol. III, he delineates "Aristotelis et veterum philosophorum discordia" on various topics, and he analyzes several issues on which Aristotle explicitly states that he differs from his predecessors. Patrizi tries to show that Aristotle is invariably mistaken in his claims. In vol. IV, Patrizi refutes the Aristotelian teachings on several key concepts, such as principles, privation, matter, and form; see Deitz, "*Falsissima est ergo*," 228–229.

6 For a discusion, see Mucillo, *Platonismo, ermetismo e "prisca theologia,"* 73–193; Borsic "Patrizi and the History of Philosophy"; Deitz, "Falsissima est ergo"; Gulizia, "Francesco Patrizi."

peripateticae.[7] It is not at Aristotle's impiety that Patrizi takes aim, but his tendentious interpretations of previous doctrines. He also highlights Aristotle's dependence on earlier authors, precisely in relation to those concepts that were subsequently presented as his own original discoveries. In this vein, he analyzes the history of Greek thought from its origins, that is, from the ancient theologians up to Plato, focusing not only on Orpheus, Pythagoras, Parmenides, and Zeno, but also on Empedocles, Anaxagoras, and then, albeit quite briefly, on Democritus.

Although limited to only a few pages, in his analysis of Democritus, Patrizi tackles his most significant physical theories, such as those of the atomistic composition of matter, the void, and the swerve of atoms (even though the swerve was not a Democritean, but an Epicurean invention).[8] Eventually, he also dwells on Democritus' view of perception. For Patrizi, Democritus is basically a physicist, or more precisely, the first Greek naturalist philosopher.[9] He is also labelled one of the most important sources for Peripatetic physics, of which Patrizi presents an original interpretation, by deriving it in fact from Democritean views.

In the first place, he points out how Aristotle's presentation of Democritus' thought on the principles of reality is flawed by contradictions. Aristotle's *Physics* and *On the Heavens* attribute to Democritus a pluralistic conception, while on other occasions, Aristotle assigns to him either a monistic or a dualistic view based on the two principles of fullness and emptiness. Returning to his favorite theme of "plagiarism," Patrizi dwells on Aristotle's bitter polemical tone and remarks that, in his attack on Democritus' view of matter, he clearly overstates his case. As a matter of fact, Democritus viewed matter as the unqualified material substrate present in every individual mixt. Due to its characteristics of *inqualificatum corpus*, Aristotle's own "matter" seems most apt to be identified with that *commune corpus*, which, divided into different "parts" and "forms," constitutes the common material structure of all beings:

> Let us now turn to Democritus, and hear what Aristotle attributed to him in *Physics* I, ch. 2, and in *Physics* III, ch. 4, and in *On the Elements* I, ch. 4. Elsewhere he says that Democritus posited infinite principles of natural things. However, we find a passage in Aristotle, in which he relates that

7 Gulizia, "Francesco Patrizi," 562.

8 For the following reconstruction, I am profoundly indebted to the detailed analysis of Muccillo, *Platonismo, ermetismo e "prisca theologia,"* 127–132.

9 Patrizi, *Discussiones peripateticae*, vol. III, bk. 1, p. 293: "Ab Orpheo Theologia universa, a Thalete Mathematica, a Democrito physica, a Pythagora, et tres hae, et praeterea ethica. Haec communis est sententia."

> Democritus established only one principle, namely, a certain common body, that is, in the afore-mentioned *Physics* III, ch. 4, where he writes, shortly after those infinite principles: [...] *Democritus, for his part, asserts that no element arises from another element, but nevertheless for him the common body itself is a principle of all things, differing from part to part in size and in shape.*[10] Is this also an Aristotelian dogma? Is the unformed (ἄποιον) body the entire matter, the principle of all things, which differing from each other in size and shape constitute the universe? And those atoms of Democritus, what else are they than the forms of things which by their nature are indivisible and uncuttable, latent first in matter, and then adorning matter externally?[11]

In this passage, Patrizi proposes a most original interpretation of Aristotle's doctrine of substantial forms. The latter are by their nature indivisible and are therefore regarded as similar to Democritus' atoms. They are at first latent in the matter, and are then made patent, arising on its surface. Thus, a view considered for centuries to be originally Aristotelian is, for Patrizi, nothing more than a camouflage of the ancient Democritean theory of the principles of reality. In virtue of this contrasting analysis, based on the attribution of the inseparability of the Democritean atom to the Aristotelian substantial form, Patrizi debunks Aristotle's criticism of Democritus. For Aristotle, "atomical" in the proper sense – that is, simple and indivisible – is a qualification that can be attributed only to substantial forms, while the Democritean atoms, being material, do not possess this feature.

Patrizi also shows that Aristotle's reconstruction of the principles of reality as outlined by Democritus and the atomists is contradictory. Aristotle's texts either refer to infinite principles, or to only two principles (full and empty),

10 Aristotle, *Physica*, 203a33.

11 Patrizi, *Discussiones peripateticae*, vol. III, bk. 2, p. 305: "Democritus accedat, quid illi Aristoteles apponat audiamus, I Physico cap. 2 et III Physico cap. 4 et I De elementis cap. 4, alibi etiam ait Democritum infinita principia rerum naturae statuisse. Attamen invenimus locum apud Aristotelem quo narrat unum principium ab eo positum fuisse, corpus scilicet quoddam commune id est 3 Physici iam dicto cap. 4 in quo scribit, mox post illa infinita principia: [...] *Democritus vero nullum primorum aliud ex alio fieri dicit, sed nihilominus ipsum commune corpus omnium est principium rerum, magnitudine per partes et figura differens.* An hoc. Aristotelicum quoque dogma est? Materiam universam ἄποιον *inqualificatum* corpus, esse omnium principium rerum, quae inter se magnitudine et figura differentes, universum constituunt? Atomique illi Democritici quid aliud sunt, quam formae rerum sui natura indivisibiles ac insecabiles, in materia prius latitantes, deinde materiam externe exornantes?"

or even to a common *corpus* of all beings. It is impossible on such grounds to establish the genuine thought of Democritus:

> But Democritus, who is rebuked by Aristotle for having established infinite principles, and because he established the existence of one common body, which he said was the principle of all things, is praised in *Physics* I, ch. 5, because he, like all other physicists, also asserted two contrary principles. And Democritus posited solid and void, the former of which he qualified as being, the latter as non-being.[12] After reviewing the various principles of Democritus, we are unable to determine what Democritus actually believed, that is, whether he established infinite or else one or two contrary principles.[13]

In a similar vein, Patrizi dissects the Aristotelian reconstruction of the views of the atomists on the phenomenon of generation. According to this reconstruction, the atomists regarded generation as a *combination* of atoms among themselves, as some Aristotelian texts would induce us to believe was an adequate description, or else they spoke of *excretion*, a phenomenon that is contrary to combination, as Aristotle himself states elsewhere:

> But that is what has been said in *Elements* I, ch. 4 of Leucippus and Democritus […]. *For they say indeed that primary masses are infinite in number, but undivided in mass, and one cannot turn into many nor many into one; and all things are generated by their combination and coiling around each other.*[14] It is exactly the opposite of what is said in the following chapter 7: […] *Empedocles and Democritus, though without observing it themselves, reduce the generation of elements out of one another to an illusion. They make it a process of excretion as though generation required a vessel. For they say that the non-existent thing is to be separated, as if*

12 Aristotle, *Physica*, 188a23–24, in Diels Kranz (hereafter: DK), II, fr. 68A45.

13 Patrizi, *Discussiones peripateticae*, vol. III, bk. 2, p. 305: "At is Democritus, qui ab Aristotele obiurgatur, quod infinita principia statuerit, quodque unum commune corpus ut ipse fecerat, principium omnium rerum esse dixerit, I Physico cap. 5 laudatur quia ipse quoque ut ceteri omnes physici duo contraria principia posuerit. […] *Et Democritus solidum et vacuum, quorum aliud quidem ut ens, aliud ut non ens esse dicit.* […] varia Democriticorum principiorum recensione, quid nam revera Democritus, an infinita an unum an duo contraria principia constituerit, statuere nequeamus."

14 Aristotle, *De caelo*, 303a5–8.

> *generation were from a vessel.*[15] Undoubtedly, combination and involution are contraries of separation.[16]

Patrizi's defense of Democritus, which is once more based on the texts of Aristotle's objections thereto, leads him to tackle other issues of a physical-physiological nature, in particular the theory of perception. According to Aristotle, Democritus regarded the soul as composed of one kind of atom, in particular – namely, fire atoms. Perception is based on *eidōla* or images, thin layers of atoms, constantly sloughed off from the surfaces of macroscopic bodies and carried through the air to the sense organs.[17] Patrizi labels as inaccurate the Aristotelian exposition of Democritus' theory of perception and his view of vision, in particular when compared to other views, some of which were developed in the Peripatetic tradition. Theophrastus' doctrine is a case in point.

By examining the texts of both philosophers, Patrizi sums up some notable discrepancies, which serve him to highlight the depth and complexity of Democritus' authentic theory. First of all, he draws attention to the fact that the Aristotelian text, centered on the reduction of vision to *emphasis* ("impression"), constitutes an arbitrary simplification of the "true" Democritean theory, more exactly reported by Theophrastus in *De sensibus*.[18] In this work, Theophrastus emphasized instead the formation of the reflected image and its movement towards the eye:

> In *On Sense*, ch. 2: [...] *Democritus is right in his opinion that* [*the eye*] *is of water; not however, when he thought that vision is an impression.*[19] Theophrastus in his book *On Sense*, which has just recently been

---

15 Ibid., 305b1–3.

16 Cf. *Discussiones peripateticae*, vol. III, bk. 2, p. 305 (Greek text skipped): "At illud quod est, I de Elementis cap. 4 de Leucippo et Democrito. [...] Dicunt enim esse primas magnitudines, multitudine quidem infinitas, magnitudine vero indivisibiles, et nec ex uno multum fieri, nec ex multis unum, sed horum complexione, et circumplexu omnia gigni. Contrarium prorsus est illi, quod est capite 7 sequenti. [...] Empedocles et Democritus latent ipsi se ipsos, generationem non ex entibus ad invicem facientes sed apparentem generationem. Inexistens enim unum quodque secerni aiunt tanquam ex vase generatio sit. Contraria enim proculdubio sunt complexio et circumplexus secretioni."

17 For a slightly more detailed account, see below section 2.4.

18 Theophrastus' *De sensu* was first issued in a composite publication: *Quaestiones Alexandri Aphrodisiensis naturales. De anima. Morales. Metaphrasis ex Damascio in primum librum De coelo et mundo. Epitome per modum commentarij in quatuor primos, et octauum physices libros. Theophrasti liber De sensu. Prisciani Lydi metaphrasis in libros Theophrasti De sensu et phantasia* (Venice, 1536).

19 Aristotle, *De sensu*, 438a5, in DK, II, fr. 68A121.

> published, imitates his teacher Aristotle but confutes him by the doctrine of not a few of the ancients, and yet does not put forward the same opinion of Democritus, *when he says that*: [...] *seeing happens through impression, and he is right in this. However, impression does not occur immediately in the pupil, but it figures the air between the sight and the object seen.*[20]

Patrizi emphasizes the articulation of the moments in which the complicated phenomenon of sensation unfolds in order to present it as less absurd and paradoxical than it appears in the simplistic words of Aristotle:

> But it is very different to say that impression is vision, and that vision happens through impression; for the former marks the identity, the latter the diversity of the act and the nearby object; such is this Theophrastic impression, against which he seems to have railed with so many words, or rather against "typosis," the figuration molded in the air.[21]

Patrizi appeals to Theophrastus for the simple reason that it allows him to highlight doctrinal discrepancies within the Peripatetic school itself. For the recovery of what he takes to be the authentic Democritean doctrine, he prefers, however, to turn to more reliable sources such as Alexander of Aphrodisias, Plutarch, and Aetius, who agree in attributing to Democritus a theory that is not based on the obscure concept of *emphasis*, but instead on that of *eidōlon*, that is, of a material entity that passes from the object to the sense organ predisposed to receiving it, and which produces the phenomenon of vision.[22]

---

20 Patrizi, *Discussiones peripateticae*, vol. III, bk. 2, p. 305 (Greek text skipped): "Libro de sensu cap. 2: [...] Democritus vere quod quidem aquam esse dicat, bene dicit, quod vero putet visionem esse emphasim non bene. At Theophrastus libro de sensu nuper impresso quamquam praeceptorem Aristotelem imitatus, veterum non paucorum dogmate eo confutat, non tamen eandem Democriti opinionem affert, ait autem: [...] *Videre quidem facit emphasi, hanc vero proprie dicit. Emphasin enim non statim in pupilla fieri, sed aerem inter medium visus et rei visae, figurari.*"

21 Ibid.: "Longe autem aliud est dicere emphasim esse visionem, et visionem fieri emphasi; illud enim identitatem notat, hoc diversitatem actus, et proximi obiecti, qualis est emphasis haec Theophrastica contra quam multis verbis invehi videtur, vel potius contra typosim figurationem in aere factam."

22 Patrizi, *Discussiones peripateticae*, vol. III, bk. 2, pp. 305–306: "Cum tamen nihil aliud eam esse Democritus voluerit, quam speciem illam, non aristotelicam (nihil enim Aristoteles de ea) sed Alexandream. Testes pro Democrito sunt Plutarchus atque Aetius unanimiter ita scribentes: [...] *Democritus, Epicurus per idolorum ingressus putarunt visivum evenire.* Idola autem haec non aliud sonant, quam species illas quae per medium diaphanum in oculos incurrere ab Alexandro contenduntur."

Against Aristotle and the ensuing doxographic tradition, Patrizi opposes other traditions and doxographies, which he views as more objective for the assessment of the genuine outlook of ancient thought. In the reconstruction of the "authentic" doctrine of vision, Patrizi individuates a basis for a correct interpretation of other ancient physiological concepts. Of particular interest to him is the case of the explanation offered by the ancient naturalists for the phenomenon of sensory perception. Aristotle had judged as "downright absurd" the reduction of any kind of sensation to the mechanism of contact. For Patrizi, by contrast, this conception is plausible. Every sensation may in fact be explained by the contact of the air with the sense organ. Smell, taste, touch, and vision can only be perceived by the collision of material particles, and Aristotle himself sometimes seems to admit this explanation:

> What he says in ch. 4 [...] *Democritus and most of the natural philosophers who treat of sense perception proceed quite irrationally; for they represent all objects of sense as objects of touch.*[23] His disciple Theophrastus does not at all admit this. Summarizing without dissimulation (*ex professo*) in the afore-mentioned booklet the views of the ancients on the senses, that is, of Parmenides, Empedocles, Heraclitus, Anaxagoras, Alcmeon, and Clidemus (whom Aristotle never mentions), Diogenes, Democritus, and Plato, he did not write such a thing either with respect to them all, or to Democritus [alone]. But even if he had mentioned this, if it were true that they had entertained his opinion, would this have been an error? Of the objects of touch and of taste Aristoteles admits that they are sensed through touch. What are the smells and sounds? It is not necessary that the air affected or moved by them touches the sense organ, if the sensation must follow, is it? What is that: visible species? Is it not just in the air which stretches out between the object and the eye? And on the very surface of that air which touches the pupil does sight arise by means of the contact of the air affected by the species?[24]

23 Aristotle, *De sensu*, 4, 442a29.

24 Patrizi, *Discussiones peripateticae*, vol. III, bk. 2, p. 306: "Quod cap. 4 ait. [...] Democritus et plurimi physiologorum, quicunque loquuntur de sensu, absurdissimum quiddam faciunt; omnia enim sensibilia tactilia faciunt. Id nequaquam discipulus eius fatetur Theophrastus, qui eo quem diximus libello ex professo antiquorum, Parmenidis, Empedoclis, Heracliti, Anaxagorae, Alcmeonis, Clidemi (huius nusquam Aristoteles meminit), Diogenis, Democriti, Platonis opiniones de sensibus recensendas sumens, nihil tale vel de his omnibus, vel de Democrito, scripsit. At etiam si scripsisset, si verum id esset, eos eius sententiae fuisse, an error hic esset? de tactus obiectis, de saporibus, Aristoteles fatetur, tactu sentiri. Quid odores, soni? Nonne aerem iis affectum, vel commotum tangere sensorium necesse est, si sensatio sequi debeat? Quid illa species visibilis,

With this final note, Patrizi's "defense" of Democritus and other Greek naturalists comes to a close.[25] To be sure, the views of Anaxagoras, Empedocles, and Democritus are called into question, especially regarding physical-physiological issues. Articulating his "defense," Patrizi set out to show the "discordia" with Aristotelian positions. There is still another passage in the *Discussiones* where the Greek naturalists are mentioned. An entire book of the second volume is devoted to the central theme of the *Discussiones*, that of Aristotle's "plagiarism" and "debts" towards the previous Greek naturalistic speculation.[26]

2 Bruno's Transformation of Democritean Lore

While Democritean atomism was materialistic, mechanistic, and deterministic, many Renaissance proponents of atomism reconciled the ancient particle theory with a vitalistic theory of the universe. Giordano Bruno is a good example. For Democritus and Lucretius, atoms aggregate in a mechanistic way, through casual encounters and accumulations. By contrast, Bruno prefers to explain these aggregations by attributing an internal force or energy to the atoms: he views the atom as a center of a 'powerful' force which irradiates vital energy and attracts other atoms, resulting in bodily aggregates whose parts are organized with the purpose of conserving the compounds. The vital cycle of natural entities is the outcome of the double motion of expansion and contraction of this universal energy, which arises from – and ends in – the atom. The idea of living atoms is closely connected with his theory of the infinite universe. Bruno conceives of an infinitely vast cosmos of which the parts do

---

nonne in omni eo aere, qui interiacet ab obiecto usque ad oculum, inque illa eius aeris superficie quae pupillam tangit, fitque visio per aeris specie affecti contactum?"

25 Ibid.: "Haec tribus physiologis celeberrimis obiecta ab Aristotele crimina theologica partim, partim naturalia sunt, quibus vel ex Aristotele ipso, vel illorum fragmentis, vel aliorum authorum testimoniis dissolvere potuimus. Quae vero a nobis non sunt in discussionem vocata, ea sunt quae et Aristotelis, et ipsorum philosophorum, vel aliorum scriptorum testimoniis caruerunt. Quae tamen omnia, uti ex discussis atque ex discutiendis post hac nobis persuadere possumus, eadem lance Aristotelem perpendisse, rationi videtur consentaneum."

26 In volume II of the *Discussiones*, Patrizi draws up an inventory of what he believed to be the vast range of themes typical of pre-Aristotelian physics. The huge number of fragments cited in Greek that appear there are mostly taken from Aristotle's works. Moreover, Patrizi does not directly discuss any theory, limiting himself to pursuing its main objective, that is, to demolish the myth of Aristotle's originality and, without pretending to make any original interpretative contribution, he lists only a variety of issues that are present in ancient naturalistic research. Cf. *Discussiones peripateticae*, vol. II, bk. 5: "Conformia cum antiquis physiologiae capita complectens."

not exert any absolute pressure on each other. Rather, each part obeys the law of its own being and its own intrinsic impulses. However, these parts are connected by the presence, in each of them, of the universal spirit by which they are merged into a single and universal "whole."[27]

It is not my primary concern here to assess Bruno's highly idiosyncratic vitalistic atomism in any detail; the focus is rather on the picture and status of Democritus in his Latin and Italian works. As a rule, Bruno's judgment on Democritus is positive, especially by contrast with Aristotle and his followers.[28] In the early works, Democritus appears first as a Pythagorean philosopher and subsequently as the proponent of innumerable worlds. It is only in the Frankfurt Trilogy, published in 1591, that Democritus assumes a prominent role in Bruno's atomism, in particular in the imagery he develops in these works. In the subsections below, I focus on the topics of infinite worlds, materialism, shape and the composition of atoms, and the relationship between perception and knowledge.

2.1 *Infinite Worlds*

Leucippus and Democritus, the founders of the Greek atomist school in the fifth century BC, were most likely the first to espouse a belief in the plurality of coexistent worlds, one century before Aristotle would so strenuously argue for a single highly structured cosmos.[29] In Epicurus and Lucretius, the connection between atomism and the infinity of worlds became explicit when they argued that the plurality of worlds derived from the assumption of an infinite number of atoms and the finitude of our world.[30]

The Aristotelian rejection of the plurality of worlds inspired critical discussion of the issue as early as the beginning of the thirteenth century. And although medieval philosophers inevitably avoided affirming the reality of other worlds, their discussions played an important role in later Renaissance debates, as many of them, including John Buridan and William Ockham, held that a plurality of worlds was not theologically impossible. In the fifteenth

27 For a summary view of Bruno's atomism, see Amato, "Atomo," and the literature mentioned in note 2.

28 See, for example, Bruno, *De la causa, principio et uno*, in *Opere italiane*, I: 639–640, where he qualifies Democritus as superior to the pedant philosophers.

29 Diogenes Laertius, *Lives of Eminent Philosophers*, IX.31: "Leucippus holds that the whole is infinite [...]; part of it is full and part void [...]. Hence arise innumerable worlds, and are resolved again into these elements."

30 For Epicurus, see Diogenes Laertius, *Lives*, X.89: "That there is an infinite number of such worlds can be perceived, and that such a world may arise in a world or in one of the intermundia (by which term we mean the spaces between worlds) in a tolerably empty space and not, as some maintain, in a vast space perfectly clear and void"; see Lucretius, *De rerum natura*, II, 1052–1066.

century, Nicholas of Cusa extended the scale of living beings beyond the Earth, asserting that none of the other stellar areas lacked inhabitants.[31]

The belief that the Earth is not the sole oasis of life in a deserted universe, and the notion that the Sun is like any other star, and our planet like any other planet, returned after the rise of Copernicanism. Although the doctrine of the plurality of worlds may seem to be implied by the heliocentric theory, the two ideas are by no means coextensive. Even though Bruno was a professed Copernican, he proudly placed himself above Copernicus, stressing that it had been him who had fully taken down the traditional, medieval view of the universe with its solid orbs and qualitatively differentiated celestial and sublunar spheres. He reasoned that because God was infinite, his creation would similarly be infinite. Each star was another sun, like our own, about which other worlds revolved. Although such a view was in principle compatible with Copernicus, it went far beyond the technical system of the latter's *De revolutionibus orbium*.

In his cosmological dialogues (*Cena de le ceneri*; *De la causa, principio et uno*; *De l'infinito universo e mondi*, published in London in the years 1584–1585), Bruno argued for an eternal and infinite universe that contained a plurality of worlds. In this context, he mentioned Democritus several times. In the *Cena de le ceneri*, Democritus is mentioned alongside Presocratic or later non-Aristotelian Greek philosophers, such as Heraclitus, Pythagoras, Parmenides, Melissus and Epicurus.[32] This broad group is narrowed in *De l'infinito* to only Democritus and the later atomist Epicurus. In the dialogue here presented, Bruno also refuted Aristotle's arguments against an infinite universe and the plurality of worlds.[33] The last reference to Democritus for this focal point in Bruno's cosmology is found in the posthumuously published *Theses de magia*.[34]

---

31 For a discussion on the possibility of other worlds, see Dick, *Plurality of Worlds*, 23–43; cf. Grant, *Planets, Stars, and Orbs*, 150–168, Del Prete, *Universo infinito e pluralità dei mondi*, and Fabbri, "Looking at an Earth-like Moon."

32 Bruno, *La cena de le ceneri*, in *Opere italiane*, I: 547: "Or questa distinzion di corpi ne la eterea reggione l'ha conosciuta Eraclito, Democrito, Epicuro, Pitagora, Parmenide, Melisso, come ne fan manifesto que' stracci che n'abbiamo: onde si vede, che conobbero un spacio infinito, regione infinita, selva infinita, capacità infinita di mondi innumerabili simili a questo, i quali cossí compiscono i lor circoli, come la terra il suo; e però anticamente si chiamavano ethera, cioè corridori, corrieri, ambasciadori, nuncii della magnificenza de l'unico altissimo, che con musicale armonia contemprano l'ordine della constituzion della natura, vivo specchio dell'infinita deità."

33 Bruno, *De l'infinito, universo e mondi*, in *Opere italiane*, II: 21, 26, 71, 161.

34 Bruno, *Theses de magia*, in *Opera latine conscripta*, III: 458: "Tertio modo mundus sumitur sicut apud Democritum, qui innumerabiles et infinitos intellexit, mundum unum accipiendo terram, alium lunam, alium singula astrorum, quorum non est definitus numerus."

Conspicuously enough, the plurality of worlds was only a minor issue in Bruno's inquisitorial trial. On 24 March 1597, he was admonished to abandon the "vanitates diversorum mundorum," and in March of the following year, this view is mentioned among the suspect theses in the summary of his trial.[35] However, at no point was it among the central heresies for which he was punished.

2.2 *Beyond Materialism*

The relevant aspects of Democritus' atomism can be summarized in the following basic list: (1) the natural world is composed of atoms and void; (2) the former are unchanging material principles (indivisible particles), which persist and merely rearrange themselves to form the changing world of appearances; (3) all changes in the visible objects of the world of appearance are brought about by relocations of these atoms. Attending this list, there are also some uncertainties: the properties of the atoms as to weight, shape, size, and motion are not well defined; while the causes of things are ascribed to necessity, there is a role attributed also to chance.

In Bruno's works, we encounter several references to Democritus as the spiritual father of atomism and materialism. The early references don't betray a particular affinity, as they dish out praise as well as opprobrium. For example, in *De umbris idearum*, Democritus is simply cited for his atomistic view of natural reality.[36] But in *De la causa*, Bruno distances himself from Democritus and the Epicureans, insofar as they reduce the substance of things to matter alone, while

> it is necessary to recognize in nature two kinds of substance, of which one is form and the other matter, because it is necessary that there is a most substantial act, in which resides the active power of everything; and also a power and a subject, in which there is no less passive power of everything.[37]

Judging by this quote, matter cannot, for Bruno, be the universal substance of reality. By contrast, in the final dialogue of *De la causa*, Bruno subscribes to the

35 Baldini and Spruit, *Catholic Church and Modern Science*, 882, 908 and 924.

36 Bruno, *De umbris*, ed. Sturlese, 194.

37 Bruno, *De la causa*, dial. III, in *Opere italiane*, I: 678–679: "[…] è necessario conoscere nella natura doi geni di sustanza, l'uno che è forma, e l'altro che è materia; perché è necessario che sia un atto sustanzialissimo, nel quale è la potenza attiva di tutto; et ancora una potenza et un soggetto, nel quale non sia minor potenza passiva di tutto."

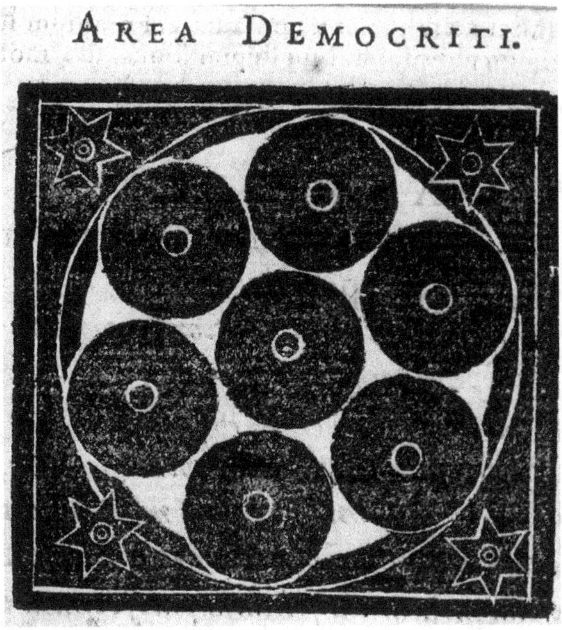

FIGURE 7.1
'Area Democriti'. Giordano Bruno, *De triplici minimo et mensura* (Frankfurt, 1591), 50

view, which he once more attributes to Democritus, that matter is the source of all forms.[38]

Another central issue of ancient atomism present in several of Bruno's works is the existence of "indivisibilia" or "impenetrabilia."[39] Remarkably enough, Bruno attributes to Democritus the notion that all atoms are spherical, a view that he then adopts and reproduces in the engravings that illustrate his *De minimo*.[40] As must be evident, this is a major departure from Epicurean and Democritean theories, according to which the shapes of atoms are indefinite and must, in any case, be different so as to constitute the diverse bodies that surround us. Bruno, by contrast, discerns in the form of atoms the representation of the perfect circle that is conceivable only by reason. His remarkable spherical atom, which is absent from the ancient corpuscularian texts on which his views at least relied, is particularly prominent in a figure entitled *Area Democriti* (see Fig. 7.1), to which we now turn.

38 Ibid., 7.

39 Bruno, *Figuratio physici auditus Aritotelis*, in *Opera latine conscripta*, I.4: 200; *Articuli adversus mathematicos*, in *Opera latine conscripta*, I.3: 24; *De minimo*, in *Opera latine conscripta*, I.3: 140, 176, 240; *De rerum principiis*, in *Opera latine conscripta*, III: 534.

40 Bruno, *Figuratio physici auditus Aritotelis*, in *Opera latine conscripta*, I.4: 145; *De minimo*, in *Opera latine conscripta*, I.3: 140; *Libri physicorum Aristotelis explanati*, in *Opera latine conscripta*, III: 272–273.

2.3 Area Democriti: *Atoms and Worlds*

The figure of the *Area Democriti* occurs repeatedly in Bruno's work, but each time in a different disciplinary context and with a different purpose, underpinning – respectively – astronomical, mathematical, metaphysical, and physical views.

From Bruno's introduction to *De minimo*, we may infer that the *Area Democriti* is generated when one original minimum is surrounded by a further six minima of the same diameter.[41] The figure is not to be related to Democritus' extant fragments, however, but rather to Bruno's medieval and early modern numerological predecessors – from Boethius to Renaissance mathematicians – who claimed that the unity contains in its being all dimensions. This central view led Bruno to define the monad as a point, then as a sphere, and again as a circle, thus designating not only the beginning of a sequence of numbers, but also its end, and hence the totality of the cosmos.[42]

Bruno applied the image of seven circles in the proof of the possibility of infinite worlds.[43] Adding to Pythagorean numerology Cusanus' idea of a *coincidentia oppositorum* and assuming both ends of the scale of being to be defined by the same figure of unity, Bruno's point – circle – sphere can be understood as a representation of both minimum and maximum.[44]

Elsewhere in *De minimo*, referring again to the figure of the *Area Democriti*, Bruno establishes a conceptual link between atoms and worlds, arguing that the void space between the spherical atoms corresponds to that between the worlds.[45] And where Aristotle had chastised the atomists for their inability to define "true mixture" and had called their atomic structures mere "heaps" (acervi), *coacervare* is precisely the verb Bruno uses to describe the formation of bodies out of atoms.[46] Spherical atoms touch each other only in one point, and this creates the vacuum in between the individual atoms.[47] As Christoph Lüthy has persuasively argued, Bruno projected Pythagorean images into the sky and then transformed them down to the level of the physical minimum.[48]

---

41 Bruno, *De minimo*, in *Opera latine conscripta*, I.3: 183.

42 Lüthy, "Bruno's *Area Democriti*," 63–70.

43 Bruno, *De l'infinito*, in *Opere italiane*, II: 145; *De immenso*, in *Opera latine conscripta*, I.2: 249–250; see Lüthy, "Bruno's *Area Democriti*," 72f.

44 *De minimo*, in *Opera latine conscripta*, I.3: 196–197.

45 Ibid., 176–177.

46 Bruno, *De minimo*, in *Opera latine conscripta*, I.3: 222; the opposition between "true mixture" and "heaps" is found in Aristotle, *De generatione et corruptione*, 328a13f.

47 Bruno, *De minimo*, in *Opera latine conscripta*, I.3, 223, 225, 241; cf. *Praelectiones*, ed. Aquilecchia, 61 and 75.

48 Lüthy, "Bruno's *Area Democriti*."

2.4 *Perception and Knowledge*

Democritus was probably the first Greek author to undertake a detailed analysis of perception as a process: the object of perception is not the sensible thing itself, but it should be seen as the outcome of the interaction between the sensible thing and sense organ. Thus, the medium in which the interaction takes place has an essential role to play in perception.[49] Things transmit strings of atoms that represent them. In turn, these strings communicate their structure to the medium. In the specific case of vision, for example, this structure is communicated to the air. In this modified medium, the atom-strings and a light-beam departing from the eye meet. Only after this contact, the sun and the daylight communicate atom-strings to the eye. Visual perception is thus based on the emission of luminosity. The representational atom-strings are called *eidōla* or "images."[50] Clearly, things are not perceived as they really are, but only as objects of perception determined by our sense organs.

Brief reference is made to Democritus' psychology of perception and knowledge in some of Bruno's early and later Latin works. Surprisingly, in his *Sigillus sigillorum*, Bruno states that Democritus qualified all knowledge as "intellect."[51] In the same book, he refers to the story of Democritus blinding himself, believing that his mind's thoughts and reflections in considering the laws of nature would be clearer and more precise if he freed them from the incitements of sight and the eye's mistakes.[52] In the *Summa terminorum*, he confirms his understanding that, in Democritus' view, all knowledge should be qualified as sense.[53]

Bruno's psychology of perception differed from that of Democritus on several points. In his physical theories, such as those concerning the relativity of the perception of motion or the physical homogeneity of the universe, Bruno attempted to set up criteria to distinguish the apparent from the real. To overlook the processes that are responsible for perception means to deny

49 *Die Vorsokratiker*, vol. II, Fragm. 88 (= DK 68 A 135).

50 Ibid., Fragm. 103 (=DK 68 A 135) and 106 (=DK 68 A 77).

51 Bruno, *Sigillus sigillorum*, in *Opera latine conscripta*, I.2: 174: "Est porro et tertius quidam modus, quo sensus significat. Epicurus enim cognitionem omnem appellat sensum, Democritus et Empedocles intellectum, Pythagorici mentem et spiritum altiorem, ipsumque intelligunt esse in omnibus pro sua ratione."

52 Ibid., 187: "Hinc Democritus mentem a sensibus avocare percupiens, sibi oculos, a quibus impedimentum patiebatur, effodit." The origin of this story is probably to be found in Aulus Gellius' *Attic Nights*, 10.17.

53 Bruno, *Summa terminorum metaphysicorum*, in *Opera latine conscripta*, I.4: 108: "Cognoscitur ergo in omnibus et cognoscit in omnibus, neque urgendum est sub tituli virtute. Placuit enim Epicuro et Democrito et aliis pluribus, omnem cognitionem appellare sensum."

the distinction between physical and phenomenal object. Bruno argued that knowledge cannot be built up from perceptual structures free from interpretation; the nature of things is something to be apprehended by rational inference. In turn, cognition is not the discernment of regularities in an unadulterated stream of experience. Cognition should intervene in experience, because the sensory data are trustworthy only when they are properly interpreted.[54]

3 Conclusion

Patrizi's collection and systematic investigation of early sources of Greek philosophy and science is part of his attempt to determine Aristotle's place in the history of thought. Thus, his concern with early sources was not strictly scholarly or historical, as can readily be seen by examining his works and the ways in which those texts were employed. In the doctrines of the Presocratic philosophers, Patrizi found topics and issues echoed by Greek authors such as Plato and his followers, and this added additional support to their appeal. Evidently, in particular in his *De nova universis philosophia*, Patrizi heavily criticizes all materialistic positions, specifically that of the atomists (Democritus, Leucippus), but his most important target remained Aristotle. Indeed, his refutation of Aristotle's criticism of Democritus is fundamental to his overall program of substituting Peripatetic natural philosophy with a basically Platonic alternative worldview.

Giordano Bruno's cosmology was marked by infinitude, homogeneity, and isotropy, with planetary systems distributed evenly throughout. For Bruno, the existence of a smallest, indivisible unit, such as the point in geometry, the atom in physics, and the monad in metaphysics, was the matrix of reality. Atoms were the units out of which nature builds her fabric, and into which bodies dissolve again. In support of his views, he frequently referred to Democritus. However, unlike Democritus, Bruno thought that atoms and void did not suffice; for him, it was necessary that atoms be endowed with a vital force.

54 For discussion, see Spruit, "Giordano Bruno's Use of Experience in Context."

CHAPTER 8

Nicholas Hill, an English Atomist

Sandra Plastina

1 Introduction

There were no orthodox early modern atomists in the sense that there was no unmitigated revival of the theories of Epicurus or Lucretius. This caveat notwithstanding, there was a host of authors who – for widely different reasons – subscribed to some of the theories connected with the names of Epicurus and Democritus. Among them was the protagonist of our essay, Nicholas Hill.[1] Like other self-proclaimed followers of Democritus and Epicurus, he offered a mixture of old and new arguments in favor of his theory.

The influence of the Epicurean doctrine – and of Lucretius' work in particular – is easy to trace in the 509 aphorisms of his *Philosophia Epicurea Democritiana Theophrastica*, which was first published in Paris in 1601 and then in Geneva in 1619. In it, Hill offers us a unique and idiosyncratic synthesis of elements of ancient atomism and of new insights from contemporary fields of natural philosophy, medicine, and chymistry.

In fact, as his subtitle indicates, Hill presented his work as the expression of a "philosophy neither new nor old" (*Philosophia nec nova nec vetus*), that is, as a somehow timeless philosophy that did not wish to force a break with the past, but gleaned from ancient philosophical thinking, notably from ancient atomism, ideas that could be applied to the explanation of natural phenomena. The amnesty towards ancient philosophical thinking did not extend to Aristotelianism, however, the scholastic terminology of which Hill was very critical. He advocated a return to a kind of pre-Aristotelian natural philosophy that could capture the wisdom of the ancient atomists. He felt that in so doing, he could more easily accommodate both the cosmological models proposed by Giordano Bruno and the chymists' theory of matter ("loquar cum chymicis") as

1 Cf. Lüthy, "The Fourfold Democritus," 446–447, which shows that Democritus, with whom atomism is associated, was a figure to whom multiple identities were attributed in the early modern period. According to Lüthy, there were in fact no fewer than four distinct Democriti of Abdera, alongside the several literary traditions that were inspired by them: the atomist, the laughing philosopher, the moralizing anatomist, and the alchemist.

 | DOI:10.1163/9789004528925_009

presented by Petrus Severinus (Peder Sørensen).[2] The terminology of the latter's *Idea medicinae philosophicae* (1571) – a synthesis of Parcelsus's doctrines – is clearly present in Hill's *Philosophia*. These various atomistic, cosmological, and chymical influences converge in a Renaissance anthropology that stresses the link between humans and the cosmos. Aphorism 63, for example, states:

> Man is a microcosm participating in all natures, and there is no force in the world, no generation and no substance to which there is no correspondence in man, something from the heaven, something from the elements, something taken from the single parts of nature and placed in his structure, in whose visual exterior and interior description shine forth all figures and the universal geometry of nature.[3]

As for the man himself, we know only little about Nicholas Hill's life. He was born in London in 1570. He attended the Merchant Taylors' School, one of the most renowned among the city's grammar schools, which since the first half of the sixteenth century had replaced the former monastic schools. In 1587 he enrolled at St John's College, Oxford, where he obtained his bachelor's degree and of which he subsequently became a fellow. He may, however, have been forced to step down from that position because of his conversion to Catholicism.[4] However, according to John Aubrey (1626–1697), a member of the Royal Society and author of *Brief Lives*, Hill was famous in Oxford as "a great Llullianist," as an expert in the *ars memoriae* and a great connoisseur of the secrets of alchemy.[5] For a while he was in the employment of Henry Percy, ninth Earl of Northumberland, and lived through some turbulent years because of the chaotic events following the failure of the Gunpowder Plot, and

2 Hill, *Philosophia Epicurea*, Aph. 128, 101. All page references will be to the 2007 edition, ed. Plastina.

3 Ibid., Aph. 63, 92: "Homo microcosmus est omnium naturarum particeps, nulla in mundo virtus, nulla generatio, nulla substantia cui aliquid in homine non respondat, aliquid a caelo, aliquid ab elementis, aliquid a singulis naturae partibus detractum, et in illius fabricam collatum, in cuius exteriori, et interiori descriptione perspecta omnes figurae et universa naturae geometria elucent."

4 See the entry "Nicolas Hill" by Trevor-Roper in the Oxford Dictionary of National Biography, art. 13287: "On 30 June 1590 he was admitted as a fellow of the college, but within a year he had been deprived (*amotus*) [of this status], probably on account of his conversion to Roman Catholicism." St. John's College was considered a stronghold of the Anglo-Catholic movement; see Mallet, *A History*, II, 236.

5 Aubrey, *Brief Lives* (1949), 253–260; there is a second edition of *Brief Lives* by Dick (Boston, MA, 1999). Aubrey's biographical reconstruction lies at the basis of Wood's *Athenae Oxonienses* (1691–1692); see Beddard, *The Source*, 1–11.

the retributions that flowed therefrom, which cost Henry Percy his freedom. Hill, suspected of participation in the conspiracy and of being pro-Catholic, was forced to leave England and went to the Dutch city of Rotterdam, where he "practiced as a physician," as Aubrey wrote. The philosopher is said by his biographer to have died there in 1610.

In this essay, we propose to analyze specifically the philosophical components that constitute the structure of the *Philosophia Epicurea*, and in particular the influence of Democritus' and Epicurus' views on atomism as well as of the philosophy of nature developed by Giordano Bruno and Francesco Patrizi. Greek atomism was known to Renaissance readers notably through the *Lives of Eminent Philosophers* by Diogenes Laertius, which contains Epicurus' *Letter to Herodotus*, which is considered the most extensive presentation of Epicurus' genuine views on atomism.[6]

Besides the genuine documents and the more or less reliable doctrinal summaries presented in doxographical collections, there were also spurious sources. Notably, the early modern understanding of the figure of Democritus was influenced by pseudonymous chymical and magical works.[7] Early modern readers were interested in the various pseudo-Democriti as they reinforced the philosophical tradition linked, on the one hand, to the development of a lay morality, and on the other to the chymists' and physicians' debates over the composition of matter. In the process, not even the staunchest supporters of ancient atomism embraced wholesale the explanations proposed by Epicurus and Lucretius, but at best adapted them to their own needs. These adaptations involved religion – the neo-atomists worked mostly in a theistic framework – as well as physics. Examples of the involvement of the latter included the identification of atoms with the traditional four elements or alternatively with the *tria prima* of the chymists, the linking the natural macrocosm to the human microcosm, or the identification of Lucretius' *semina rerum* with Anaxagoras' homoeomeries.[8] The resulting neo-atomism could not in any way resemble a homogenous movement; doctrinal incoherence and what Lasswitz has called

6 A Latin translation by Ambrogio Traversari had appeared in 1472.

7 Pseudo-Democritus, *De arte magna*. For the magical works by pseudo-Democritus, see Berthelot, *Les origines*. As for the connection between Democritus' philosophy and chymistry, see Hershbell, *Democritus*. In the first half of the seventeenth century, chymists came to explain empirical findings by an atomistic theory of matter. But as Newman argues (*Experimental Corpuscular Theory*, 291), the classical studies on corpuscular and atomistic matter theories have ignored the fact that chymists had, even before the end of the sixteenth century, adopted such ideas.

8 See Lüthy, "Thoughts," esp. 4–5, where he proposes to follow Stephen Clucas ("The Atomism") in labeling as neo-atomists those who named themselves atomists in the seventeenth century, so as to avoid a mistaken doctrinal identification between ancient and modern atomism.

a "general anarchy" prevailed, with the "atom" figuring in a variety of incompatible systems of natural philosophy.[9] These systems were often developed as alternatives to the Aristotelianism of the schools. It is therefore always interesting to watch out for the role played by the concepts of atoms, corpuscles or minima or by the historical figures of Democritus, Epicurus, and Lucretius in these alternative natural philosophies of the *novatores*, these inventors of new systems, whose canonical list traditionally starts with Bernardino Telesio's *De rerum natura* (1586) and Francesco Patrizi's *Discussiones peripateticae* (1581).[10]

The utterly dissimilar, alternative models of natural philosophy that were formulated between the end of the sixteenth and the first half of the seventeenth century all share one element: they all return, in one way or another, to pre-Socratic Greek naturalism: "*philosophi qui ante Platonem et Aristotelem scripserunt*," as Patrizi wrote in 1571 to the scholar Gian Vincenzo Pinelli.[11] More specifically, many of the early "*Novateurs modernes en la Philosophie*" – to use Charles Sorel's phrase from some decades later – drew inspiration from the atomistic philosophy they pieced together with fragments of Leucippus, Democritus, Epicurus, as well as from Lucretius' poem.[12]

One of the earliest philosophers to embrace Democritus explicitly was Giordano Bruno, who may be seen in various respects as having attempted to revive "the doctrine of Democritus."[13] In Bruno's dynamic philosophy, an

In this spirit, we ought to refer to "neo-atomism" in the same way that we refer to "neo-Stoicism" or "neo-Platonism."

9 Lüthy "The Fourfold Democritus," 446, compares the two ways of dealing with the messy landscape of seventeenth-century atomism (Lasswitz's "general anarchy" in *Geschichte*, 520). The first way, suggested by Clucas, has been explained in our previous note; the second, suggested by Joy, *Gassendi*, starts from the inconsistency of early modern atomism and tries to retrace its complex cultural origins. Whether the second approach works is unclear. As Lüthy points out, summarizing Clericuzio's argument in "L'atomisme," 227–235: "Clericuzio demonstrates the profound conceptual differences separating ancient atomism from its alleged revival by Gassendi, but he also believes that developing a linear conception of the history of atomism is a futile enterprise because seventeenth-century corpuscularism derived its inspiration from a number of divergent traditions."

10 Patrizi, *Discussiones*,l. III, ch. 1, 293, for example, praises Democritus as the inventor of physics; see Leen Spruit's chapter in this book.

11 Patrizi, *Lettere ed opuscoli inediti*, 7.

12 The treatise *Des Novateurs*, published by Sorel, firstly in his *Perfection de l'Homme* (1655), was to be included in the fourth volume of *Science universelle* in 1668.

13 As for the representation of material substances in terms of aggregates of spherical bodies, there is no agreement among scholars over the role of ancient atomism in the early modern physical and visual models of the constitution of matter; cf. Lüthy, "Bruno's Area Democriti," which stipulates that the origin of the model of the spherical atom can be traced to the tradition of neo-Pythagorean numerology, for which the numerical unities (including monads) are essential. Bruno first applied the spherical unit to astronomy,

infinite cosmology and a Copernican astronomy are wedded to a conception of atoms that pulsate with life and grow from within, being the units of which the cosmos is in the last instance made up. In this respect, Bruno's dialogue *De l'infinito, universo e mondi* is emblematic, as it views cosmology from an atomistic perspective. Life and death are based on the elementary structure of the atoms and their double movement of influx and efflux, as they join and separate and give birth to infinite variations, transmutations, and vicissitudes of forms. Atoms, the ultimate subjects of that infinite local motion, produce at a constant rate innumerable compositions and figurations in keeping with the rhythm of infinite matter. At a gnoseological level, as well as at the ontological and cosmological one, Bruno's theory relies on a model of life that in turn requires an identification of the mathematical point, the physical atom, and the monadic soul.[14] In fact, Bruno's atomism is a far cry from the anti-finalistic or anti-vitalistic models that were produced after Descartes. For Bruno, the Lucretian term for atoms, *semina rerum* ("seed of things") retained its vitalistic significance.[15]

Those who embraced the atomistic intuitions of the ancient philosophers hoped that by recognizing the existence of an immanent principle of motion, some *vis* that moved the elements, and by explaining every transformation in

---

before including it in a more complex theory of matter. His "Area Democriti" and similar graphic representations of the atomistic structure of matter as crafted by Bruno for his *Articuli adversus mathematicos* (1588) and *De minimo* (1591) influenced the development of atomistic models, first in the domain of crystallography and then more generally in a corpuscular theory of matter, starting with Kepler's analysis in his *Strena* (1611) of the structure of snowflakes.

14 Bruno, *De l'infinito*, 43. Bruno states in the "Proemiale epistola" of *De l'infinito*, "dall'infinito sempre nova copia di materia sottonasce. Di maniera che megliormente intese Democrito ed Epicuro che vogliono tutto per infinito rinnovarsi e restituirsi, che chi si forza di salvare eterno la costanza de l'universo, perché medesimo numero a medesimo numero sempre succeda e medesime parti di materia con le medesime sempre si convertano." ("For from infinity is born an ever fresh abundance of matter. Thus Democritus and Epicurus, who maintained that everything throughout infinity suffereth renewal and restoration, understood these matters more truly than those who would at all costs maintain belief in the immutability of the universe, alleging a constant and unchanging number of particles of identical material that perpetually undergo transformation, one into another.") The translation is Waley Singer's, from Bruno, *On the Universe, Infinite and Worlds*.

15 For interesting considerations on Bruno and his epistemological theories on the atomistic idea of matter, see Meinel, "Early Seventeenth-Century," 74, which gives an assessment of Bruno's role in the contested link between seventeenth-century atomism and empirical research: "For Bruno the existence of a smallest, indivisible unit, such as the point in geometry, the atom in physics, and the monad in metaphysics, was the matrix of reality, the measure and prerequisite of cognition."

terms of a material recomposition, they could render Aristotle's substantial forms superfluous. The early years of the sixteenth century saw a resurgence of this expectation also in England.[16] To those dissatisfied by scholastic learning, the new philosophers subscribing to the *secta Epicurea* were innovators, *novatores*; to most others, however, they looked like 'schemers' of whom to be wary.

In 1621, Robert Burton signed his *Anatomy of Melancholy* as "Democritus Junior." But, while appreciating and praising Democritus, he treated the followers of Democritus' atomistic philosophy with irony, distancing himself from a series of new and paradoxical doctrines circulating in those years under the name Democritus. Addressing the reader, "Democritus Junior" explains that he will not provide

> a pasquil, a satire, some ridiculous treatise (as I myself should have done), some prodigious tenet, or paradox of the Earth's motion, of infinite Worlds in an infinite waste, so caused by an accidental collision of Motes in the Sun, all which Democritus held, Epicurus and their Master Leucippus of old maintained, and are lately revived by Copernicus, Brunus, and some others.[17]

It is not difficult to identify Nicholas Hill as belonging to the group of "some others," for in the second part of *The Anatomy*, entitled "Digression of the Ayre," Burton names Hill among those who subscribe to the motion of the Earth, to the infinity of space and the plurality of worlds.[18] Burton was not alone in this judgement. As Anthony Wood was to explain later in the century, Hill "adopted the notions of old Democritus about atoms and was a great patron of corpuscular philosophy."[19] And Ben Jonson referred to Hill as "an English man who maintained Democritus's opinions"; in an ironic epigram, he addressed Hill as the English atomist par excellence: "Than all ridiculous / Where of Democrite and Hill Nicholas / one said, the other swore, the world consists."[20] In his polemics against the new science, Jonson specifically criticized Hill for the obscurity of his prose.[21]

16 On the question of the relation between corpuscular philosophy and chemistry in the seventeenth century, see Clericuzio, *Elements*, 75–102; with regard to Hill, see esp. 75–81.

17 Burton, *The Anatomy*, I, 1.

18 Ibid., II, 52–53.

19 Wood, *Athenae*, II, 86–87; see also Aubrey, *Brief Lives* (1949), 256.

20 Jonson, *Works*, vol. 145, 198.

21 See McPherson, "Ben Jonson's Library," 17–20. On his copy of *Philosophia Epicurea*, Jonson, quoting Martial 10.21.3, noted: "Non lectore tuis opus est, sed Apolline libris" ("Your works

Three years after the publication of the second edition of the *Philosophia Epicurea*, Tobias Adami in the 1622 preface to Tommaso Campanella's *Apologia pro Galileo* mentioned Hill, next to William Gilbert, among the supporters of the Copernican theory.[22]

2 The Philosophia Epicurea

In contrast to Giordano Bruno, Hill embraced a type of atomism that proposed a whole series of different atomic forms. Right from the start, he stressed that atoms manage to combine and create larger structures thanks to their hooked, curved or rounded shapes.[23] This is in keeping with Lucretius' doctrine or the description that Cicero furnished in *De natura deorum* (1.24.66):

> As for the outrageous doctrines of Democritus, or perhaps of his predecessor Leucippus, there are certain minute particles, some smooth, others rough, some round, some angular, some curved or hook-shaped, and that heaven and earth were created from these, not by compulsion of any natural law but by a sort of accidental colliding.[24]

As the title of his book, *Philosophia Epicurea Democritiana Theophrastica*, indicates, Hill lumped together the doctrines of Democritus and Epicurus as proponents of the very same doctrine. In Aphorism 138, the followers of Epicurus and Democritus rise jointly to challenge the supporters of the abstract and metaphysical principles of Aristotle's *Physics*. Hill rebuffs as inconsistent the view that an infinite God has created a finite world.[25]

According to Hill, Democritus and Epicurus proposed a physics that was based on the movement of atoms and on their sudden deviation from a trajectory of rectilinear motion, a model Hill was convinced could explain all natural phenomena:

---

need no reader, but rather (an) Apollo (to understand them"), implicitly complaining about the text's obscurity.

22 Campanella, *Apologia pro Galileo*, 1622, 4.

23 Hill, *Philosophia Epicurea*, Aph. 12, 79: "Continuitatis causam ab atomis non deducibilem, respondeo unceatas, hamatas, annulares, atomos posse convenienter dari regularibus et primis irregularibus non sufficientibus."

24 Cicero, *On the Nature of the Gods*, transl. Rackham, I. 24, 66.

25 Hill, *Philosophia Epicurea*, Aph. 138, 103: "Sic neque natura dignitatem tuetur suam si in principiis actionem suam fundet actum suum trascendentibus: Pessime ergo aiunt qui infinitum Deum ad finiti mundi existentiam supponunt."

> To put it more boldly and confidently, the Epicureans and Democriteans have indeed approached the explanation of nature excellently by beginning from the inclination of unstable atoms, if they merely seek, and clearly derive, the solution to physical questions thence.[26]

The "inclination" mentioned here refers to the sudden deviation or swerve of atoms from a vertical fall, which Lucretius explains as *clinamen* in *De rerum natura* 2.216–250 and Cicero in *De fato*, 10.22 and other works. Besides their ability to swerve, Hill considered solidity (*soliditas*), eternity (*aeternitas*), and simplicity (*simplicitas*) as the further distinguishing features of atoms. The swerve fulfills a role in the formation of physical bodies, as it allows the otherwise vertically descending atoms to collide and get hooked up in new combinations, whereby it is the diversity of their basic shapes that allows for both their entanglement and for the secondary qualities and other properties of the compound bodies. In the *Philosophia Epicurea,* just as in Epicurus and Lucretius, Hill describes the number of atomic shapes as extensive, but not infinite: "The principles are infinite in number [...] and there are many figures of atoms."[27] As this sentence indicates, Hill used a range of terms to refer to atoms: he alternated the Greek term *atomi,* which Cicero had introduced into Latin, with *principia, primordia rerum,* and *semina,* which are Lucretius' words. The term *semina* also features in several passages of the Latin translation of Epicurus' *Letter to Herodotus,* where it refers however not to single atoms, but to aggregations of several of them.

In his *Philosophia Epicurea,* Hill, like Lucretius, puts less emphasis on the indivisibility of the *semina* or *genitalia,* but instead on their generative potential and on the power contained within them.[28] In Aphorism 8, he explains: "The generative seed is the [result of the] confluence of many primordial bodies which strengthen each other."[29] Aphorism 169 explains further: "The individual seed is not so much the generative principle as a birth in itself."[30] But these atoms as seeds are not only regarded as the kernels of life, but – and here Hill Christianizes ancient atomism – they are also coeternal with God

26 Ibid., Aph. 138, 103–104: "immo ut audacius, et confidentius loquar, Epicurei, et Democritiani optime naturae explicationem aggressi sunt ab atomorum instabilium inclinatione exorsi, si modo physicorum dubiorum solutionem illinc petent, et derivent apertam."

27 Ibid., Aph. 473, 179: "Principia sunt numero infinita [...] figurae atomorum multae."

28 See Sedley, "Lucretius' Use," 230–231; Beretta, *La rivoluzione culturale,* 138–145.

29 Hill, *Philosophia Epicurea,* Aph. 8,102: "Semen generativum est multorum primordialium confluentia se invicem confortantium."

30 Ibid., Aph.169, 111: "Semen individuum non tam generativum principium, quam ipsa genitura."

and bring God's work to fruition in physical terms. In his Aphorism 133, Hill explains that "the material atoms give substance to the species [of things], they conclude God's physical action, impenetrable and truly solid."[31]

The idea that there is a limit to the division of natural bodies – that there are "impenetrable and truly solid" principles – is a fundamental tenet of Epicurean physics, as Lucretius had stressed (1.561–2). For Lucretius, the essential property of atoms was not so much their indivisibility as their simplicity (*simplicitas*); in 1.548, he explains that "elements are entirely solid. Since otherwise there is no way they could have been preserved through ages of infinite time till now, in order to restore things one again."[32] Lucretius regarded this simplicity as both eternal and also linked to solidity.

For Hill, it was not only the atoms themselves that were indivisible, but also their "first compositions" (*primae compositurae*). This is made clear from the beginning, in Aphorism 2, where he states that "the imperceptible seeds of things are indissoluble and in no way subject to destruction."[33] As for *compositura*, it is a Lucretian term, taken from *De rerum natura* 4:328. Hill maintained that it is not the indivisibility of atoms that must provide the foundation for matter theory, but their simplicity: "That the first principles are simple is sufficiently clear by the very word" (Aph. 47).[34] In Aphorism 116, he further declares that "the first corpuscles are truly solid, impenetrable, unchangeable, multiform, and define the ends of divine action in nature."[35] As nuclei of resistance and the ultimate units of solidity, if such first corpuscles – did not exist, everything would be destined to destruction.

The Greek atomists were the first, according to Hill, to introduce an independent domain of physics, as they set out to study the structure of the world with a specific methodology. In a bid to insert himself into that tradition, Hill explains his intentions in the first of the sixteen objections that precede the aphorisms: "The first [elements], just as in nature so also in science, are coordinated [with one another], not subordinated [to one another]."[36] In his eyes, the entire order of natural phenomena must be viewed within a horizontal

31 Ibid., Aph. 133, 102: "Atomi materiales substantiant species, divinam actionem physicam terminant, impenetrabiles et vere solidae."

32 Lucretius, *On the Nature of Things* (transl. Watson).

33 Hill, *Philosophia Epicurea*, Aph. 2, 81: "sunt insensibilia rerum semina indissolubilia, et interitui nullatenus obnoxia."

34 Ibid., Aph. 47, 89: "Prima principia simplicia cum sint verbo satis declarantur."

35 Ibid., Aph. 116, 100: "Prima corpuscolasunt vere solida, impenetrabilia, intalterabilia, multiformia, divinae actioni in natura terminos ponentia."

36 Hill, *Philosophia Epicurea*, 77: "Prima ut in natura, sic in scientia esse coordinata, non subordinata."

framework of mutations, which are explained by the varied ways in which atoms position themselves through the *clinamen*, the "inclination" or deviation that the atoms need to perform so as to come into contact with one another. His proposed "coordination" of invisible and indivisible particles which combine to engender life in matter contrasts with the Aristotelian subordination of physical processes under abstract and metaphysical principles.

Interestingly, the term *clinamen*, a neologism derived from the Greek word παρέγκλισις, is also linked to grammatical expressions meaning "derivation," "conjugation," and "declension." Lucretius had described the process as follows:

> When particles are borne by their own weight, on a downward path straight through empty space, at undetermined times and random places, they swerve a little – not much, just enough, so you can say they have changed direction.[37]

Hill uses terms like "even more audaciously" or "more confidently" to describe the idea that the encounters of invisible and immutable bodies should generate the macroscopic world of phenomena, but he was aware that the idea of atoms was to most people shocking. As he wrote, in Aphorism 268: "Atoms move around each other in a circle with a perpetual and never-ending motion, and they are the manifest causes of our restlessness."[38] In the minds of most readers, the atomists' ideas and their new cosmological theories will have generated confusion and bewilderment. Aware of this, Hill anticipates their reactions, almost as though he was able to conjure up their perplexed expressions, and hastens to justify his views.

In objection 7, for example, Hill responds to the distrust that readers may have felt when confronted with new words and philosophical concepts. In *De rerum natura* (1.138, and elsewhere), Lucretius had addressed the same problem, that is, the difficulty of expressing new philosophical ideas effectively when the language at the author's disposal was not fit for the purpose: "Above

37 Lucretius, *De rerum natura*, 2.217–224: "Corpora cum deorsum rectum per inane feruntur / ponderibus propriis, incerto tempore ferme / incertisque locis spatio depellere paulum, / tantum quod momen mutatum dicere possis. / Quod nisi declinare solerent, omnia deorsum, / imbris uti guttae, caderent per inane profundum, / nec foret offensus natus nec plaga creata / principiis: ita nil umquam natura creasset." Translation by Watson from Lucretius, *On the Nature of Things*.

38 Hill, *Philosophia Epicurea*, Aph. 268, 129: "Atomi perpetuo, et indesinente motu circumrotatae seinvicem circumeunt, et inquietudinis nostrae sunt causae manifestae".

all since we must deal with many things, employing new words, because our language is impoverished and the subject new."[39]

In fact, Hill's philosophy is characterized not only by the recovery of terms typical of the Greek philosophical tradition, especially the Epicurean and Stoic traditions.[40] He also introduced his own terminological and conceptual innovations. He preferred Greek terms to scholastic, Latin ones, and was not averse to coining neologisms, as when he referred to matter's ability to assume a certain shape with the word *formicapax*, in imitation of a Greek term that characterizes matter's capacity to assume all forms – "παμμορφία, vel omnis formae capacitas" (Aph. 463). Hill also often utilized compound words that are combinations of Greek and Latin terms. One example to which he frequently resorted is *συστάσις*, also in the Latinized form *systasis* and in the mixed form *syστάσις*, a term that Hill used to indicate nature's substance, which in his vocabulary refers to its overall stability in the face of the incessant motion of the *minima*. It is Hill's way of expressing the concept of a general equilibrium resulting from the unavoidable cycle of birth and death to which all entities formed by the random aggregation of *solida primordia*, the 'modules' of life, are subject (for example, the term *syστάσις*is used in Aphorisms 263 and 264 to refer to stable atomic aggregates).[41]

The question of stability is addressed repeatedly in Hill's *Philosophia Epicurea*. After all, if the *clinamen* occurs in unforeseeable ways and at unforeseeable times, our knowledge of nature can never be secure, but can at best be probabilistic. According to Epicurus and Lucretius, only emptiness and atoms exist of their own accord. This is all we can know with certainty. Emptiness, the

39 Lucretius, *De rerum natura*, 1.138–139: "'multa novis verbis praesertim cum sit agendum/ propter egestatem linguae et rerum novitatem." The translation is again by Watson.

40 On the influence of Stoicism in Jacobean England, see Clucas, "Noble virtues in extremes," 267–291. The reading of Stoic texts was an excellent example of the way in which humanistic reading practices could help to shape both aristocratic self-image and political conduct in the Jacobean period (ibid., 291). When they were denied the possibility of acting in the political sphere, philosophers adopted Stoic consolation as both a therapy and as a language of political resistance: in the Aphorisms 485bis, 487, 489, 490 (*Philosophia Epicurea*, 182–183), Hill provides a brief and disconsolate analysis of the political-religious conspiracy, of which Trevor-Roper accuses him of having been one of the promoters.

41 Giuliana Leone's recent edition of the second volume of Epicurus' περὶφύσεως makes it possible to analyze in great depth the composition of this lexical family, especially the words σύστασις and ἀπόστασις; see *Epicuro. Sulla natura, libro II*, comments to column 14, pp. 514–518, and to column 76, p. 555.

vacuum, is in fact the second important element of Hill's natural philosophy: all natural phenomena are possible only if a vacuum is assumed.[42]

Laws are created at the precise time when prime particles combine. As soon as a phenomenon is manifest, as soon as a body assumes a form, a law can be enunciated. The laws of nature are laws of conjunction, and nature consists only of combinations of *mixta*:

> In nature there are truly minima parts that exist in reality, and the mixture is formed and defined by their corpuscularity for otherwise excess in mixed bodies would not have been defined by any quantity.[43]

Drawing inspiration from Bruno's dialogues and in particular from *De immenso*, Hill considers nature the repository of the true elements of the just laws.[44] "Laborious earth" (*daedala tellus*, taken from *De rerum natura* 1.7) assembles and puts together each of its component pieces in succession, according to the logic of construction that regulates nature's generative processes. According to this view, the earth is the origin of all species, as the earth and nature alone are capable of giving life to beings.

In a series of aphorisms, Hill explicitly embraces the tenets of Epicurean atomism. There he refers, for example, to the aggregation of atoms and the way in which these combine in a variety of ways and in different quantities, change their position and merge with other atoms, thereby creating the various physical bodies. In Aphorism 35, he stresses the importance of combinations and connections between particles: "Form is the state and condition of a thing, the result of material principles joined together, a principle that constitutes and does not operate."[45] Again, in Aphorism 331, he states: "In fact, Nature performs

---

42 Hill, *Philosophia Epicurea*, Aph. 289, 134: "Effluxus, et influxus, rarefactio, condensatio, corruptio, generatio, gravitas, et levitas, motus, et alteratio, non imaginabilis absque vacuo."

43 Ibid., Aph. 209, 118: "Sunt vera in natura minima actu existentia, in quorum existentia, et minimitate mixtio fundatur, et terminatur, aliter enim excessus in mixtis aliqua quantitate non prescriberetur."

44 Bruno, *De immenso*, lib. I, cap. IX, 433 (ed. Canone): "Cum vere natura manus sit cunctipotentis, Vis, Actus, Ratio, Verbum, Vox, Ordo, Voluntas. Naturae cursum quicumque ergo impedit, ille est Excors, insanus, calomastix, impius, exlex, Numinis intemerata etenim natura ministra est, Legum suppeditans iustarum vera elementa." Lib. III, cap. V, 548 (ed. Canone): "Naturam ipsius moderari lege novantem Omnia, ut e quovis quidvis disponat."

45 Hill, *Philosophia Epicurea*, Aph. 35, 86: "Forma est status, et conditio rei, resultantia principiorum materialium connexorum, principium constituens, non operans".

nothing with accuracy in whatever species without the encounter of many principles of the same kind."[46]

Almost at the end his *Philosophia Epicurea,* Hill lists the salient elements of his own atomistic vision of nature. To begin with, we must know that his *principia* are innumerable.[47] But whereas the number of atoms is countless, they possess only a limited number of atomic shapes. Still, the number of combinations into which they can enter is once again limitless. Whenever the number of atoms or their respective positions change, the compound of which they are parts will change into a different substance. Moreover, the component atoms of compounds are always in motion. But even though all prime corpuscles are constantly moving, the entirety of matter in the universe appears to be at rest. This should come as no surprise, for the motions of the corpuscles take place at a material level that is way below the threshold of our senses. Motion in things is not perceivable, because, while all atoms are constantly moving, their general combination appears quiet. As Lucretius had already pointed out, just as atoms themselves are invisible to us, so too are their movements.[48]

But Lucretius also compares atoms to dust motes that are visible in a shaft of sunlight. Their dancing motion provides Lucretius with the concrete model for the invisible movements of prime particles.[49] Moreover, the eternal struggle in which they appear to be engaged is an exact representation of the life-generating *aeternum certamen.* Thanks to this constant atomic dance, Lucretius concludes that everything flows: *omnia fluere constat.* In this context, it should be noted that the motion that drags atoms through space is constant and stable, like that of a river that never stops flowing. Here, Lucretius offers us another macroscopic analogy to explain the behavior of the invisibly small. By analogy with this theory of ever-moving atoms, death can also be viewed as sort of creation, because when a body dissolves, its corpuscles will find a new collocation at an unspecified time and place, generating new temporarily stable

46 Ibid., Aph. 331, 143: "Nihil enim natura praestat exacte in quacunque specie sine multorum eiusdem generis principiorum concursu."

47 Ibid., Aph. 473, 179: "[…] atomi sunt innumerae, figurae atomorum multae […], nihilominus in communi ratione conveniunt, materia, figuris primis, efficiente physico omnem compositionem ingredientibus, et a seinvicem nihil mutuantibus ad suam entitatem, qumvis cooperantibus, et conferentibus vires mutuas."

48 Lucretius, *De rerum natura* 2.308–14: "Illud in his rebus non est mirabile, quare,/ omnia cum rerum primordia sint in motu,/ summa tamen videatur stare quiete,/ praeterquam siquid proprio dat corpore motus./ Omnis enim longe nostris ab sensibus infra/ primorum natura iacet; quapropter, ubi ipsa/ cernere iam nequeas, motus quoque surpere debent."

49 Ibid. 2. 123–124: "Dumtaxat rerum magnarum parva potest res/ exemplare dare et vestigia notitiai."

constellations. Meanwhile, the sum total of the universe remains unchanged, while being is renewed endlessly – a scenario in which mortal creatures lead their lives as temporal compounds. But change takes place not only at the individual level: some species grow, while others decline, and animal species may change in a brief time span.[50] Hill, who follows Lucretius' account of the origin of life and the emergence of humankind as given in *De rerum natura* 5. 795–800, considers the latter "the last germ of the blooming mother Earth."[51]

The theory of the ancient atomists had the further advantage of explaining also the world of feelings in terms of their causes. In this respect too, Hill agrees with Epicurus' preference for sensory knowledge, which guarantees its truthfulness: "The proper objects of sensation require and presuppose a harmony of sensation, good eyesight, a fixed distance, and a medium fit for this purpose, and cannot be further than a distance at which they can be perceived."[52]

In fact, the *Philosophia Epicurea* would not have done justice to its title if it had not reacted to Epicurus' theory of vision and of the *simulacra*, these prime generators of vision.[53] Epicurus had explained in the "Epistle to Herodotus" how *simulacra* emanate from physical bodies as images that have the same configuration as the solid entities from which they emanate but, due to their lightness, are yet very different from the way things manifest themselves to the senses. In book IV of *De rerum natura*, Lucretius elaborated on this idea, insisting that *simulacra*, like atoms, are imperceptible and invisible and that they assume their shapes very quickly. In fact, Lucretius spoke of "a point in time" ("temporis in puncto"; verses 164, 193 and 214), emphasizing the extreme speed with which the *simulacra* were formed. To render the same idea, Hill borrowed the adjective "punctual" from Bruno's *De minimo*.[54] Bruno had there invoked Lucretius when insisting on the subtlety of images.[55]

Hill adopts a series of synonyms to render the Greek term *eidola* (εἴδωλα) in Latin: *simulacrum, imago, effigies, species*, and *figura*. The *eidola* are exhaled, emanated and effused into the air as the rarefied, invisible atomic images of

50 Ibid. 2. 71–79.

51 Hill, *Philosophia Epicurea*, Aph. 64, 92: "ultimum genitalis terrae efflorescentis, et omnen possibilem combinationem molientis germen."

52 Ibid., Aph. 50, 89: "Propria sensuum obiecta sensuum compositionem, valetudinem, distantiam certam, medium taliter dispositum exigunt, et praesupponunt, nec longius sunt quam sentiuntur."

53 Ibid., Aph. 17, 83: "Species et simulacra rerum sunt nullius quantitatis, nihilominus coalescentibus punctualibus formis proportionalem induunt quantitatem."

54 Bruno, *De minimo*, 1, 9, 47: "Dicimus non punctualem lucem videri (quam neque sensu neque ratione a sensu desumpta possumus pro punctuali definire), sed lucis diffusionem."

55 Ibid., 46: "A minimo nostris obiecto sensibus, altam accipito docti rationem mente Lucreti, indicat ut tenui natura constet imago."

things. This notion, first introduced by Democritus, had been elaborated upon by Epicurus, who spoke of the constant flux of *eidola*, a term that Lucretius Latinized as *simulacra*. For Lucretius, such *simulacra* also explained not just sensation, but also feelings and dreams. The elegance of the explanation consisted in the fact that the *simulacra* are made of atoms just like the sensorial organs that they subsequently penetrate and affect. Lucretius in fact compares *simulacra* to *vapors*, *fumes*, and *odors*. As such, they differ from the luminous, colorful but immaterial Aristotelian visual species. Hill concurs, stating in Aphorism 203 that: "The appearances of sensible things are simple emanations, not substantial effusions; even though they depart from the object from every side, nothing perishes."[56]

As mentioned in our introduction, Hill however was not simply reintroducing an unchanged Epicurean natural philosophy, but was modifying it according to more recent influences. In his theory of *simulacra*, for example, he is influenced by Girolamo Fracastoro, to whom he refers with regard to the concept of *subnotio*.[57] In Fracastoro's opinion, *species spirituales* issue from natural forms. Owing to a "latent principle," they are the cause of the mutual attraction and repulsion of bodies. The "spiritual" nature of these species sometimes seems to lie in their thinness. In other words, they are material, but imperceptible. However, some species propagate across such huge distances that "cannot be reduced to atoms and corpuscles," though they are *simulacra materialium*, such as the light emitted by stars, or flavors, smells and sounds, or the species issued by magnetic iron protuberances at the Earth's poles, which cause compass needles to move.[58]

With regard to the explanation of remote actions, Hill generally owes much to Fracastoro. While admitting that all action takes place through contact ("omnis actio fiat per contactum"), he allows for two kinds of contact: corporeal and virtual. In the case of virtual contact, the connection with the cause – which is remote – is slight: what moves from one body to another, causing a specific effect, are in fact mere *effluvia* or *species spirituales*, emissions of rarefied matter, which are imperceptible but generate phenomena that are usually attributed to sympathy and antipathy. In Aphorism 388, Hill embraces this view explicitly: "Those things that are internally ὁμογενή, that is homogeneous,

56 Hill, *Philosophia Epicurea*, Aph. 203, 117: "Species rerum sensibiles sunt emanationes simplices, non substantiales effluxus, quibus ab omni parte exeuntibus obiecto nihil deperit."

57 Ibid. Aph. 348,145: "Est quaedam animae operatio innominata ab Aristotele, a Fracastorio subnotio dicta, quae est confusa apprehensi obiecti successive et per partes intensissima inspectura, in qua neque veritas, neque facultas, sensu vel actuu existente, vel non."

58 Fracastoro, *De sympathia et antipathia rerum*, ed. Pennuto, 54: "Attractio autem haec ad atomos et corpuscola reduci posse non videtur."

[…] reciprocally attached to one another through bodily or virtual contact, strengthen one another […]; of these, we have evidence in the arousal through one another of magnets and magnetic objects."[59]

But Hill does not simply copy Fracastoro. While his *Philosophia Epicurea* deploys the same vocabulary – *spiritus*, *eidolon*, and *effluvium* are used interchangeably – he attributes a different meaning to them than does Fracastoro. For Hill, spirits and *effluvia* function as bodily emanations, and less so as the carriers of hidden forces, sympathies, and antipathies. They thereby take on a much more material meaning, as they refer to corpuscles that act by penetrating the porous textures of bodies. Hill's *Philosophia Epicurea* in fact testifies to the way in which the meaning of *effluvium* gradually veered away from the Renaissance meaning of spirit and towards a worldview that thought in terms of atoms, pores, and bodily textures.

Still, it is not a mechanistic philosophy that Hill is here proposing. His combination of the concept of "texture" together with that of "effluvium" inserts itself rather into the context of a theory of *materia activa* –a theory in which matter is endowed with properties that cannot simply be attributed to the quantitative features of the particles or to an assembly of interlocking parts, but in which the corpuscles themselves possess a perennial, independent source of movement.

The concept of *semina* also defies a mechanical interpretation. To begin with, the *semina*, those first aggregations of atomic particles, are the result of God's creative act (a thought that Epicurus would have rejected).[60] Hill in fact follows cues provided in Severinus' *Idea medicinae philosophicae*: accordingly, *semina* are to be understood as particles of matter that are able to preserve the fecundity of their species, that is, genuine *rationes seminales*, which act as intermediaries between the divinity's creative idea and their instantiation in Creation. Hill attributes to the *semina* the germinal properties of all that exists in nature. In accordance with Paracelsian thinking, the true principles are, for

59 Hill, *Philosophia Epicurea*, Aph. 388, 155: "Quae sunt inter se ὁμογενή, i. homogenea, […] sibiinvicem applicita per contactum corporalem, aut virtualem, seinvicem corroborant […] cuius manifestam habemus in magnetum, magneticorumque per seinvicem expergefactione."

60 Hill, *Philosophia Epicurea*, Aph. 2, 81: "Primae compositurae sunt insensibilia rerum semina indissolubilia, et interitui nullatenus obnoxia"; Aphorism 3, 81: "Semen primarium est certarum atomorum in certam figuram conformatio, et coincidentia"; Aphorism 4, 81: "Semen primarium est materiae atomicae, figurae, virtutisque divinae prima complexio"; Aphorism 8, 81: "Semen generativum est multorum primordialium confluentia seinvicem confortantium."

him, forces that are physically minuscule and hidden within natural objects, and which imprint upon them a certain "signature" (*signatura*).

Note that for Severinus, a *semen* is essentially a spiritual entity, invisible but able to morph into a physical form; in fact, the boundaries between spirit and body blur within the *semen*:

> Seeds are the chains between two natures, which connect visible bodies to invisible ones: in these are contained the laws of motions, the predestinations of times, the lithurgies of generations and transplantations, and the managements of the universal, mundane anatomy.[61]

For Severinus, *semina* are moreover particles of matter that are able to maintain the fecundity of their species. They are the intrinsic vital principles of natural objects, the bearers of the latter's mechanical spirits, and the *quinta essentia* of the entire anatomy peculiar to each of their species. They provide the means by which a natural body takes shape, with colors, flavor, dimensions, forms, and other intrinsic features (*signaturae*) appropriate to its functions and purpose.[62] At the same time, the so-called "seminaries" (*semina*) are also bodies, albeit spiritual bodies, and this explains how even a minimal fraction of a *semen* ("minima gutta seminis") contains the species' entire anatomy ("universam anatomiam totius speciei continet").[63]

Hill takes over much of Severinus's Paracelsian vocabulary, which explains the title word "Theophrastica," which refers to Philippus Aureolus Theophrastus Bombastus von Hohenheim, more widely known as Paracelsus. In Aphorism 331, for example, Hill appropriates Severinus's statement that the smallest seminal droplet contains the whole anatomy of the semen itself, but that for a complete generation, you needs a greater quantity than such a droplet.[64] It is significant, in this respect, that he introduces his borrowing from Severinus's *Idea medicinae* with the statement that "primary atoms have

61 Severinus, *Idea medicinae*, 58: "Semina sunt vincula utriusque naturae, visibilia invisibilibus coniungentia: in quibus motuum leges, temporum praedestinationes, generationum et transplantationum *lithurgiae*, et universae mundanae anatomiae dispensationes continentur." On *semen* as intrinsic vital principle of natural bodies, cf. Hirai, *Le concept de semence*, 217–265. On this passage, and specifically on the notion of "lithurgia" in Severinus, see Jole Shackelford's chapter in this book.

62 Severinus, *Idea medicinae*, 96.

63 Ibid., 101.

64 Hill, *Philosophia Epicurea*, Aph. 331, 143: "Nihil enim natura preaestat exacte in quacumque specie sine multorum eiusdem generis principiorum concursu, quamvis enim minima seminis guttula habeat in se totam seminis anatomiam, tamen ad generationem complendam aliqua illius copia exigitur." Compare this with Severinus, *Idea medicinae*,

motion."[65] This is noteworthy because Hill gives Severinus's seeds a much more bodily, or indeed atomistic, twist. As Antonio Clericuzio has pointed out, "Hill's *semina*, unlike those of Severinus, are units of matter, clusters of atoms, and spirit is a very subtle body."[66]

Following Severinus, a number of Paracelsians gave *semina* a central role in both medicine and cosmology, but Hill went further than most in giving these *semina* an atomistic interpretation. Importantly, a few years later, a similar combination of an atomistic conception of matter with Paracelsian theories was to set off the philosophical research program of Henry Percy, Earl of Northumberland, a patron of the arts and natural philosopher whose circumstances were linked to those of Nicholas Hill. In certain significant passages of Percy's *Advice to his Son*, published in 1609, we find him suggesting a synthesis similar to Hill's between an atomistic theory of matter, which explains generation and corruption ("The doctrine of Generation and Corruption unfoldeth to our understanding the method general of all atomical combinations possible in homogeneal substances"), and an alchemical tradition which, without the philosophical edifice of atomism, would remain "a mere mechanical broiling without this philosophical project."[67]

Bringing the discussion back to Hill, the combination of Epicurean-Democritean atomism and Paracelsian celestial influences in the *Philosophia Epicurea* leads him to postulate a connection between the force in atoms (*minima vis*), which is the efficient cause of its movements, and the divine force (*prima vis*), which is the efficient cause of the cosmos. At the end of times, when the world will have collapsed, all will return to the latter: divine "force and root," *vis et radix*.[68]

We have just drawn attention to Severinus's influence on Hill's atomism. We must now add yet another source. The notion of *prima vis*, to which Hill resorted when speaking of God's actions, and the features he attributed to the first, divine principle, that original source of energy, bear witness to the

---

58: "Semina sunt vincula utriusque naturae, visibilia invisibilibus coniungentia: in quibus [...] universae mundanae anatomiae dispensationes continentur."

65 Hill, *Philosophia Epicurea*, Aph. 331,143: "Atomi primariae motus habent."

66 Clericuzio, *Elements*, 77; cf. Hill, *Philosophia Epicurea*, Aph. 1–2, 9, 35, 41.

67 Percy, *Advice to his Son*, 70.

68 Hill, *Philosophia Epicurea*, Aph. 110, 98: "Prima vis, causa rerum efficiens, activa, universalis, simplex, absoluta essentia, materiale virtutum fundamentum Deus est, radix, ad cuius nomen omne genuflectendum, et ad quem iure postliminii omnis virtus, et energia redit, soluta mundi compage, et dissitis a se primis principiis specierum." Aph. 200, 117: "Minima vis per materiam atomicam in motum provissimam effective infinitatur materiam deificans quodammodo."

influence of Francesco Patrizi's *Nova de universis philosophia* (1591).[69] In fact, Hill followed Patrizi in basing his cosmology on four principles, that is to say, on God, time, space, and matter (Aphorism 352). As was the case with Severinus, Patrizi provided Hill with notions that could be reinterpreted into his atomistic framework. Take the following description contained in Patrizi's *Pancosmia* of the primordial chaos, referring explicitly to the empty space before the creation of the world, through which atoms might have flown:

> Before the world we inhabit had been forged by God, a void existed, in which either the atoms flew about or chaos or formless matter spread, with irregular motions; therefore, there existed space before the formation of the world.[70]

Here, atoms are presented as only one among several possible entities that swarmed about in the primeval void. But it is evident that for someone committed to an atomistic conception of matter, like Hill, this provided a fruitful hypothesis.

As must have become obvious by now, Hill's work provided a synthesis of several of the great philosophical themes of the Renaissance, which he borrowed, *inter alia*, from the works of Paracelsus, Severinus, and Francesco Patrizi. We will return to Patrizi in section 4, below. But first, we have to add to our list of influences the name Giordano Bruno, whose influence we examine in section 3.

Hill's place in the history of philosophy, in fact, must be defined with respect to these thinkers. It is therefore curious that studies on Hill have often focused primarily on his critical position vis-à-vis Aristotelian philosophy. Scholars have stressed how, in many passages of *Philosophia Epicurea*, Hill passes a damning judgement on scholastic theology, expressing the view that the combination of Aristotelian philosophy and Christian religion had compromised the understanding of both the essence of God and the natural world. But as we

69 The manuscript notes on animal organism by Walter Warner (ca.1557–1643), another English atomist, display a comparable combination of tenets from Patrizi's light-metaphysics with atomism. Jan Prins, *Walter Warner*, 50, has pointed out the influence of Patrizi on both Hill and Warner: "Warner's arguments about the nature of space, matter and 'vis' are prefigured in Patrizi *Nova de universis philosophia*, and some of Hill's aphorisms show a strong similarity to Warner's views on the principles of nature in general and on the 'vis radiativa' in particular."

70 Patrizi, *Nova de universis philosophia, Pancosmia* I, "De spacio physico," 65: "Ante quam hic quem incolimus mundus a Deo esset fabrefactus inane erat, in quo vel atomi volitabant, vel chaos volutabatur, vel materia informis, motibus inordinatis, spacium ergo ibi erat ante mundi formationem."

shall see, even that anti-Aristotelianism is indebted to Renaissance predecessors, and notably to Giordano Bruno, to whom we now turn.

3 Nicholas Hill and Giordano Bruno

The *Philosophia Epicurea* refers repeatedly to Bruno. Hill took up several ideas contained in Bruno's *De minimo*, including the doctrine of the triple minimum of point, atom, and monad; the atom being the physical expression of the minimum.[71]

Bruno's influence is also evident in Hill's cosmology. To begin with, Hill subscribes to Copernicus's heliocentrism, which he combines with Bruno's idea of an infinite universe containing a plurality of worlds.[72] Hill is convinced, in fact, that the infinity of the world helps to clarify many mysteries of the Christian religion. For if it were accepted that the world was merely finite, the infinity of the supreme being would in no way be intelligible. Nor would God's incarnation and the coexistence of the divine with human nature stand out for its extraordinary significance and uniqueness, but could rather be explained "secundum proportionem." Moreover, if a finite world were given (*dato mundo finito*), the apparent potential infinity of all individuals would be superior with respect to God's single act of Creation.[73]Since each agent acts according to its

71 On the minimum, compare the following passages. Hill, *Philosophia Epicurea*, Aph. 140, 104: "Atomus pivativa absoluta punctus, actualis minimum materiale, respectiva primum in speciali." Cf. Bruno, *De minimo*, "Liber de minimi existentia," II, 10: "Minimum est substantia rerum quatenus videlicet aliud a quantitatis genere significatur, corporearum vero magnitudinem prout est quantitatis principium. Est, inquam, materia su elementum, efficiens, finis et totum, punctum in magnitudine unius et duarum dimensionum, atomus privative in corporibus quae sunt primae partes, atomus negative in iisce quae sunt tota in toto atque singulis, ut in voce, anima et huiusmodi gens, monas rationaliter in numeris, essentialiter in omnibus." Aph. 138, 103: "Quemadmodum in particulari productione a minimo insensibili ad infinite maius, et sensibiliter magnum, sic in universali productione natura a minimo intelligibili exorditur, et in maximum intellegibile desinit, et quemadmodum omnes nostrae laudatae actiones apparatum excedunt."

72 A similar denial of the existence of the sphere of fixed stars, which is a necessary step towards the notion of an infinite university, is also found in Gilbert's *De magnete*. Gilbert's theories are accepted by Hill, *Philosophia Epicurea* (2007) in Aph. 435, 167, where he writes: "Magneticam terrae naturam satis probant." Cf. Gilbert, *De magnete,* 215. On Gilbert's ideas, see Jones, *Ancients*.

73 Hill, *Philosophia Epicurea,* Aph. 499, 184: "Infinitas mundi christianae religionis mysteria multum dilucidat et exaltat, nam mundi terminatione data Dei infinitas non est intellegibilis, et consequenter θεοτοκία, id est Dei genitura, et θεανθρωπία, id est divinitas et humanitas non sunt tam stupenda, aut magna nisi secundum proportionem, quae prerogativa

nature, if an infinite God had produced only a world that was finite in magnitude, virtue, and essence, He could not have expressed Himself fully, nor could He have given to each rational creature a first indication of His presence, and to each man that of the Trinity and of the nature of angels.[74] Like Bruno, Hill thinks about the infinity of God having to express itself in a cosmic immensity; Hill insists that God must have expressed Himself entirely in the infinite production of things and in the correspondence between Himself as an infinite cause and a corresponding effect.

Evidently, the assumption of an infinite universe erases the notions of spatial "superiority" and "inferiority" that had characterized the closed, geocentric system, and with it the traditional notions of gravity and levity.[75] In an infinite space, longitude, latitude, and depth coincide, and in movement "rest and the simply quickest motion over a finite magnitude coincide; in a circle, the longest and most extended chord and the line coincide."[76]

Hill bravely addresses the question of whether an infinite universe does not reduce our Earth to a speck of dust and annul the eminent place of humanity.[77] His equally brave answer is that once you assume that the entire universe is made of the same matter, by analogy, the noble race of humans may be found anywhere in the cosmos.[78] The men that Hill imagines on other globes might be giants one hundred times taller than we are and perhaps sixty times taller than those living on the Sun. Countless worlds swarming with life abound in

est cuiusque seminuli, quinetiam dato mundo finito omnium individuorum apparens infinitas potentialis Dei instantaneo actui praevalebit."

74 Ibid., Aph. 123, 100–101: "Omne agens agit secundum modum suae naturae, infinitus ergo Deus si finita solummodo haec mole, virtute, et essentia produxerit, nullatenus se expressit, atque hinc rationali creaturae prima suspicio et multi πολυπροσώπου θεοῦ, multi personalis et incommunicabilis nobis Christi, et angelicae naturae."

75 Hill, *Philosophia Epicurea*, Aph. 205, 118: "Infinitas mundi tollit realitatem superioritatis et inferioritatis, illasque respectivis entibus annumerat, respectivitas earum doctrinam gravitatis, levitatisque funditus evertit, sublata gravitate, et levitate, proiectorum a terra corporum ad terram reditus magneticae attractioni tanquam parienti famosissimo, et apertissimo coitu ad imputationem subeundam parato obtruditur."

76 Ibid., Aph. 52, 89–90: "[…] simpliciter velocissime supra finitam magnitudinem motus et quies, in circulo extensissima maxima corda et recta coincidunt."

77 Ibid. Aph. 482, 180–181: "Dubitatur utrum opinio concepta de infinito mundo hominem non reddat homunculum, aut de gradu non deturbet, nam mundo id est universo infinito dato nec terra, nec sol, nec omnes nobis conspicui globi notabilem totius partem constituunt, et formiculae quantitas nobis videbitur obtigisse."

78 Ibid., Aph. 278, 132: "Globi superiores (oportet loqui cum vulgo) eiusdem sunt materiae cum globa quam incolimus, in quibus etiam ea sunt omnia quae apud nos secundum analogiam […] tum sibiinvicem hominibus gyganteis, mole exuperantibus nostrates centimes sexagesis forsan qui in sole sunt, ad quos comparati homunciones nos sumus."

Hill's cosmological vision, which he describes in his *Philosophia Epicurea* with no show of any theological qualms.

Hill's aphorisms regarding comets are preceded by an interesting proposition regarding the distance between the Sun, the Moon, and the other planets. He reverses the traditional image of the world as being kept intact by a chain of being and interconnected through a principle of dependence or subordination. Instead, he subscribes to Bruno's statement according to which God is everywhere at all times. In *De infinito*, Bruno had stated: "My opinion is that God is in all places at all times." For Bruno, the divine essence, which is everything in all things, fills everything and is even more intrinsic to things than their own essence, since it is the essence of essences, the life of lives, the soul of souls.[79]

Hill fully concurs. In this world, where old orders and ranks are abolished, where all divine powers are everywhere actualized, and where there is no longer for the sublunar world any hierarchical dependence on the celestial world, one divine substance in its infinity embraces everything, whether small or large. As Hill tells us, "the infinity of God is intensive objectively in the biggest thing, while it is established subjectively in the smallest."[80] In his cosmology and the theology that is associated with it, Hill follows Bruno closely. And, like Bruno (and before him Nicholas of Cusa), he believes that the notion of an infinite God in an infinite universe strengthens rather than weakens Christian theology and religion. We recall Hill's statement, cited before, that the doctrine of "the infinity of the world much clarifies and exalts the mysteries of the Christian religion."[81]

Moreover, in an infinite universe coincident with God's divine substance, freedom and necessity coincide thoroughly. Hill offers us a concise synthesis of Bruno's reasons for postulating this coincidence: "In things, God perceives, understands, foresees, and carefully superintends, coexisting with individuals, investing with being what does not exist as well as created things."[82] In Aphorism 376, he adds: "God has in the souls of all the things, of spheres, and

---

79 Bruno, *De infinito*, 101; see Bruno, *De immenso* (1884), I, 11, 312: "Deum esse infinitum in infinito, ubique in omnibus, non supra, non extra, sed praesentissimum, sicut entitatis non est extra et supra entia, non est natura extra naturalia, bonitas extra bonum nulla est."

80 Hill, *Philosophia Epicurea*, Aph. 205, 118 and Aph. 275, 131: "[...] intensiva Dei infinitas est objective in maximo, fundatur autem subjective in minimo."

81 Ibid., Aph. 499, 184: "Infinitas mundi christianae religionis mysteria multum dilucidat et exaltat."

82 Ibid., Aph. 370, 150: "Deus in rebus sentit, intelligit, providet, et sollicite administrat singulis coexistens, inexistens et res creatas substantians."

of animals, the first active hypostasis."[83] He also subscribes to the metaphor of Bruno's *De infinito* of God as the "center which amplifies in a sphere."[84]

But how can we know such an infinite God? Hill suggests that we have to exclude intellectual knowledge, as God cannot be placed among the abstract objects of thought, irrespective of what theologians may have believed. Thanks to our dianoetic activity, we can perceive space, bodies, and substances through experience. Accordingly, God should be studied and honored in nature as a whole. Hill finds it blasphemous to consider God merely an efficient cause: since He is active everywhere, he must be revered above all in his ubiquitous action.[85]

Hill would probably have agreed with Bruno's characterization in the proem to *De infinito* of his own doctrine as a philosophy that opens the senses, satisfies the spirit, magnifies the intellect, and that can propel one toward the real beatitude to which a man can attain.[86] After all, Hill denies the possibility that a finite idea can represent the infinite. Rejecting a rational theology, he states that the finite intellect can only grasp something of the divine if it receives divine gifts: "Doubtlessly nothing can be said or understood about God if not by that person to whom God supplied an additional new force of judgment, a new spirit, and the divine faculty of seeing."[87] It must be evident that there is a mystical streak in Hill, just as there had been in Bruno. This comes to the fore when he recommends contemplation, which he describes as a holiday and sabbath of the senses and faculties.[88]

The other path by which we can reach God is through nature, His only offspring, as it were, and a force that assists Him in the process of generation.[89] God shows and expresses Himself in nature and through nature – a profound

83 Ibid., Aph. 376, 152: "Deum in animis totius, globorum et animalium habere primam hypostasin energeticam."

84 Bruno, *De l'infinito*, 83. Hill, *Philosophia Epicurea*, Aph. 275, 131: "Itaque intensiva Dei infinitas est obiective in maximo, fundatur autem subiective in minimo."

85 Hill, *Philosophia Epicurea*, Aph. 291, 134–135: "Abstracta primi efficientis consideratio est blasphema, Deum enim actione spoliat, nec aliter cogitandus est, adorandus, et colendus, quam in naturae intemerabili lege, in bene ad eandem legem instituti animi religione, in rerum specie, in mundi vultu, in animalium innumerabilitate."

86 See the "Proemiale Epistola" of the Italian dialogue *De l'infinito*, in Bruno, *Dialoghi filosofici italiani*, 317.

87 Hill, *Philosophia Epicurea*, Aph. 151, 107: "Nihil sane de Deo potest eloqui, aut intelligere, nisi cui Deus novum κριτήριον, animum, et Dei visivam facultatem superaddidit."

88 Ibid., Aph.146, 106 and Aph. 148, 106: "Quos contemplatio deitatis sublevat, eos captives abducit."

89 Ibid.,162, 109: "Deus genuit naturam προτοτοκόν, και μονογενή primogenitam et unigenam, i. unigenitam filiam, cuius dignitatem vult salvam, et quam singulis occupationibus auxiliatricem accersit ministram."

truth of which the ancients had been well aware, according to Bruno's argument in the *Spaccio*:

> That God therefore, in his absoluteness, has no relationship with us; but it is in close relation with the effects of nature, more intimately so than nature itself; so that, if it isn't nature itself, it is nature's very nature; and it is the soul of the world's soul, if not the soul itself.[90]

In Hill's *Philosophia Epicurea*, just as in Bruno's *De l'infinito*, the atomistic structures are repeated at each and every level: just as, in terms of the universe as a whole, every planet is a complex unity surrounded by empty space, so each body – including the bodies of the very planets – is constituted by minute atoms surrounded by emptiness. Obviously, the difference is that planets and other macroscopic objects are complex, articulated structures with specific configurations, while atoms are simple, indivisible bodies, which by virtue of their shapes and the mobility afforded by empty space can hook up to form larger entities. So, while at the microscopic level, there exists a kind of composite homogeneity which consists of material parts that match up while being numerically distinct and yet are entirely identical as far as their nature is concerned, at the level of the infinite universe – intended as a single entity and organism – there exists an infinite fabric made up of suns and planets and of the ether between them. Planets, despite their differences, play a role that is similar to that of atoms, in giving what we might call a "macro-corpuscular shape" to the universe.

This image of an infinite number of finite worlds dotted across the immensity of space provides an altogether new conception of the cosmos and of nature. As for the species, or categories that define entities, which populate these worlds, they are finite in number and are constrained by the reproductive rhythm of the 'macrocontext' in which they exist. According to Bruno, the planets and space are like infinite lands, infinite suns, infinite ether; according to Democritus and Epicurus, this infinity is both full and empty, possessed of a dual nature. At the same time, in each planet, different finite species, each comprised in the others and each subordinate to the others, give rise to a multiplicity of possibilities, contributing to making the infinite universe single and whole, so that from an infinite number of finite worlds similar to ours there descend infinite worlds, not like a single continuous entity but like

90 Bruno, *Spaccio* in *Dialoghi filosofici italiani*, 636: "Talmente dunque quel dio, come assoluto, non ha che far con noi, ma per quanto si comunica alli effetti della natura, et più intimo a quelli che la natura istessa: di maniera che se lui non è la natura istessa, certo è la natura de la natura; et è l'anima de l'anima del mondo, se non è l'anima istessa."

a composition of the innumerable multiplicity of such worlds. The universe's extension results from the endless aggregation of its minimal constituent parts into increasingly complex structures, into a multiplicity of entities and, eventually, into planets which – like huge atoms and infinite in number – make up the coarse fabric of the universal entity. That Hill's philosophy is indebted to that of Bruno is evident, for instance, in Aphorism 138:

> Just as, in the individual production, from an imperceptible minimal body to an infinitely bigger one, and to a sensibly big, so too, in the universal production, Nature begins from the minimal intelligible and ends at the biggest intelligible; just as all our praised actions go beyond their preparation.[91]

Hill, who here takes his cue from Bruno, created a model that works simultaneously at the macro- and micro-levels: the one, overarching unity of the universe consists of an infinite number of world-atoms, which are separated by the ethereal space that fills the cosmos. Analogously, an infinite number of atoms, equally separated by the void so that they can change their position and enter into new compositions, gives corporeal substance to the entities that structure the planets. In the course of its perennial existence, the cosmos spawns a succession of all the possible entities that are implied by the divine form. The infinite worlds do not in fact function as static or closed systems, but constantly assume new configurations.[92] When the compositional possibilities engendered by atoms are exhausted in any given world, the vital cycle of that specific macro-context does not necessarily come to an end. According to Hill's argument in Aphorism 249, it may receive from other planets and stars, or cede to them, the elements that are needed to generate further combinations:

> Worlds do not exhaust or empty themselves, but through joint forces, they restore themselves, and, joined together, they strengthen each other with an alternating exchange of qualities, and one restores the weakened forces of the other.[93]

91 Hill, *Philosophia Epicurea*, Aph. 138, 103: "Quemadmodum in particulari productione a minimo insensibili ad infinite maius, et sensibiliter magnum, sic in universali productione natura a minimo intellegibili exorditur, et in maximum intellegibile desinit, et quemadmodum omnes nostrae laudatae actiones apparatum excedunt."

92 Cf. Bruno *De l'infinito*, "Proemiale Epistola," 317: "Perché dall'infinito sempre nuova copia di materia sotto nasce. Di maniera che migliormente intese Democrito et Epicuro, che vogliono tutto per infinito rinovarsi e restituirsi."

93 Hill, *Philosophia Epicurea*, Aph. 249, 124: "Orbes non se exhauriunt, aut exinaniunt, sed mutuis viribus se refocillant, et confoederati alterna idiomatum communicatione seinvicem corroborant, et redintegrant fractas vires alter alterius."

4 Nicholas Hill and Francesco Patrizi da Cherso

I have tried to show in section 3 the ways in which Hill's *Philosophia Epicurea* may be considered an instrument for the diffusion of Giordano Bruno's ideas in England.[94] But, as mentioned before, Hill was also much taken by the thoughts expressed in Francesco Patrizi' *Nova de universis philosophia* (1591).

Patrizi's work was known to various English intellectuals at the time.[95] To give some pertinent examples: Henry Percy's library contained copies of his work;[96] Walter Raleigh, eminent intellectual and politician of the Elizabethan Age, appreciated Patrizi's dialogues in the *Historia*;[97] and the father of the magnetic philosophy, William Gilbert, was influenced by certain of Patrizi's ideas.[98] The *Nova de universis philosophia* was particularly esteemed because of its criticism of Aristotle's philosophy and its support for a *sapientia vetus*, that is, for the "philosophers who had written before Plato and Aristotle" ("philosophi qui ante Platonem et Aristotelem scripserunt"), a list that included, besides the

---

94 Cf. Ricci, *La fortuna*, 56–63, who compares some of Hill's aphorisms with the corresponding sources of Bruno. The most evident points of contact are Hill's support of Copernicanism and his acceptance of an infinite universe, which Hill accepts unaltered from Bruno's *De immenso*. For a study of Bruno's influence on the intellectuals connected with the Earl of Northumberland, see Gatti, *Giordano Bruno*. The Earl of Northumberland commented a copy of *Furori* and owned *De progressu et lampade venatoria logicorum* (Wittenberg, 1587) and *De specierum scrutinio et lampade combinatoria Raymundi Lullii* (Prague, 1588) in his Alnwick Library, and *De imaginum, signorum et idearum compositione* (Frankfurt, 1591) at Petworth House.

95 Jean Prins, in his study on the mathematician Walter Warner, underlines Patrizi's influence on Warner and Hill: Warner's arguments about the nature of space, matter and *vis* are all prefigured in Patrizi's *Nova de universis philosophia*; see Prins, *Walter Warner*, 47–50.

96 On the books owned by Henry Percy, see Batho, "The Library." The library also contained Patrizi's *La militia* (1584) and *De paralleli* (1594). It is significant that the Earl of Northumberland, a proud opponent of James I, was inspired in his personal resistance to Stuart politics by Patrizi's military treatises, which in turn were clearly linked to Machiavelli's teaching. On the theme of Machiavelli's influence on Patrizi, see Vasoli, "Il 'platonico machiavellico'."

97 The influence of Patrizi's ideas during the Elizabethan Age, above all of those expressed in *Dialoghi della Retorica* and in *Dialoghi della Historia*, was partly due to Jacopo Aconcio, an Italian religious exile in England. On the presence of Patrizi's work in England, see also Jacquot, "Les Idées," 340. Raleigh, writing about the philosophy of nature and natural magic in his work *The History of the World*, follows Patrizi also with regard to cosmological ideas; see Rattansi, *Alchemy*, 122–138, esp. 127.

98 Gilbert changed a number of the theses expressed in *Nova de universis philosophia*; others, such as Patrizi's hypotheses on primary matter and on the generation of Earth, were criticized through the references from *De magnete* (V, 12) and *De mundo*, as Zilsel, "The Origins," has pointed out.

Presocratic philosophers, also Hermes Trismegistus and Zoroaster, according to the historiographical chronology first devised by Marsilio Ficino.[99]

Hill does not mention Patrizi explicitly, but the latter's influence is obvious, not only in the terminology Hill uses but also in his thematic choices. His specific theory of light, for example, echoes that of Patrizi in the first treatise of the *Nova de universis philosophia*, entitled "Panaugia." This work, which in turn echoes Philo of Alexandria's *De opificio mundi*, describes "panaugia" as an infinite divine light.[100] For his theory of light, Patrizi combined optical observations, astronomical considerations, and metaphysical speculations. For Patrizi, light is a phenomenon that lies in between corporeal and incorporeal matter: "Therefore, light is the *simulacrum* and image equally of incorporeal and corporeal things, and a kind of medium between divine immaterial entities and the nature of the bodies."[101]

Patrizi's terminology follows the teaching of his day: he distinguishes *lux*, which is light in its own essence, from *lumen*, which is diffused light detached from its source.[102] Hill follows Patrizi by distinguishing corporeal and incorporeal light, but adds a third category, which contains items that seem to be both corporeal and incorporeal. He maintains that *prima lux* is purely incorporeal, while the other *luces* fall into the mixed category.

Patrizi proposed nine grades of diffusion of light, inspired as he was by Ficino's analogy between planets and the choirs of angels. The titles in

99 On Patrizi's views about the *prisca theologia*, see Muccillo, *Platonismo*. Patrizi republished the treatises of the *Corpus Hermeticum* and added to the corpus established by Ficino and Pico della Mirandola by including further documents related to the *prisca theologia* and by publishing less known Neoplatonic works, like Proclus' *Elementa theologica et physica* or the *Mystica Aegyptiorum theologia*, which he published as an appendix in his *Nova de universis philosophia*.

100 In a 1589 letter to Baccio Valori, Patrizi explained his reasons for composing his *Nova philosophia:* "I draw your attention to the fact that in my *Philosophia*, just as Aristotle had discovered the first mover by way of [an analysis of] motion, so in the *Panaugia* I find it by way of *lumen* and light, and subsequently in the *Pancosmos* with a Platonic method I retrace the production of light." ("[...] avvertendo a Vostra Signoria che nella *Filosofia* siccome Aristotele per via del moto trovò il primo motore, così nella *Panaugia* io lo trovo per via del lume e della luce, e poi nel *Pancosmo* con metodo platonico discendo alla produzione della luce"); see Patrizi, *Lettere*, 70.

101 Patrizi, *Nova de universis philosophia*, *Panaugia*, l. 1, "De luce," fol. 2v: "Lux ergo et incorporeorum, et corporum aeque, simulacrum et imago, et medium quoddam inter divina incorporea, et corporum naturam."

102 The distinction between *lux* and *lumen* is drawn by several authors and, among Patrizi's contemporaries, by Agostino Steuco in his *De perenni philosophia*;see Muccillo, "Marsilio Ficino," 668–669. On the general structure of the *Panaugia*, see Puliafito Bleuel, "Per uno studio."

Panaugia reflect these distinctions: *De luce*, *De diaphano*, *De radiis*, *De lumine*, *De opaco*, *De aereo lumine*, *De celesti luce et lumine*, *De luce et lumine supercoelesti*, *De lumine incorporeo*, and *De fonte et patre luminum*. He starts with *lux*, which is the first grade and the origin of the phenomenon of light, and then turns to examining *radii* and *lumen*, which play an important role in his world system. On the one hand, *lux* is "some kind of universal form of all that is visible, whether translucent, opaque, or of all colors"; on the other, it is "going through everything as if a single life, grants life and nourishes with the smoothest pleasantness."[103] Beyond the role that Patrizi attributes to light for the act of seeing, *lumen* is the instrument through which heat and the hidden virtues of celestial bodies are spread. For this reason, it constitutes the real "bond of the universe, between the superior and the inferior" in a world in which there is no place for pure darkness.[104] Hill follows Patrizi in this: for him, too, darkness is not a phenomenon of light deprivation, just as there is no genuine absence of light. Darkness is not the opposite of light, but the minimum level at which light is manifest and perceptible, even when it is unable to illuminate other objects.[105]

In his attempt to provide an elemental order distinct from that of Aristotle, Patrizi attributes to light a role of primary importance. It is one of his four fundamental principles, together with space, heat, and fluidity or *fluor*. Like Patrizi, Hill bases his cosmology on four principles: God, time, space, and matter: all four of them are required for the generation of the physical universe.[106] He refers to God as the universal principle and the primary source of energy. His concept of *vis*, which he employs to render God's action evident, is related to Patrizi's light.[107] In both cases, we have an explanation of natural phenomena in terms of the propagation and diffusion of light, which is viewed as the

---

103 Patrizi, *Nova de universis philosophia*, *Panaugia*, 1. 4, "De lumine," fol. 11v: "universalis quaedam forma totius visibilis, tum diaphani, tum opaci, tum colorum omnium." and "quasi vita una per omnia means, blandissimaque suavitate sua et intus et extra vitam donat et nutrit."

104 Cf. Patrizi, *Nova de universis philosophia*, *Panaugia*, l. 4, "De lumine," fol. 11r. Compare this with Hill, *Philosophia Epicurea*, Aph. 57, 91: "Umbra est minima lux"; Aph. 59, 91: "Tenebrae non sunt privatio lucis, non luci contraria affectio, sed lucis minimum gradus in se perceptibilis, sed alia non revelans"; and Aph. 385, 153: "Umbra non est ens privativum, sed positivum."

105 Patrizi, *Nova de universis philosophia*, *Panaugia*, l. 6, "De aereo lumine," fol. 14v.

106 Hill, *Philosophia Epicurea*, Aph. 352, 146: "Non est unum in natura primum, sed deus, materia, spatium, tempus, Tetrarchae sunt."

107 Ibid., Aph. 125, 101: "Quemadmodum lux per vitrum refracta multiplicatur, sic vis prima sua in materiales, et impenetrabiles atomos incidentia intenditur, et maiorificatur."

active principle *par excellence* and the universal cause of local motion and of all quantitative change.

But the action of light goes beyond the realm of the senses: it also affects the mind, to which it appears at the same time extremely clear and extremely obscure, notably concerning the image and *simulacrum* of God. *Lumen* is in fact defined as "visible divinity and the image of God."[108] Curiously, in the *Panaugia*, Patrizi refers to *spiritus* as an instrument of the soul – "the spirit by which the soul perceives" – but the *Nova de universis philosophia* does not develop an independent doctrine of *spiritus*, for example, in Ficinian terms. Instead, some of the features of *spiritus* are attached to *lumen*. Through light, we are told, we may reach the divinity and grasp some of His essence. Also, all of nature may be described in analogy with light "in a triple mode of seeing: direct, refracted, and reflected."[109]

Hill, inspired by what he found in Patrizi, distinguishes between *fulgor*, an immaterial but intelligible light; *lux*, the light of a luminous source; and *lumen*, diffused light. *Lux* is at the same time also the image and expression of *fulgor*: light, as the intelligible image of the mind's brilliance, is a physical, universal, primary, unconcealed and uninterrupted substantial form which, through the rarefaction and condensation of matter, acquires consistency. The primogenital light shapes matter by emanation and penetrates the material parts of the world.[110] Since "the principles of causation and intellection are the same," light insinuates itself into all material parts of the world and shapes them; it also reaches the eye and illuminates it with its intense rays. It also reveals things to us, thereby permitting knowledge.[111]

---

108 Patrizi, *Nova de universis philosophia*, *Panaugia*, l. 2, "De diaphano," fol. 4v: "[…] numen visibile, Deique simulachrum."

109 Hill, *Philosophia Epicurea*, Aph. 159, 108: "Actio naturalium formarum et totius naturae processus secundum analogiam descriptissime proponitur in triplici videndi ratione, directa, refracta, reflexa, numerorum generatione, et figuram geometrica constructione multum lucis afferente."

110 Ibid., Aph. 136, 103: "Lux primogenita. Lux formatrix individuorum. Lux essentialis coloritius rei status. Lux effluvialis radialis. Lux umbratilis, representativa lucis forma, et species."

111 Ibid., Aph. 299, 136: "Eadem sunt principia efficiendi, et ut intelligendi, et lux materialibus mundi partibus se insinuans omnia format, ita oculum pulsans, et radio foelicitans beatissimo, singula nobis expandit exploratoris functa officio." Clucas, "Corpuscular," has studied the influence of medieval theories of light on Hill's and Walter Warner's corpuscular philosophies, including Robert Grosseteste's treatise *De luce* and Roger Bacon's writings on optics. Both of these medieval authors reasoned on the theoretical basis of al-Kindi's *De radiis* and developed laws concerning the diffusion of force that were modeled on the laws of geometrical optics.

Of all of this exposition, the starting point is the easiest to grasp. Hill, like Patrizi, starts with the observation of the world around us and there empirically finds *lux* and *lumen*, which he takes to be the most clearly observable phenomena in the physical world, the world of our senses. As these become known to us in bodies, it is necessary to consider bodies in their relationship to *lux* and *lumen*. When we do, we see there are three kinds of bodies: (1) bodies that give light (*corpora lucida*), such as the Sun, stars, fire; (2) bodies that resist light (*corpora opaca*), such as the Moon, the Earth or clouds; (3) bodies that receive light (*corpora diaphana*).

In Patrizi's *Panaugia*, light is that which "permeating all, forms and produces" ("omnia permeando format et efficit").[112] Its understanding takes us to ever higher levels of understanding:

> All cognition [has] its first origin from the mind: it has its first beginning from the senses. Because of the nobleness of nature, superiority of forces, and dignity of actions, sight is first in rank among the senses. By sight, brightness and light are first and known in the first place.[113]

Lumen brings about sight, sight leads to wonder, wonder moves to contemplation, contemplation accomplishes knowledge. In agreement, Hill adopts Patrizi's definition of philosophy as "the descendant of light, brightness, sight, admiration, contemplation, and ἀκμή, that is, energy and acme."[114] In this philosophy of light, the role of *lux* is all-embracing: it spreads in the air, shines in the sky, pours out of our eyes, gleams in our senses, glows in our imagination, dominates our mind, and generally provides evidence of the universal divine presence, as Hill explains in Aphorism 245. It is incorporeal, indivisible, and without dimension; and it can be perceived only when it mixes with a certain quantity of matter.[115]

Lumen, by contrast, which is, as it were, an image of light and almost its shadow ("sic lumen lucis umbra est solummodo") spreads its goodness and

112 Patrizi, *Nova de universis philosophia, Panaugia*, l. 1, "De luce," 1v.

113 Patrizi, *Nova de universis philosophia, Panaugia*, l. 1, "De luce," 1r–v: "Cognitio omnis, a mente primam originem: a sensibus exordium habet primum. Inter sensus, et naturae nobilitate, et virium praestantia, et actionum dignitate, visus est primarius. Visui prima, et primo cognita, sunt lux, et lumen."

114 Hill, *Philosophia Epicurea*, Aph. 300, 136: "lucis, luminis, visus, admirationis, contemplationis proles, et ἀκμή, i. vigor, acies." Cf. also Patrizi, *Nova, Panaugia*, l. 1, "De luce," 1v.: "Philosophia ergo, lucis, luminis, admirationis, contemplationis proles est verissima. Lux ergo, et lumen primaria eius proles, ante omnia sunt nobis cognoscenda. Per ea, ad primam lucem, patremque luminum ascendendum."

115 Ibid., Aph. 56, 91.

perfection over the entire world and to the individual bodies of which the world is comprised: "Indeed, while it penetrates all things with its tepidity, it nourishes all things within and fertilizes within and without. For, through the participation of *lumen*, all things are generated."[116] Like *lux*, *lumen* is also incorporeal and imperceptible, becoming visible only when reflected in solid bodies. In the process, the "first *lumen*" (*lumen primum*) gets increasingly degraded and is gradually obscured, even while remaining active and in motion, acting incessantly on what is alive.

Hill's aphorisms also reflect notions that are found in the second part of the *Nova de universis philosophia*, which is entitled *Panarchia*. There, Patrizi addresses the hierarchical cosmic order, which is divided into the nine gradations that depend on the eternal source of *Unomnia*, the ineffable foundation of the unity of everything.[117] In one of his aphorisms, Hill states that the only real science is "*panarchia*" – Patrizi's term – which might alternatively be called "*pamphysicia*."[118] The proper object of this science is a transcendent entity, whose logical underpinnings are the features of the four primary principles: God, time, space, and matter.[119] The relevant aphorisms of the *Philosophia Epicurea* offer an argument in support of the principle that every multitude is from some unity, and indeed from a one that is proper to itself. Like unity with regard to numbers, so "*unomnia*" is, in Neoplatonic terms, the absolute origin, the ultimate unity, the "all-governing condition" (*panarchica conditio*), and the primary principle from which every multiplicity proceeds.[120]

5 Conclusion

The *Philosophia Epicurea* maintains that impenetrable and solid atoms bring about the species and thereby complete the divine act from a physical point of view.[121] Thanks to their unchanging identity, they confer unity to the world.

116 Patrizi, *Nova de universis philosophia, Panaugia*, l. 4, "De lumine," 11r: "Dum enim cuncta penetrat tepore suo cuncta alit intus, et foecundat intus, et extra. Luminis namque partecipatione omnia gignuntur."

117 Ibid., *Nova de universis philosophia, Panarchia*, l. 11, "De entium ordinibus," 23v.

118 Hill, *Philosophia Epicurea*, Aph. 175, 112: "Unica est sola et scientia Pamphysicia, Panarchica, quae subiectum habet adaequatum Ens trascendentissime acceptum. Cuius fondamenta sunt passiones primorum, Dei, temporis, spacii, materiae passione logice accepta."

119 Ibid., Aph. 124, 10: "Quemadmodum numerus nihil aliud est quam reassumpta unitas, ita species rerum numeri physici nihil sunt aliud quam reiterates, et reimpressus Dei actus."

120 Patrizi, *Nova de universis philosophia, Panarchia*, 7v.

121 Hill, *Philosophia Epicurea.*, Aph. 133, 102: "Atomi materiales substantiant species, divinam actionem physicam terminant, impenetrabiles et vere solidae"; and Aph. 116, 100: "Prima

And thanks to their solidity and resistance, they are at the basis of the process of generation. Without their resistance, Hill argues, the divine force could not establish itself; for where there is no resistance, there can be no alteration of the subject, and without alteration, generation would be unrealizable. The solidity of atoms is the primary *lumen* and the principle of every protoplastic production, of living matter, and of the physical basis of life.[122] Just as refracted light spreads through glass, so the primary *vis*, which gives life to everything, reinforces itself and increases its intensity by colliding with the impenetrable material atoms. In Aphorism 428, Hill attributes the rarefaction and condensation of light, and the *contractio* that follows from it, to the separation of its parts and to the effects caused by two fundamental physical agents: namely, fluidity, or *fluor*, and the atoms that govern the circular movement of souls, since "each soul is [...] mobile by nature and full of very thin atoms and moves spherically [...] and the very movement of the limbs of animals and men is the consequence of the spherical movement of the soul."[123]

The cause of the continuity and coherence of the world's physical structure must be attributed, above all, to that primogenital *fluor*, something thin, spiritual and ethereal, which has, according to Hill, been rightly linked to that vital power and energy with which every living body is endowed.

In *Pancosmia*, the fourth part of his *Nova de universis philosophia*, Patrizi explains that the active force that generates and produces beings in nature, defined *calor*, needs a principle to be able to act. This principle is *fluor*, which replaces matter as the active principle from Aristotelian philosophy. For this doctrine, Patrizi had drawn inspiration once more from the ancient wisdom of Hermes, Moses and the *prisci theologi*; but the word *fluor* also refers to the flowing of all things in ancient Greek philosophy and to the water in Homer and Thales. *Resistentia* or *renitentia*, impenetrability is due to this *fluor* or

---

corpuscula sunt vere solida, impenetrabilia, inalterabilia, multiformia, divinae actioni in natura terminos ponentia."

122 Ibid., Aph. 134, 102: "Absque resistentia non est mora, aut residentia virtutis activae, absque mora nulla alteratio subiecti, absque alteratione generatio impossibilis, itaque soliditas atomorum productionis protoplasticae primum, et praecipuum lumen est." The word "protoplasm" indicates those substances that are in general necessary for the manifestation of life; it can be defined as living matter or as the physical basis of life. For an analysis of the word as used by Hill, which is close to Presocratic naturalism, and for a reconsideration of the relationship between the concept of life and the doctrine of matter in Presocratic Greek science, see Hall, "Origini scientifiche."

123 Cf. *Philosophia Epicurea*, Aph. 375, 151, where Hill gives a definition of spherical movement: "Motus vero sphaericus est ab omni puncto ad omnem profluxus, refluxus, confluxus, interfluxus, et transfluxus (animae materialitate, et commensuratione data) intellegibilis satis."

humor ("fluor, seu humor"), which spreads in space in different degrees of density, and is regarded as the primary feature of bodies.[124] In Latin, it is also called *antitypia*.

And what does Hill do with this? For him, the physical causes of generation are due to both atoms, which are coeternal with the efficient cause in their incessant movement, and to humidity/fluidity, Patrizi's *fluor primaevus*, which acts as the bond and link of that continuity that unifies things.[125] The "humidity" (*humiditas*) that the *Philosophia Epicurea* promotes – spiritual and material at the same time – shapes things by tempering, agglutinating, dissolving and separating them.[126] It may strike us as odd that the creation of species and their destruction realizes itself through the same action of primogenital humidity or fluidity, of original heat and of "mechanical spirits," but then, as Paracelsus maintains, the world is the universal manifestation of a unique vital principle that acts through the varied instrumentality of forms.

In the eyes of Hill, the truth of this worldview is demonstrated, for instance, by the various, sometimes contradictory effects of "primeval heat" ("calor primaevus") and fluidity in the flowing, corporeal condition of substances such as air, ether, or water. Hill, by assuming fluidity to be the essential feature of original humidity, revises the thesis of Patrizi's *Nova de universis philosophia* and redefines the unifying element of the material universe. Patrizi defines the elements of phenomenal reality as corporeal/incorporeal entities from whose progressive condensation stem the various components of the sensitive world. This process of condensation explains the genesis of the four elements of the Aristotelian tradition and of ether itself, which is no longer perceived as an unchanging substance which differs qualitatively from sublunar matter, but is now one that is assimilated in its essence to the air surrounding the Earth. This idea of development, a dynamic concept of being, gives new meaning and new substance to the concept of matter. Matter is not shaped by form from the outside, but develops its shape and size from the inside. Nor does form envelop and compress matter, but matter in itself tends to constantly assume new shapes. In other words, matter is not a naked power without active faculty, but it must be viewed as the vivifying seed of all things. By the same token,

124 See Patrizi, *Nova de universis philosophia, Pancosmia*, l. 7, 78r–v.

125 According to Patrizi, the *minimum* presents itself as a unit and plays an important role, being *primum* in space, prior to creation itself; see Patrizi, *Nova de universis philosophia, Pancosmia*, l. 1, "De spacio physico," 65r.

126 Hill, *Philosophia Epicurea*, Aph. 256, 126: "Humiditas est fluida, materialis, et corporea substantiae condition, aeri, aetheri, aquae, et ceteris ad haec analogiam habentibus cognate qualitas."

nature maybe represented as the immanent force within things. These central themes in Patrizi's *Nova de universis philosophia* also recur in Hill's work.

In any reading of Nicholas Hill's work, the scope and variety of the many philosophical influences to which he was subject are easy to discern. The obvious term to use would be "eclectic": atomistic concepts, Paracelsian doctrines, Bruno's theory of an infinite universe, and Patrizi's metaphysics of light are all essential components of Hill's *Philosophia Epicurea*. The aphoristic presentation furthermore allows for a manner of placing ideas of different extraction next to each other, without a complete integration or synthesis. As a consequence, it is not evident that Hill manages to present a consistent and unitary philosophy.

Instead, he seems awkwardly poised between a variety of disparate philosophical traditions.

From the outset, indeed, Hill presents his *Philosophia Epicurea* as being manifestly "without method" and without subordinating one set of views to another, but placing them next to one another.[127] It is therefore an unsystematic work, which lacks methodology and which through its structure subverts the ideal of a rigorous and systematic argumentative framework. In his choice of the aphorism as the form for the expression of his ideas – a device that lends itself to the evocative rather than the systematic and the analytical, tending towards a philosophy that asserts the existence of a multiplicity of explanations for natural phenomena and their interpretations – Hill appears to encourage his readers to embrace the manifold cognitive possibilities afforded by an infinite, pluralistic universe that has no center.

127 Hill, *Philosophia Epicurea*, 77: "Obiicienti": "Tractatum istum esse ἀμέθοδον, i. sine methodo, seu via aut ordine, respondeo prima ut in natura, sic in scientia esse coordinata, non subordinata."

CHAPTER 9

Finite God and Infinite Space: Conrad Vorstius and David Gorlaeus

Kuni Sakamoto

1 Introduction

We usually associate 'atoms' with the domain of physics; we think of Democritus's cosmos as made of atoms and void, or Lucretius's particles whose shapes explain the properties of elements; or, again, we think of the atoms in Newton's *Opticks*, in Dalton's chemistry, or in modern physics. But the revival of atomism in the early modern period was not driven only by natural-philosophical considerations. There were also metaphysical and even theological reasons that could lead to atomistic conceptions. Some of these will be documented in this chapter, in which I will trace a line of thought that led from the theologian Conrad Vorstius to the young theology student David Gorlaeus.

To us, David Gorlaeus (1591–1612) is known as an early modern atomist. His life was tragically short. He died at the age of twenty-one while studying theology at Leiden University. He left behind two books which were published posthumously: the *Exercitationes philosophicae* (Leiden, 1620) and the *Idea physicae* (Utrecht, 1651). While the latter attracted almost no attention, the *Exercitationes* circulated widely among French and Dutch intellectuals by the middle of the seventeenth century.[1]

The *Exercitationes* came under fierce debate during the so-called "Utrecht Crisis."[2] The crisis broke out towards the end of 1641 when the Cartesian physician Henricus Regius (1598–1679) held a series of disputations at the University of Utrecht. During these disputations, he declared that the human being is one not by itself (*per se*) but only accidentally (*per accidens*), and he revealed that he had adopted this thesis from Gorlaeus's *Exercitationes*.[3] This and other heterodox ideas of Regius provoked a refutation from the conservative theologian Gisbertus Voetius (1589–1676). Remarkably, Voetius's criticism identifies

1 On the reception of Gorlaeus's works, see Lüthy, *Gorlaeus*, 133–161.

2 On the crisis, see especially Verbeek, *La querelle d'Utrecht*.

3 Regius, *Disputatio medica tertia*, thesis 9; Voetius, *Testimonium*, 23; Verbeek, *La querelle d'Utrecht*, 94.

 | DOI:10.1163/9789004528925_010

an earlier provenance for the idea of the human being as an accidental union, noting that Gorlaeus had taken this idea from the *Philosophiae triumphus* (1573) by Nicolaus Taurellus (1547–1606). To this, Voetius added the historical detail that Taurellus had been denounced in 1610 as an "atheist physician" by the Heidelberg theologians in their unfavorable report on the theology of Conrad Vorstius (1569–1622).[4]

Voetius did not develop his observation further. It is Christoph Lüthy who has shown that the association of the three – Taurellus, Vorstius, and Gorlaeus – has a historical substance which goes far beyond Gorlaeus's borrowing of a single idea from Taurellus. Lüthy has found it particularly noteworthy that the three shared an idiosyncratic ontology on which to base their entire philosophy and theology. They all interpreted "being" (*ens*) as applicable to both God and created things; they accordingly allocated the same attributes of "being" to both created things and God. Quantity (*quantitas*), as one of those attributes, was therefore ascribed also to God. According to Lüthy, by attributing quantity to God, they physicalized the supreme being. The resultant thesis is highly provocative: combined with the ontology of Taurellus and Vorstius, Gorlaeus's atomism opened up the possibility of counting God as an indivisible physical entity, an atom.[5]

I agree with Lüthy's claim that Taurellus, Vorstius, and Gorlaeus shared the same ontology and attributed quantity to God. But I disagree with his assumption that attributing quantity to God would necessarily lead to the physicalization of God. As Andreas Blank has shown, Taurellus understood quantity as the amount of causal power, so that it did not designate anything physical.[6] My contention in this chapter is that Gorlaeus likewise did not physicalize God, and that it was his atomism that prevented him from so doing.

With this purpose in mind, the present chapter focuses on the issue of God's omnipresence, because, at the time, it was commonly recognized that

---

4 Voetius, *Testimonium*, 28, trans. Verbeek, *La querelle d'Utrecht*, 98; Taurellus, *Philosophiae triumphus*, sig. δ6r = Wels, ed., 110; Van Ruler, *Crisis of Causality*, 186–192; Goudriaan, *Reformed Orthodoxy and Philosophy*, 234–237. For the remarks of the Heidelberg theologians, see *Bedenckingen*, sig. B1r. Their report has been translated in a slightly abbreviated form in Schweizer, "Vorstius," 446–447. See also Spijker, "Heidelberger Gutachten," 209–212, 223–225.

5 Lüthy, *Gorlaeus*, 119–131. For other studies by Lüthy on Gorlaeus, see Lüthy, "David Gorlaeus' Atomism"; Leijenhorst and Lüthy, "Erosion of Aristotelianism"; Lüthy and Spruit, "Henricus de Veno." Lüthy's contributions have made older studies somewhat obsolete. For recent studies, see Hattab, *Descartes on Forms and Mechanisms*, 160–185; Van Ruler, "Substantiële entiteiten zonder vorm"; Goudriaan, "Gorlaeus en de Arminiaanse kwestie"; Hattab, "Aristotelianism and Atomism Combined."

6 Blank, "Taurellus," 670. See also Petersen, *Geschichte der aristotelischen Philosophie*, 234; Leinsle, *Das Ding und die Methode*, 1:162–163.

the materialization of God resulted from a heterodox understanding of divine omnipresence. I concentrate here especially on the relationship between Vorstius and Gorlaeus.[7] As we shall see, it is highly likely that Gorlaeus was indebted to Vorstius on the issue of divine omnipresence, and it is undeniable that Vorstius physicalized God (or at least, he is widely understood – and not without reason – to have done so). To the question of whether Gorlaeus's debt to Vorstius extended to the point of materializing the Creator, however, my contention is that it did not.

In what follows, I first briefly summarize what the sixteenth-century theologian Fausto Sozzini (1539–1604) said about God's omnipresence, because Vorstius developed his theology based on that of Sozzini. The main body of the article is devoted to an examination of Vorstius's theology and its relationship with Gorlaeus's atomism. It shows that, while his ontology was greatly indebted to that of Vorstius, Gorlaeus did not physicalize God. In the concluding part, I reflect on how Gorlaeus's atomism helped him to avoid lapsing into the Vorstian heresy.

2 Sozzini – The Anti-trinitarian Denial of Omnipresence

In order to assess Sozzini's views, let us first take a look at some of the doctrinal basics of the issue of divine omnipresence. Since at least the Middle Ages, God was supposed to be omnipresent in three ways – by knowledge, by power, and by essence. Omnipresence by knowledge means that God knows all things happening everywhere. He is omnipresent by power in that his providence and operation extend to all the places of the world. Finally, essential omnipresence signifies, quite literally, that his essence (or substance) is everywhere. Both in the Catholic and Protestant churches, orthodox theologians allowed for all three types of omnipresence.[8]

It was commonly held that essential omnipresence was possible under the condition that God was infinite. When understood in terms of place, divine infinity was conceived as immensity (*immensitas*). A standard theology textbook of the time defined immensity as signifying that God "is measured by and circumscribed to no place but penetrates and fills each and every place without the multiplication, extension, inclusion, and division of His own essence."[9]

7 Aza Goudriaan has already compared Vorstius and Gorlaeus on the issue of God's locality, albeit very briefly: see Goudriaan, "Gorlaeus en de Arminiaanse kwestie," 243–244.

8 See, for example, Turretin, *Institutes*, III.9.4, 1:197.

9 Gerhard, *Theological Commonplaces*, II.171, 164.

Thanks to his immensity, God could penetrate and fill everything so that his essence could be ubiquitous.

The doctrine of omnipresence became a focal point of debate in the early modern period. This was partly because of its denial by Sozzini, in the scandalous context of his heretical attack on the doctrine of the Trinity which appeared in his posthumous *Christianae religionis brevissima institutio* (Raków, 1618).[10] In this discussion, Sozzini begins by pointing out that the notion of divine omnipresence and immensity was introduced to justify the doctrine of the Trinity. According to this teaching, God is one in his essence, but three in his persons. To avoid this obvious contradiction – that a thing is one and plural at the same time – theologians appealed to divine infinity and immensity: God's immense essence enables him to inhere in three persons without undergoing any division or multiplication. For Sozzini, however, this argument was untenable; nobody would fail to recognize its absurdity. If it were valid, God's immensity "would be able to render what is necessarily true to be false and the necessary natures of things would change. These results implicate not a little contradiction and are obviously very much absurd and impossible."[11]

Sozzini anticipates one possible objection: if one accepts the standard view that God's power is immense and infinite, one must conceive of his essence in the same way. Sozzini responds that this reasoning is wrong: "In fact, the latter kind of immensity [of essence] belongs to quantity (if it is possible to speak in this way when the discourse is about God). The former [the immensity of power], however, belongs to quality."[12] It is noteworthy that, albeit with reservations, Sozzini regards the essence of God as something quantifiable. The gist of his argument is that quality and quantity are two different things, so that one cannot infer the immensity of the one from that of the other. For instance,

---

10 The most comprehensive treatment of Sozzini and Socinianism so far is Daugirda, *Die Anfänge des Sozinianismus*. See also Mulsow and Rohls, *Socinianism and Arminianism*; Mortimer, "Socinianism and Unitarianism." For an overview of *Brevisissima institutio*, see Pioli, *Fausto Socino*, 263–272.

11 Sozzini, *Brevissima institutio*, 155–156 = *Opera omnia*, 1:685a: "Itaque frustra penitus laborant, qui, dogma istud unius essentiae in deo et pluralitatis personarum, rationi non repugnare sibi demonstrandum suscipiunt; et ad dei essentiae infinitatem sive immensitatem recurrere, ut manifesta ista contradictio tollatur, vanum prorsus esse, nemo est qui non videat. Cum ista divinae essentiae immensitas efficere non possit, quin idem plane sit in deo essentia, quod persona, alioqui per eam id, quod necessario verum est, falsum redderetur, ipsaeque necessariae rerum naturae mutarentur; quae non minorem contradictionem implicent, et porro plane absurdissima atque impossibilia sunt." See also Scribano, *Da Descartes a Spinoza*, 159–160.

12 Sozzini, *Brevissima institutio*, 157 = *Opera*, 1:685a: "Haec enim immensitas ad quantitatem (si ita, cum de deo est sermo, loqui licet) illa vero ad qualitatem pertinet."

one can never recognize the other properties of an animal from its size; such inferences would make one a laughing-stock. Therefore, the immensity of God's essence cannot be deduced from that of his power.

There may be a further objection: what is finite never produces anything infinite because there can be no "proportion" (*proportio*) between them. Therefore, if God's essence were finite, his infinite power could not be the result of his finite essence. It follows that the immensity of his power demands his essence to be immense. Sozzini's answer was as follows:

> To this I reply: the divine power (with which we are here particularly concerned as a topic adduced as an example) is infinite in a way that, as we have already said, it is not circumscribed to any limits. Consequently, it extends without any single exception to all things at all that can happen. For certain things cannot happen in any way, as all those things that contain a contradiction, as we have mentioned shortly before. Such as the one repeatedly taught in theological schools: undoing a thing already done. Everyone admits that even the power of God does not extend to such things. From this, it is apparent that the power of God is not absolutely infinite in such a way that there would not be any proportion between it and his essence. Therefore, his power comes out of his essence even if that essence is not immense.[13]

It was a commonplace among theologians that even God could not cause contradictory events. On this basis, Sozzini claims that divine power is not infinite in an absolute sense and can therefore arise from a finite essence.

Finally, Sozzini examines a biblical passage that was adduced as evidence for God's essential omnipresence. In the Book of Jeremiah, God is presented as asking: "Can a man hide himself in secret places so that I cannot see him? [...] Do I not fill heaven and earth?" (23:24). This can be interpreted as affirming God's omnipresence through which he can be omniscient of all human behaviors. However, Sozzini cautions that this must be read in conjunction with the

13 Sozzini, *Brevissima institutio*, 157 = *Opera*, 1:685a: "Ad hoc respondeo, ipsam divinam potentiam, de qua, ut in exemplum allata, hic nominatim agimus, ita esse infinitam, sive ut antea nos ipsi locuti sumus, nullis limitibus circumscriptam, ut ad omnia penitus, ac sine ulla prorsus exceptione pertingat, quae fieri possunt. Quaedam enim sunt, quae nullo modo possunt fieri, ut sunt ea omnia, quae, ut paullo ante attigimus, contradictionem implicant. Quale est illud in ipsis theologicis scholis decantatum, ut factum infectum fiat. Ad quae omnes fatentur ipsam dei potentiam non pertingere. Ex quo constat, non ita absolute infinitam esse dei potentiam, ut nulla sit futura proportio inter ipsam et dei essentiam, unde illa proficiscitur, etiamsi dei essentia non sit immensa."

immediately preceding sentence: "'Am I a God at hand,' says the LORD, 'and not a God afar off?'" (23:23).[14] Sozzini holds that this "seems to suggest that God's substance is present everywhere not equally and therefore it is not immense or infinite, though God himself is ubiquitously present through his power and providence."[15] If God can be conceived of as "afar off," his substance must be distant from the earth, so that his essence is not omnipresent.

With the above argument, Sozzini arrived at a surprising view: God's essence is finite and somewhat quantifiable. Based on this understanding, he ultimately tries to draw the conclusion that God is not omnipresent, but instead exists in a specific place – heaven.[16] We will see in the next section how this heretical view was further developed by an Arminian theologian.

3 Vorstius – Finite God in Heaven

Vorstius started his academic career as a professor of theology at the Academy of Steinfurt and then moved to Leiden University to take over the theology professorship from Jacobus Arminius (1560–1609). However, he was immediately expelled from the city because his theological views came under fierce attack from his opponents. Especially damaging was the charge that he was committed to the Socinian heresy. Vorstius pleaded that this was a groundless criticism, because, as an Arminian theologian, he had never denied the Trinity. Nonetheless, as Pierre Bayle has perceptively remarked, "the disputes raised against him, were doubtless intermixed with a great deal of passion; but after all, those who suspected him to be strongly inclined to Socinianism, were not much in the wrong."[17] In fact, Vorstius communicated with Sozzini and his followers, and closely examined Sozzini's works such as *Praelectiones theologicae*, which he possessed in manuscript form. Consequently, his theology showed a great deal of similarity with that of Sozzini.[18]

14 I have used the Revised Standard Version when quoting from the Bible.

15 Sozzini, *Brevissima institutio*, 159 = *Opera*, 1:685b: "Immo superiora verba, Numquid deus ego e propinquo, et non deus e longinquo? suadere potius videntur, ipsius dei substantiam non ubique pariter praesentem esse, et sic eam non esse immensam sive infinitam, quamvis ubique tamen sua virtute et providentia sit ipse deus praesens."

16 Sozzini also takes Christ to be a human being, not a God. This meant that Christ first had to ascend to heaven in order to learn what to teach on the earth. This view also presupposes God's presence in heaven; see Gomes, "The Rapture of the Christ."

17 Bayle, *Dictionary*, 512.

18 On Vorstius's connections with the Socinians, see Daugirda, *Die Anfänge des Sozinianismus*, 392–402.

Here, I focus on Vorstius's *Tractatus theologicus de Deo, sive de natura et attributis Dei* (Steinfurt, 1st ed., 1606; 2nd ed., 1610). This book is a collection of disputations over which Vorstius presided. He added extensive notes to the texts of the disputations, in which he also addressed the issue of omnipresence.[19] He starts his discussion by summarizing what Girolamo Zanchi (1516–1590) had written about omnipresence in his *De natura Dei, seu de divinis attributis* (Heidelberg, 1577). Vorstius considers Zanchi's views to be representative but finds within them two problematic theses. The first is as follows:

> God is infinite actually, of himself, and simply, obviously because he has the simplest essence, which is not limited to, or included by, any boundaries. He is therefore present everywhere. Indeed, he is present even where there is no place, such as in the uppermost parts of the heavens.[20]

Included in the premise of this thesis is Aristotle's definition of place, according to which "the place of a thing is the innermost motionless boundary of what contains it."[21] It was widely recognized that this definition could not ascribe any place to the outermost sphere of the heavens, which is surrounded by nothing. However, even in that sphere, Zanchi notes, God can be present because he is essentially infinite, and therefore, not contained by anything in the first place. Thus, his essence is everywhere.[22]

The second thesis can be understood as giving details of how God is essentially omnipresent. Vorstius outlines this thesis as follows:

> Next, the second question [of Zanchi] teaches that God is present ubiquitously in his essence, but without magnitude or quantity. Therefore, he is present ubiquitously also in his power or potency. He is so not only in so far as he gives existence to things (as is claimed by Aquinas, who was criticized by Zanchi on this ground), but also in so far as he is in all things

19 For Vorstius's discussion on divine omnipresence, see also Scribano, *Da Descartes a Spinoza*, 162–167; Salatowsky, "God in Time and Space," 67–69.

20 Vorstius, *Tractatus de Deo*, notes to disputation III, 231: "… denique [deum] infinitum esse actu, per se, et simpliciter: quia nimirum essentiam habeat simplicissimam, nullis omnino finibus descriptam aut inclusam; eoque totam ubique locorum praesentem, imo etiam ubi nullus sit locus, v.g. in supremis coelis."

21 Aristotle, *Physics*, 4.4.212a20. On the place of the outermost sphere of the heavens, see Grant, "Medieval Doctrine of Place."

22 Note that Vorstius's summary of Zanchi's argument is inaccurate. Zanchi does not speak of the outermost sphere but the region beyond it. He claims that God is present even there; see Zanchi, *De natura Dei*, II.6.1, 110a–b.

essentially. This is, Zanchi says, just like the soul is in the whole body and in every part of it.[23]

Two points are noteworthy. First, the essence of God is accompanied by neither magnitude nor quantity. Second, the way God is omnipresent is said to be analogous to the way the soul is in the body.

Vorstius sets out to examine this standard view. He divides his discussion into three parts. The first is based on the Bible. In the Book of Isaiah, God is presented as saying, "Heaven is my throne and the earth is my footstool" (66:1). This clearly situates God in heaven.[24] The Gospel of Matthew describes the angels as beholding God (18:10). If this is the case, the essence of God cannot be absolutely infinite because there must be a certain "proportion" (*proportio*) between the one who beholds and the one who is seen.[25] Following Sozzini, Vorstius also takes Jeremiah 23:23–24 as evidence against the essential omnipresence of God. Rather, it teaches that divine providence extends through all things.[26]

The second part of the discussion appeals to the fact that "even pagans have admitted the following by natural instinct (as Aristotle testifies in *De mundo*): God lives in heaven, namely, the supreme region of the universe, just as he finds himself in the acropolis or on the royal throne."[27] As it happens, *De mundo* is not by Aristotle, and its authenticity had been questioned since Antiquity. Many sixteenth-century scholars had begun to consider it spurious, but Vorstius was amongst those who still attributed it to Aristotle. This is understandable given the similarity between his view and that of *De mundo*: the latter presents God as situated in the heavens and preserving the cosmic order by his power.[28]

23 Vorstius, *Tractatus de Deo*, notes to disputation III, 231: "Deinde quaest. 2. tradit, deum tota sua essentia, sed absque magnitudine seu quantitate, ubique praesentem esse, eoque etiam virtute seu potentia: non tantum quatenus omnibus rebus esse dat (ut Aquinas exponit, qui ab ipso Zanchio propterea taxatur) sed etiam quatenus essentialiter in omnibus rebus adest: sic ut tota, inquit, anima est in toto corpore, et tota in qualibet ejus parte." Cf. Zanchi, *De natura Dei*, II.6.2, 115a–b; Gründler, *Die Gotteslehre Girolami Zanchis*, 85.

24 Vorstius, *Tractatus de Deo*, notes to disputation III, 232.

25 Ibid., 237.

26 Ibid., 232–233.

27 Ibid., 236: "Ipsi etiam ethnici hoc naturae instinctu (teste Aristotele in lib. de mundo) deum in coelo, utpote suprema mundi regione, et quasi acropoli seu regia sede, habitare confessi sunt."

28 Thom, "Cosmotheology of *De mundo*." For various interpretations of *De mundo* in the Renaissance, see Kraye, "Daniel Heinsius"; eadem, "Aristotle's God"; Hirai, *Le concept de*

The third part of the discussion is based on the nature of God. Vorstius first enumerates eight arguments for God's essential infinity and omnipresence and then refutes them one by one. The fifth position argues for the infinity of God's essence based on that of his power.[29] To this, Vorstius replies as follows:

> The power of God, although it is rightly said to be infinite with respect both to us and to possible things, is yet not thoroughly infinite because it does not extend to impossible things. Rather, in God himself, his power remains within the limit of his natural qualities. Therefore, his essence is also not infinite, because just as an infinite power requires an infinite essence, a finite power requires a finite essence.[30]

Vorstius's argument is identical to that of Sozzini. He concludes that God's power (and hence, his essence) is finite from the premise that even God cannot bring about mutually contradictory things simultaneously.[31]

Vorstius finally takes up the issue of whether one can apply the categories of place and magnitude (and quantity) to God. He first affirms that the Bible explicitly attributes these properties to the Creator. However, there have been some scholars who have interpreted the relevant biblical passages only figuratively, taking them as similar to those descriptions that attribute to God a human-like body. Vorstius responds as follows:

> [I]f we talk about the coarse and sensible magnitude of bodies and the purely physical place (such as the one described by Aristotle in book IV of the *Physics*), I am very much willing to admit that these things should be taken figuratively about God. But what forbids the spirit and even God to

---

semence, 83–103; Hirai, *Medical Humanism*, 46–79; Kraye, "Disputes over the Authorship of *De mundo*," 181–197.

29 Vorstius, *Tractatus de Deo*, notes to disputation III, 234: "Habet infinitam potentiam. Ergo et essentiam."

30 Ibid., 238: "Potentia dei, etsi respectu nostri, et rerum possibilium, recte immensa dicitur, non tamen prorsus infinita est: quia non extenditur ad impossibilia: immo in ipso deo consistit intra naturales dei qualitates. Ergo nec essentia infinita est: quia ut infinita potentia requirit essentiam infinitam, ita etiam finita finitam."

31 Unlike Sozzini and Vorstius, Taurellus never failed to argue for God's infinity, both in terms of his substance and power. He also maintained that God fulfills everything by his substance because he is substantially infinite. It is therefore misleading to follow the Heidelberg theologians in associating Taurellus and Vorstius on the issues of God's infinity and omnipresence; see Taurellus, *Philosophiae triumphus*, III, 264 = Wels, ed., 444; Petersen, *Geschichte der aristotelischen Philosophie*, 231–233.

> have spiritual magnitude of their own, which is certainly not sensible but intelligible in some way? Similarly, what forbids us to accept by the term "place" (namely, when it is considered generally and metaphysically) not the surface of some surrounding body, but any void region or any space empty of any genuinely subsisting thing? At this point, we oppose Plato and others to Aristotle, particularly since the latter's definition of place is defective in many ways, and therefore, cannot be properly applied to all physical places, as has been shown not just once by the most erudite Scaliger and other Christian philosophers. See also Taurellus's *De rerum aeternitate*, especially pp. 64, 71, 88, 126.[32]

Decisive here is the new definition of place introduced by Julius Caesar Scaliger (1484–1558). In his *Exotericae exercitationes* (Paris, 1557), Scaliger rejects the Aristotelian notion of place. His contention is that place cannot be properly understood in relation to the surrounding body; rather, it should be defined independently of the relation between surrounding and surrounded. Thus, he concludes that place is a three-dimensional space, which he also calls "void."[33]

Scaliger's new conception plays two roles in Vorstius's theology. The first concerns the place of the outermost sphere. As already mentioned, Aristotle's definition makes it difficult to attribute place to that part of the universe. The desire to find a solution to this difficulty was one of the motivating factors that led Scaliger to introduce his new definition; it would be unproblematic to say that the outermost sphere occupies a certain three-dimensional space, even if it is surrounded by nothing.[34] Therefore, the new definition can be applied to "all physical places." For Vorstius, the attribution of place to the last sphere

---

32 Vorstius, *Tractatus de Deo*, notes to disputation III, 239: "Respondeo, si de crassa et sensibili corporum magnitudine, deque loco mere physico (qualem Aristoteles lib. 4. physicorum descripsit) loquamur, omnino me fateri, haec de deo figurate intelligenda esse. Sed quid vetat, etiam spiritus, imo et deum, suam habere spiritualem magnitudinem; non sensilem quidem, sed aliquo tamen modo intelligibilem? Quid item vetat, nos per locum, scilicet in genere et metaphysice consideratum, non superficiem ambientis alicujus corporis; sed vacuam quamlibet regionem, seu quodvis inane spacium cujusvis rei vere subsistentis, accipere? Aristoteli Platonem et alios hac in parte opponimus: praesertim quum illius definitio loci multis modis vitiosa sit: ut quae non omnibus quidem physicis locis recte accommodari possit: sicut a doctissimo Scaligero, et aliis Christianis philosophis, non semel hactenus ostensum est. Vide etiam Taurellum de rerum aeternitate, praesertim pag. 46, et 64, et 71, et 88, et 126 et d." Part of this passage is translated in Leijenhorst and Lüthy, "Erosion of Aristotelianism," 394.

33 Scaliger, *Exercitationes*, 5.2, 6v–7r; Sakamoto, *Scaliger*, 82.

34 Scaliger, *Exercitationes*, 5.3, 7r; Sakamoto, *Scaliger*, 84.

has another significance: it allowed him to claim that God occupies that region with his finite essence.[35]

Secondly, Scaliger's definition made it easier for Vorstius to speak about the place of God. The traditional Aristotelian definition had rendered it nonsensical to say that God occupies a place: as God is wholly incorporeal, he is not surrounded by any body. However, now that this place is reconceived as a space independently of bodies, its occupation by God does not give rise to the same kind of controversy. In this connection, Vorstius writes as follows:

> As for place and local presence, it certainly appears that one should have no doubt that they can truly and correctly be attributed to God. It is so if place is considered generally and abstractly, as well as if local presence is taken not as circumscriptive or definitive (as they say), but as repletive and substantial, by which the divine substance fills its divine place in a divine manner.[36]

Here, Vorstius refers to three modes of presence – circumscriptive, definitive, and repletive – a distinction that goes back to the Middle Ages. The circumscriptive presence means that something occupies a place by surrounding it. Corporeal things exist in this way. The definitive presence signifies that a thing exists in a specific place and not in others. This can be applied to the way angels occupy places. Finally, the notion of repletive presence has been used to explain how God fills every place by his immense essence. In applying this last category to God, Vorstius follows the tradition, but he substantially modifies its meaning: God no longer fills all places; his repletive presence is now limited to heaven.[37]

Vorstius goes on to argue that the Bible clearly attributes magnitude (and quantity) to God. To this claim, an objection could be raised that God is incorporeal, so that he should have no quantity. However, Vorstius reminds the reader that angels, though endowed with no matter, are generally considered to have finite quantities. If this is the case, why should God – like angels, a

---

35 See the passage quoted in n. 20.

36 Vorstius, *Tractatus de Deo*, notes to disputation III, 239: "Sane de loco, et locali praesentia, quin deo vere et recte attribui possit, minime dubitandum videtur: si modo locus generatim et abstracte consideretur; et localis praesentia, non pro circumscriptiva vel definitiva (ut loquuntur) sed pro repletiva et substantiali, qua scilicet divina substantia, locum suum divinum, modo divino impleat, accipiatur." Part of this passage has been translated in Gerhard, *Theological Commonplaces*, II.176, 170.

37 On the three modes of presence, see Turretin, *Institutes*, III.9.5, 1:197.

spiritual entity – not have quantity? Vorstius concludes, therefore, that we must follow the Bible in attributing quantity to God.

Vorstius argues that depriving God of quantity would lead to absurd conclusions:

> It appears to be a strange paradox that God lacks all quantity in such a way that he is substantially whole in any thing, even in the tiniest things like the burrow of a beetle or the tip of a spear. This would mean to force the biggest and the most ample essence of God into a small space, or to make of the biggest and immense God a very small and tiny thing.[38]

If the entire essence of God were omnipresent, he would be inherent even in the filthiest and lowliest of things. Vorstius repeatedly notes that such a mixing of God and creatures would seriously compromise divine dignity.[39] To avoid such blasphemy, God must be understood as quantifiable.

Vorstius concludes his discussion by criticizing one key element in the analogy drawn in the standard view of the scholastics – namely, that the omnipresence of God can be understood analogously to the soul's presence in the body. He points out that the deployment of this analogy in support of God's omnipresence rests on a mistaken view of the soul; contrary to this misconception, Vorstius insists that the essence of the soul by no means extends to the entirety of the body. This is obvious from the principle that an essence cannot exist simultaneously in different places. Generally, the essence of an entity has a limit that coincides with its place. Thus, occupying several places at once would mean that a thing transcends its limit, or "it would be outside of itself." Avoiding this absurd consequence, Vorstius states that: "To be sure, the soul is diffused through all parts of the body by its power and operation, but its essence is contained in a certain bodily place (for instance, in the heart or in the brain)."[40] From this, he concludes that the soul also has a certain quantity.

---

38 Vorstius, *Tractatus de Deo*, notes to disputation III, 240: "Et sane mirum videtur esse paradoxon, deum sic omnis quantitatis expertem esse, ut totus substantialiter sit in re qualibet, etiam minima, v.g. in antro scarabei, aut in apice unius pili, etc. Hoc enim maximam et amplissimam dei essentiam esset in angustum cogere, sive ex maximo et immenso deo minimum et pusillum efficere." Part of this passage has been translated in Gerhard, *Theological Commonplaces*, II.176, 169.

39 Vorstius, *Tractatus de Deo*, notes to disputation III, 236–37; Vorstius, *Oratio apologetica*, 18, trans. Schweizer, "Vorstius," 460; Scribano, *Da Descartes a Spinoza*, 166.

40 Vorstius, *Tractatus de Deo*, notes to disputation III, 242: "Virtute quidem et operatione per omnes corporis partes anima diffusa est: sed essentia ejus uno aliquo certo loco (puta vel corde, vel cerebro) continetur."

He then supports his view by referring to Book IV of *Universae naturae theatrum* (Lyon, 1596) by Jean Bodin (1530–1596), in which the soul is conceived as corporeal and endowed with quantity.[41]

To the extent that Vorstius always maintains that God is spiritual and incorporeal, he cannot be said to physicalize the Creator. It is also undeniable, however, that he sometimes approaches materialism, especially when he holds that God has a finite quantity. His assertion that the soul has a three-dimensional quantity has the same implication, because it is underpinned by an analogy between the soul's presence in the body and God's existence in the universe (his objections to the use of the analogy for divine omnipresence notwithstanding).

Vorstius would have denied that he was materializing God – the quantity he ascribes to God is spiritual, not physical. Such protestations are not particularly persuasive, however. This is especially evident when we turn to a critique of Vorstius's *Tractatus* in the *Tractatus de Deo et attributis divinis* (Mainz, 1611) by the Jesuit Martin Becanus (1563–1624). Becanus portrays Vorstius as making a rather subtle distinction: while Vorstius denies that God's whole substance (or essence) is omnipresent, at the same time, he does not explicitly oppose the idea of God's substantial omnipresence. Becanus concludes that what is omnipresent, in Vorstius's view, is not God's whole substance but only the parts thereof: "Therefore, according to Vorstius, God's substance has its parts and is divisible. But what are these parts? This is completely unheard of." Vorstius's view, which renders God divisible and therefore material, is described by Becanus as "the new gospel of a new heretic."[42]

Even if Vorstius had denied the substantial omnipresence of God, by Becanus's reckoning, his theology would not have been the better for it: "For if, as he says, God is not present according to his substance elsewhere than in heaven, it clearly follows that, when Christ was active on Earth, he was not truly God according to his substance."[43] This implies a denial of the Trinity and is thus tantamount to a revival of Arianism. Vorstius's theology therefore ends up either with the physicalization of God or with anti-trinitarian Socinianism. For Becanus, either of these options implies atheism.

41 Bodin, *Universae naturae theatrum*, IV, 515–516; Blair, *Theater of Nature*, 140.

42 Becanus, *Tractatus de Deo*, 36: "... ergo substantia dei secundum Vorstium partes habet, et divisibilis est. At, quae illae partes sunt? Hactenus inauditum est"; 37: "Novum novi haeretici evangelium."

43 Ibid., 35: "Nam si deus secundum substantiam suam, ut ipse ait, non est alibi, quam in coelo; plane sequitur, Christum, cum in terris versaretur, non fuisse verum deum secundum substantiam."

4 Gorlaeus – The Infinite Locality of God

Having examined the positions of Sozzini and Vorstius on divine omnipresence, we are now ready to turn to Gorlaeus. On one of the central tenets of Vorstius's doctrine – that God and created things occupy the same kind of place – Gorlaeus was in agreement, as seen in his treatment of the notion of "attributes" (*attributa*). According to Gorlaeus, one cannot – in reality – separate a thing's attributes from its "being" (*ens*); the separation is possible only in thought. Gorlaeus gives the following reasoning for this:

> Indeed, since being is common to both God and created things, it can include nothing in its conception that is not appropriate to both God and created things. Therefore, in being there is not this distinction between its attributes [and its being] because also these attributes are common to God, in whom there is no distinction.[44]

Because being is univocal to both God and created things, they have the same attributes. Specifically, Gorlaeus enumerates the following six attributes: unity (*unitas*), truth (*veritas*), goodness (*bonitas*), existence (*existentia*), locality (*localitas*), and durability (*durabilitas*). The inclusion of unity in this list is of particular significance for its author. For Gorlaeus, that which is truly existent necessarily has the attribute of unity. By contrast, composite things lack unity as they are reducible to their constituents, which Gorlaeus calls "atoms."

The attribute of locality also presupposes atomism: since all atoms must occupy some place, locality necessarily accompanies being. Although locality and durability are not included in the definition of any entity, it is impossible to imagine something that occupies no place or does not endure even for a moment. Therefore, these attributes are "inseparable affections that belong to an entity by itself," not accidentally.[45] Is this locality an attribute common to both atoms and God? Gorlaeus answers as follows:

> We ascribe also to each entity its locality, even to God himself, but we ascribe the infinite locality to the infinite God. Indeed, since he is everywhere, he is necessarily in all places. As he is in all places, it is necessary

44 Gorlaeus, *Exercitationes*, 2.4, 52: "Quum enim ens sit commune et deo et creaturis, nihil in suo conceptu includere potest, quod non competat et illi et his. Quare non est in eo haec attributorum distinctio; quia et haec attributa deo sunt communia, in quo nulla datur distinctio."

45 Ibid., 57: "Sed durabilitas et localitas sunt affectiones inseparabiles, quae enti per se competunt."

> that he is in a place. Those who say that God is everywhere and nowhere therefore contradict themselves. For what is in "every-where," is ubiquitous; and what is in "no-where," is nowhere.[46]

For Gorlaeus, it is contradictory to assert the omnipresence of God while denying locality to him. If one admits God's ubiquity, one should ascribe a place to him. This locality of the divine is exactly the same as the locality of atoms. The only difference is that the latter is finite whereas the former is infinite. This raises the following questions, however: What is infinite locality? If infinity means not being limited to any place, is the phrase "infinite place" not a contradiction?

To clear up these doubts, we have to look at how Gorlaeus conceives of place.[47] He first rejects the Aristotelian definition. Suppose that body A surrounds body B, and then A begins to move. If one adopts Aristotle's definition, one would have to say that not only the place of A but also that of B would change. Gorlaeus regarded this as absurd. Moreover, since one body can be surrounded by plural bodies and plural bodies can be surrounded by one body, the Aristotelian definition allows one body to have plural places and plural bodies to occupy one place. These were also unacceptable consequences for Gorlaeus.[48]

Abandoning the traditional definition, Gorlaeus proposes that "place is not something real, but a pure nothing."[49] It is therefore neither substance nor accident. If it were a substance, it would have to be either a spirit or a body. If it were a spirit, place would be an "intelligent being" (*res intelligens*) and it is inconceivable how such a being could contain bodies. By contrast, if place were a body, to say that a body occupies a place would mean that the body is contained in another body. This, for Gorlaeus, is also impossible.[50]

Place cannot be an accident either. If place were to be regarded an accident, it would have to have a substratum in which it inheres. Is this substratum the surrounding body? If this were the case, when the surrounding body moves,

---

46 Ibid., 58: "Ascribimus et unicuique suam locationem, etiam ipsi deo, sed infinito hinc infinitam. Quum enim ille sit ubique, necesse est, ut sit in omni loco. Quod est in omni loco, necesse est, ut sit in loco. Nam illi sibi ipsis contradicunt, qui ajunt deum esse ubique et nullibi. Est enim ubique, qui in omni ubi, nullibi qui in nullo ubi."

47 A brief summary of Gorlaeus's argument can be found in Donahue, *Dissolution*, 173–174; Grant, *Much Ado*, 392.

48 Gorlaeus, *Exercitationes*, 10.2, 214.

49 Ibid.: "Dicam quod res est, locus non est aliquid reale, sed purum putum nihil," modified translation from that given in Lüthy, "Gorlaeus's Atomism," 252.

50 Gorlaeus, *Exercitationes*, 10.2, 214–15.

the place would simultaneously change even if the surrounded body remained unmoved. This is because the accident of place would move together with its substratum. Or should one think that the accident of place migrates from the first substratum to the next (namely, the body that is to surround the contained body after the movement)? This would mean that an accident migrates from one substratum to another, a phenomenon that no Aristotelian would admit. Even if one thinks that the accident of place inheres in the contained body, Gorlaeus adds, the same kind of absurdity would arise.[51]

From the above, place is neither a substance nor an accident but a pure nothing. What is this nothing? Gorlaeus describes it as follows:

> Rather, before the world was made, there was a pure nothing. For there was no other entity than the one and only God; and this inane space is infinite, located, long, and deep, without limit. In this, the world has been placed.[52]

The nothing he refers to here is the space that was already existent even before the creation of the world. It has no limit and is therefore infinite. At first it contained nothing, so that it was a vast void (*vacuum*). From this, Gorlaeus gives a more precise definition of place: whereas a void is a space devoid of bodies, place is a space filled up with bodies.[53]

Gorlaeus's conception of place using such terms as *void* and *space*, reminds the reader of Scaliger's *Exercitationes*. Indeed, his consultation of this work can be confirmed from the following remark:

> A space that is actually void [of any bodies] is admitted outside of this world. Whether it should be admitted within this world, I do not know. But a space [filled up with bodies] is admitted [within this world]. For when a body makes way for another, void arises in order to be filled. If the entry of other bodies were then prohibited, an actual void would arise.[54]

---

51 Ibid., 215. But Gorlaeus admits that some accidents migrate from one substratum to another. See ibid., 5.2, 92–95; Pasnau, *Metaphysical Themes*, 505.

52 Gorlaeus, *Exercitationes*, 10.2, 215: "Verum ante conditum mundum erat putum nihil. Non enim aliud ens, quam unus solus deus. Et hoc inane spatium est infinitum, locatum, longum et profundum, sine terminis. In eo positus est mundus," modified from the partial translation given in Lüthy, "Gorlaeus's Atomism," 252.

53 Gorlaeus, *Exercitationes*, 10.2, 216.

54 Ibid.: "Datur hoc spatium actu vacuum extra hunc mundum, in eo an detur, nescio: spatium tamen datur. Cedente enim corpore corpori vacuum fit, ut impleatur. Quodsi aliorum corporum ingressus prohiberetur, fieret actu vacuum."

Gorlaeus borrows almost verbatim from Scaliger the phrase "when a body makes way for another, void arises in order to be filled".[55] For Scaliger, there is no actual void in this world because bodies move so as not to produce a void. Gorlaeus, however, reserves judgment on this point.

Gorlaeus's reservation stems from his atomism. When dealing with the shape of atoms – which he suspects are spherical – he writes the following:

> In that case, if two or three atoms touch each other, it is necessary that between them there be a void. If two or three spherical balls touch each other alternately, the space that lies between them cannot be filled by a spherical body. Therefore, neither can the space that lies between atoms be filled by an atom.[56]

If atoms are spherical, there must be a void. This discussion of contiguous atoms is one that Gorlaeus has once again borrowed from Scaliger. Their difference consists in this: that while Scaliger combines the argument with his denial of the existence of an actual void – in support of his conclusion that there were no atoms–, Gorlaeus, the atomist, uses it to allow the possibility of an actual void existing within the world.[57] Gorlaeus maintains that atoms are too small for humans to identify their shape.[58] While he could not rule out the possibility that they might be spherical, he therefore had to conceive of space as capable of being devoid of any body.

Gorlaeus also differs from Scaliger in admitting an extra-cosmic infinite space. Where does this idea come from? One might first think of ancient atomism; Gorlaeus might have come across Cicero's remark that Epicurus considered the whole to be necessarily infinite.[59] Though this is possible, the more likely source for Gorlaeus's wording is the scholastic doctrine of "imaginary space" (*spatium imaginarium*).[60] This doctrine teaches that beyond this finite world there exists an infinite space devoid of any body; God occupied it from eternity and created the world within it. Such a supposition of the

---

55 Scaliger, *Exercitationes*, 5.2, 7v: "Est enim vacuum, spatium, in quo est corpus. Cuius natura per se talis est, ut cedente corpore corpori, fiat vacuum, ut impleatur."

56 Gorlaeus, *Exercitationes*, 13.1, 243: "Tum duae aut tres atomi, si se tangant, necesse est inter illas interjectum esse vacuum. Duo enim globi rotundi, aut tres, si se invicem tangant, quod inter illos interjacet spatium, a rotundo corpore nequit repleri. Quare nec quod interjacet inter atomos, compleri potest ab atomo."

57 Scaliger, *Exercitationes*, 5.1, 6v; Sakamoto, *Scaliger*, 81–82.

58 Gorlaeus, *Exercitationes*, 13.1, 244.

59 Cicero, *On Divination*, 2.50.103 (fr. Usener, 297).

60 On this doctrine, see Grant, *Much Ado*, 101–255.

extramundane infinite space made it possible to avoid confining divine immensity to the finite world. In the early modern period, the theory received support from the Jesuits in particular, as well as from some Reformed theologians. Gorlaeus must therefore have been familiar with this doctrine.[61] But he significantly transformed it. According to the traditional understanding, the void-space exists only outside the world; within it, there exist the Aristotelian places. Gorlaeus, by contrast, admits the void-space both inside and outside the world, by adopting Scaliger's new definition of place.

Gorlaeus has thus arrived at an atomism that postulates the void-space both within and outside the created world. With this new conception of space, he can easily make sense of his thesis that the infinite God has the same attribute of locality as the locality attributable to atoms. Both God and atoms fill three-dimensional space. The difference is that atoms occupy the finite space of this world, whereas God fills the infinite space, both within and beyond this world.

5 Conclusion

The present chapter has documented some remarkable agreements between Vorstius and Gorlaeus. They both thought that we could talk about the place of God in the same way as that of created things. They both opposed the view that we could attribute locality to God only figuratively. This required them to abandon the traditional, Aristotelian definition of place, since it could not be applied to God. Instead, they adopted the alternative definition given by Scaliger. It is likely that Gorlaeus learned from Vorstius that Scaliger's

61 Voetius, *Selectarum disputationum parts quinta*, 576–579; Goudriaan, *Philosophische Gotteserkenntnis*, 93–108; Beck, *Gisbertus Voetius*, 255–260. It may be noted that Gorlaeus's wording is somewhat similar to that of Lambert Daneau (ca. 1530–1595), a reformed theologian who was a professor of theology at Leiden University from 1581 to 1582. Like Gorlaeus, Daneau admitted the pre-creation void in which God would place the world. Unlike Gorlaeus, he did not qualify this void as infinite. But he maintained that beyond the extremity of the universe there exists an infinite space for the seat of the blessed and of Christ. See Daneau, *Physica christiana*, 213: "Etsi enim Deus ante mundum hunc conditum omne illud spatium, in quo nunc est mundus occupabat, replebatque vi sua nobis incognita, modoque ineffabili, sicuti etiam nunc implet omnia, postquam condita sunt, vi et modo suae praesentiae et maiestatis et potentiae … tamen huius, quae per rerum creationem facta est, impletionis respectu, locus hic, ubi nunc est positus mundus, vacuum quiddam certe erat"; and idem, *Physices christianae pars altera*, 11.5, 27v. On Daneau, see Bizer, *Frühorthodoxie und Rationalismus*, 32–50; Fatio, *Méthode et théologie*; Systma, "Calvin, Daneau, and 'Physica Mosaica.'"

definition made it possible to ascribe place to God, even if there is no conclusive evidence to confirm this at present.[62] Yet, there were also important differences between the two concerning the doctrine of omnipresence. Like Sozzini, Vorstius denied the essential omnipresence of God, a denial which led him to declare divine finitude. Becanus attacked this thesis as inadmissible because it made God divisible. Gorlaeus, by contrast, admitted that God was ubiquitous, so that God's essence was infinite. At least on this point, Gorlaeus was faithful to the orthodox doctrine and did not move toward the physicalization of God.

Sozzini and Vorstius had concluded that God was finite on the basis of the locality which was one of God's attributes. Starting from the same premise, Gorlaeus nevertheless maintained God's infinity. Where did this difference come from? It has become clear that it emerged from their contrasting understanding of the relationship between infinity and locality. For Sozzini and Vorstius, as well as for many scholastics, infinity and immensity carried meanings of not being limited to any place. This made it difficult to explain how an infinite being had locality. To solve this difficulty, medieval theologians developed subtle and complicated theories about God's location such as the doctrine of imaginary space. Sozzini and Vorstius, in turn, cut the Gordian Knot in a simple but heretical manner: they denied God's infinity in order to secure his locality. Gorlaeus, by contrast, did not admit any tension between God's being infinite and having locality. He combined the scholastic doctrine of imaginary space with Scaliger's new definition of place and arrived at the idea of the void-space spreading through our finite world to the extra-cosmic infinity. This new understanding served two purposes. First, it enabled him to allow an actual void to exist within this world, a possibility for which his atomism required admission. Secondly, it made it possible to ascribe the same attribution of locality both to atoms and to the infinite God. Thus, Gorlaeus's atomism was far from necessitating the physicalization of God. It rather helped him to avoid the heretical tendency that could be traced back to Sozzini.

Overall, this chapter has revealed the complex construction of Gorlaeus's philosophy. As Lüthy has pointed out, Gorlaeus's philosophy comprised a confluence of Vorstius's idiosyncratic Protestant theology, Scaliger's natural philosophy of Italian provenance, and the newly revived atomism. Moreover, these elements were combined in a complicated way, perhaps more complicated

62 Note that Henricus de Veno (ca. 1574–1613) also accepted Scaliger's concept of place, but he did not apply it to God. In fact, he denied that God had a place, on the grounds that his essence was infinite; see Lüthy and Spruit, "Henricus de Veno," 1133. As his position contradicts Gorlaeus's view in this respect, his teaching could not be the sole source of inspiration for Gorlaeus.

than Lüthy has assumed. Vorstius's theology was purged of Socinian implications, while Scaliger's new definition of place served not only the atomistic supposition of interstitial voids, but also, by its conflation with the notion of imaginary space, it served to make sense of God's infinite locality. Such is the strategy that can be traced throughout his two books. Gorlaeus's life was surely brief, but his twenty-one years were enough for the atomist to leave us with many enigmas that still remain to be solved.

CHAPTER 10

Atomism, Mechanism, and Chymistry in the Natural Philosophy of Walter Warner

Stephen Clucas

In the first two decades of the seventeenth century, atomism was not a widely held doctrine amongst English natural philosophers. Amongst natural philosophers with interests in alchemy or chymistry, proponents of atomism were even rarer. One such proponent – together with his colleague and friend Thomas Harriot (1560–1621) – was Walter Warner (ca. 1557–1643), one of several mathematically – and philosophically – inclined clients of Henry Percy, ninth earl of Northumberland (1564–1632).[1] Although Warner never published any of his atomistic writings, his manuscripts show a natural philosopher who was seeking mechanical and atomistic explanations for chymical – and even physiological – processes well before the later and more widely-known corpuscularian hypotheses of Robert Boyle (1627–1691).[2] The origins of Warner's atomistic theories are difficult to establish, as he rarely cites the works of other philosophers. He certainly appropriated some of his key ideas from a manuscript treatise by an unidentified early sixteenth-century English philosopher (the "Generall principells or conclusions"), and worked closely with Harriot, who seems to have elaborated a similar atomistic theory.[3]

Warner's matter theory was atomistic, although he saw atomic parts not as absolutely indivisible, but as actually undivided.[4] Warner also maintained that matter – and its component parts – were "eternall [...] both in respect

1 On Henry Percy and the 'Northumberland circle', see esp. Shirley, *Thomas Harriot*. On Northumberland circle atomism, see Kargon, *Atomism in England*; Jacquot, "Harriot, Hill and Warner"; Clucas, "Corpuscular Theories," and idem, "All the Mistery of Infinites." On Warner's life and manuscript writings, see Prins, *Walter Warner*.

2 On Boyle's corpuscular theories, see Newman, "The Alchemical Sources," idem, "Robert Boyle's Debt"; Clericuzio, *Elements*, 103–148 and idem, "A Redefinition of Boyle's Chemistry"; Banchetti-Robino, *Chemical Philosophy*.

3 "Generall Principells or conclusions," Lambeth Palace Library, MS Sion L40.2/E10, fols. 82r–88v. On Warner's use of the "Generall Principells," see Clucas, "Corpuscular Theories," 184–185.

4 British Library [hereafter BL], Add. MS 4394, fol. 396v: "the prime or simple partes of matter are called atomi not for that they can suffer no further division but because they are not actually divided or discrete."

 | DOI:10.1163/9789004528925_011

of beginning and end."[5] Like the author of the "Generall principells", he saw the universe as being composed of four principles: time, space, matter, and a spherically radiated force which he calls 'vis' (or "Virtue radiative").[6] The coherence of bodies depended on the projection of this radiated force, which he describes as the "squarer and cutter of atomi."[7] Although bodies are sometimes able to resist this force, Warner says that matter is "both moveable in whole and alterable in partes by the action of vis," and thus he sees radiated force as the primary instrument of what Aristotelians would call "local motion" – the principle underlying all natural changes.[8]

Warner seems to have explored various possibilities regarding the shape of atomic particles, and the existence (or non-existence) of vacua. At one point he claims that atoms must have "plaine sides" because their composition was reliant on the "laterall cohaerance <and congruence>" imposed by vis. He argued that there were two "speciall resons" against the "conceit of rotundity of atoms":

> one *a priore* w[hi]ch is the direction <and parallell> proiection of virtue radiatiue [...] [and] an other *a posteriore* [which is] the frustrating of reflexion and annihilating of virtue by disgregation and conseque[n]tly the [...] taking away of motion.[9]

But he also seems to have been open to the possibility that atoms might have "spherikall or sphaeroidicall sides, cylindricall or conicall or conoidicall or such like sides," albeit with a recognition of the problems this posed for any explanation of the conjoining of atoms.[10] At this point, he seems to have held a plenist position, arguing that "Matter and vertue radiative do fill the universall space."[11] In another place, however, he speculates on the packing (or "constructure") of spherical atoms – interspersed with vacua – into various geometrical

5 Ibid., fol. 382r. Cf. "Generall Principells," fol. 83r: "Matter is eternall and enfenet."

6 Ibid., fol. 129v. Cf. "Generall Principells," fol. 83v: "Matter cannot bee considred with owet tieme and space <and> vis to effect anytheng for with out <all> thes foer notheng can be effected but quiet – from henc ther foer must be considred the effecte of thenges acording to the prerogatiue of thes 4 actors Tyme space matter vis."

7 Add. MS 4394, fol. 397r. On Warner's concept of *vis radiativa* in Warner's matter theory, see Clucas, "Corpuscular Theories," 182–196.

8 Ibid., fol. 386r.

9 Ibid., fol. 397r.

10 Ibid.

11 Ibid., fol. 386r.

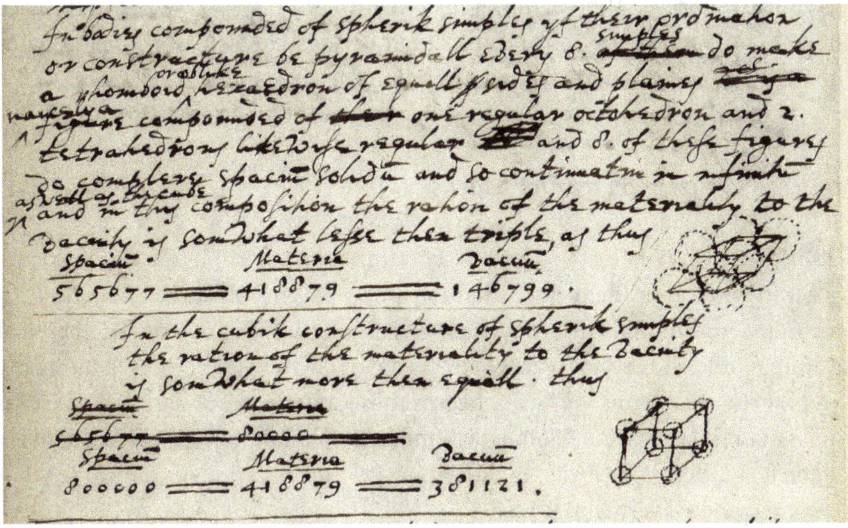

FIGURE 10.1 Spherical packing of atomic particles. Walter Warner, British Library, Add. MS 4395, f. 71v
COURTESY OF THE BRITISH LIBRARY

arrangements including the cube (see Fig. 10.1).[12] In this version of his theory the proportion between "materiality" and "vacuity" in these constructures is what accounts for the density and rarity of bodies.[13] As I have elsewhere previously discussed Warner's matter theory, and its relationship to that of other atomists in the "Northumberland circle,"[14] I want to focus here on what he refers to as the "chymicall groundes" for his atomism.[15]

In recent decades, a great deal of attention has been paid to the contributions made by chymical philosophers to the development of the corpuscularian philosophy,[16] and it should be noted that there is evidence that Warner was a reader both of Daniel Sennert (1572–1637) and Gaston DuClo (born ca. 1530), each of whom are known for the development of corpuscular

12 Giordano Bruno opted for spherical atoms in his *De triplici minimo et mensura* (1591), but Bruno's atoms are ensouled and – unlike Warner – he sees the universe as "essentially vital rather than mechanical." See Gatti, "Giordano Bruno's Soul-powered Atoms," 170–171. On Bruno's spherical atoms, see also Lüthy, "Bruno's *Area Democriti*."

13 See Add. MS 4395, fol. 71v. See esp. his statement that "in the most dense or compact bodies that are of this kinde the porosity or vacuity is more then a third part <quarter> of the whole and in the most rare, more <lesse> then half."

14 Clucas, "Corpuscular Theories," and idem, "All the Mistery of Infinites."

15 Add. MS 4396, fol. 396v.

16 See, for example, Newman, *Atoms and Alchemy*; idem, "Experimental Corpuscular Theory," 291–329, and idem, "The Significance of 'Chymical Atomism'."

chymical theories.[17] When Warner turns his attention away from the coherence of solid bodies, and begins to think about various kinds of "spiritual" activity, he draws upon his chymical reading, and makes use of chymical examples.

1 Warner and Fire Atomism

In British Library MS 4395 – almost certainly written in the 1620s – there are about twenty folios dedicated to atomistic or corpuscularian explanations of heat, fire, and humidity or "aquosity," which makes extensive use of chymical examples and shows that Warner conceived of chymical processes in atomistic and mechanical terms.[18] Warner begins these notes on heat with the question of its material causation. "Seing it is a manifest phenomene that heat passively taken is an alteration caused in the patient by the operation of some agent that is actually hot," Warner argues,

> it may be doubted whether it be an alteration only formall w[i]th out the transfiguratio[n] or participation of any thing materiall from the agent to the patient [...] or whether it be effected by the ingression of some materiall spirit that passeth from the agent and is [...] insinuated and receved into the body of the patient.[19]

In deference to the prevailing Aristotelian distinction between the physics of the celestial and sublunar realms, Warner confines his investigation to a consideration of "terrestrial or elemental" and not "aetherial or celestial heat," but his explanation of sublunar phenomena proceeds in a very un-Aristotelian fashion. He begins with a definition. "Heat," he says, "in so far as it is in the patient is nothing other than the alteration of its parts by the fiery spirit emitted by

17 On Sennert, see Newman, "Corpuscular Alchemy"; Clericuzio, *Elements*, 23–33; Michael, "Sennert's Sea Change" and Banchetti-Robino, *Chemical Philosophy*, 26–33. On Duclo, see Principe, "Diversity in Alchemy." Warner's reading notes from Sennert's *De chymicorum cum Aristotelicis et Galenicis consensu ac dissensu* (1619) can be found in Northamptonshire Record Office, Isham-Lamport Papers, MS IL3422, notebook VIII, fols. 11v–13r. His notes from Gaston Duclo's *Apologia chrysopoeiæ et argyropoeiæ* (1602) can be found in BL, Add. MS 4391, fols. 50r–51v (inscribed "Claueus de triplici praeparatione" and "Clauei processus"). Warner's colleague Thomas Harriot also took notes from Duclo's work (see BL, Add. MS 6788, fol. 230r).

18 BL, Add. MS 4395, fols. 49r–70v. The manuscript has a *terminus ante quem* of 1619 as can be seen from the reference on fol. 82r to a work by the mathematician Alexander Anderson "Mathematicoru[m] excercitationu[m] [...] 1619."

19 Ibid., fol. 49r.

the heat-producing body and received into the porosities of the patient."[20] He then proceeds to outline, at some length, fourteen "faculties or conditions" of this fiery spirit (*spiritus ignei*), many of which, in their arguments, draw on chymical examples.

But before we look at some of these chymical implications, we need to consider the first three conditions which briefly – albeit not unproblematically – outline the corpuscularian assumptions which underlie the rest of the manuscript. The first condition of the fiery spirit is the "subtility or minuteness of its single parts or indivisibles," which are "reduced not only to absolute simplicity but also to the utmost possible atomization."[21] The second condition, which (according to Warner) follows from the first, and from the spirit's discontinuity, is its "density" or the "brevity or exiguity of the interstices, or the propinquity of its indivisibles."[22] The third condition of the fiery spirit is "extensive power or impetuosity" which extends from itself in all directions, that is to say, spherically.[23] Considered together, Warner points out that the first two of these three conditions are the same as those of other spirits, whereas the third is a condition which it shares with spirits, but also with other fluid bodies such as air and water.[24] The corpuscularian tenor of these first three conditions is made problematic, however, by Warner's immediate qualification of the claims he is making. The ingression of the fiery spirit, he says, is "specious" (*in specie*), that is, as "a completely spiritual substance or consistency [which is] closely-packed or joined together and not [...] individually [...] or one at a time [...] not as an atom but as a spirit." This is because "no atom or particle is subtle enough to allow it to enter the pores of the altered bodies alone or one at a time." Only a substance with a "spiritual consistency" (*spirituosa consistentia*) can freely penetrate and enter the pores and vacuities of bodies and permeate

20 Ibid.: "Calor quatenus in patiente est nihil aliud est quam alteratio partiu[m] ipsius causata a spiritu igneo ab agente calefactiuo emisso et in porositates patientis recepto."

21 Ibid.: "Prima spiritus huius ignei conditio est simploru[m] seu singuloru[m] seu indiuiduoru[m] eius summa subtilitas seu parvitas, et quidem tanta ut [...] non solu[m] ad absolutam simplicitatem sed etia[m] ad extrema[m] possibilem atomizatione[m] redacta esse censeantur."

22 Ibid.: "Secunda [conditio] quae ex prima et discontinuitatis causa sequuntur est <densitas hoc est> interstitioru[m] brevitas <vel exiguitas> seu [...] ipsoru[m] individuorum propinquitas."

23 Ibid.: "Tertia huius spiritus conditio [...] quae etiam ex discontinuitatis causa consequitur est potentia <seu impetuositas> extensiua seu seipsu[m] omniquaque extendendi, vnde omnis ei actiuitas."

24 Ibid.: "Duae primae conditiones sunt propriae huius respectu alioru[m] spirituu[m], tertia ei communis est cu[m] aliiq[ue] non [...] solu[m] spiritibus vt aere et sequi sunt alij sed [...] suo etiam modo cu[m] continuis liquidis omnibus."

it.[25] This is problematic because Warner is speaking of *spiritus* – an extremely fine and subtle body, as understood in the medical tradition; corporeal rather than immaterial – and he has already characterised the fiery spirit as "reduced to the utmost possible atomization."[26]

This is the cue for Warner's first chymical example. "The extensive power of the spirit," he writes, "is founded on its coordination or consistency, which conserves its power and keeps it whole." This power would be lost if it worked by means of individual atoms – no matter how fine or gross – and no effect or alteration would follow in the inner parts of the patient; only "some superficial mixtion or corrosion" would occur, caused by the impact of the particles which were first emitted by the agent becoming lodged in the superficial pores and preventing other particles from entering. "We can see an example of this," Warner continues, "in the glass vessels used in distilling the spirit of salt nitre and the spirit of the salt of Orion and other similar things," which suffer "a corrupt mixture and impaction of this kind."[27]

The fourth condition (or "faculty") of the fiery spirit is the "faculty of transcursion," that is, its continually moving across the surface of the body. "This faculty of transcursion," says Warner, "is sphericall or *omniquaq[ue]*

25 Ibid., fol. 49r–v: "Primae conditionis necessitas est ad penetrationem seu ingressione[m] primae cu[m] secunda ad ingressionem in specie hoc est in tota substantia seu consistentia <spirituosa> seu confertim vel aggregatim et non [...] <in indiuiduo> [...] <vel> singulatim. [...] non vt atomus, sed vt spiritus. Non eni[m] sufficit tanta particularu[m] seu atomoru[m] subtilitas qua possint solitaria seu singulatim poros corporu[m] alterandoru[m] ingredi sed tantula[m] esse oportet vt vel densissimoru[m] in natura corporu[m] poros seu inanitates [...] in spirituosa consistentia et catervatim [...] <libere> penetrare et ingredi possi[n]t. et non tantu[m] penetrare et ingredi sed et permeare et transgredi."

26 On the medical conception of *spiritus* in the Middle Ages and the Renaissance, see Bono, "Medical Spirits"; Debus, "Chemistry and the Quest for a Material Spirit of Life"; Walker, "Medical 'Spirits' and God," and idem, "Medical Spirits in Philosophy and Theology."

27 BL, Add. MS 4395, fol. 49v: "Nam extensiua spiritus potentia hoc est tota eius actiuitas et operatio fundatur in coordinatione seu consistentia spirituosa qua salva et integra manente vim suam exercet destructa vero hac, perit illa concessa igitur <solitaria tantum> atomoru[m] propter nimia[m] eoru[m] crassitiem [...] vel raritatem [...] ingressione, nullus inde effectus seu alteratio calorifica sequeretur in interiore patientis substantia sed tantu[m] mistio <seu corrosio> quaeda[m] superficialis, particulis quae prius ab agente emittuntur, impactis et haerentibus in poris superficialibus et viam reliquis quae sequu[n]tur praecludentibus cuius exemplu[m] videmus in vasis vitreis quibus perficitur destillatio spiritus halonitri, et salis orionij et similiu[m] [...] mistione et impactione <huiusmodi> corruptis." The "spirit of salt nitre" (*spiritus halonitri*) was prepared by distilling purified salt in vinegar and was used as a diuretic. The "spirit of the salt of Orion" was a medication prepared from distilled urine used in the treatment of kidney stones and gout. See Libavius, *Syntagma*, 442–443.

[omnidirectional]," that is, it is a "continuation of the originall extension" of the spirit. This "sphere of actiuity of the igneous spirit" extends "until the cessation of the motion of the indivisibles from the origin of the fire [...] or until they become so rarefied that by this rarity or singularization they have lost all power."[28] Warner seems, then, to conceive of "spiritual consistency" as an unbroken "stream" of "indivisibles" or "particles." This is confirmed later in the manuscript, when he writes that when the "igneous spirit" is resolved "into his singles," the change is to be understood as "virtuall and not materiall for the singles that were before in the state of spirituality remaine the same <vnaltred> only [...] by distraction or separation from the sphere of spirituality singularized."[29] In any case, Warner states that "Singularity [i.e., particles in a singularized state] is the destruction of spirituality."[30]

Warner's quibbling about whether fire acts by means of individual atoms or particles in a spiritualized state is typical of the corpuscularian equivocations found throughout these notes. He refers to the particles of fire variously (in a mixture of Latin and in English) as "particles (*particulae*)," "atoms" (*atomi*), "minims" (*minima*), "singles" (*singula*), "indivisibles" (*individua*), and "simples" (*simpla*). He thereby blurs – consciously or unconsciously – the boundaries between the *minima* of the scholastics and the atom *per se*.[31] Allowances must be made for the fact that these are, of course, private manuscript notes, where internal consistency is perhaps not at such a premium, and Warner, moreover, frequently signals unresolved matters in his papers which are "further to be considered." Having insisted on the spiritual consistency in his qualification to the second condition, for example, he leaves open a further possibility, i.e., that the particles begin acting individually and, in the process, become aggregated:

> Although *densitas seu constipatio singuloru[m] vel individuoru[m]* be set downe before for the second condition yet [...] that is further to be considered of; for although their rarity be such as they can not enter the pores of solid obiects *in forma et consistentia spirituosa <vel> turmatim* but only *singulatim* as when the intervalls [...] are gretter then the amplitude of the pores, yet it may be vnderstood that by meanes of the retardation or stay w[hi]ch they necessarily finde in their transition they may be

28 BL, Add. MS 4395, fol. 49v: "the sphere of actiuity of the igneous spirit doth extend vsq[ue] ad cessationem motus [...] individuoru[m] ab originali ignario emissoru[m] [...] vel vsque ad tanta[m] eoru[m] rarescentia[m] vt ob raritatem seu singularizationem vim omnem amiserint."

29 Ibid., fol. 57r.

30 Ibid., fol. 49v: "Singularitas eni[m] est spirituositatis destructio."

31 On Aristotelian *minima*, see van Melsen, *From Atomos to Atom*, 58–72.

> so collected or aggregated that they may *ignescere* <or *spirituascere*> or attaine *gradu[m] alique[m] ignitionis* <[*vel*] *igneitatis*>.[32]

In the course of these reflections, Warner's chymical knowledge is never far from his mind. When he comes to consider the pervasive quality (or "faculty") of fire, for example, he refers his reader to the effects of calcination:

> That transmentive faculty removes that unsalted thing from the salty spirit of fire, augmenting the weight of calcined things, and various empyreumatics of this kind […] for that fiery spirit is pure and leaves nothing ponderous or corpulent behind it, it pervades everything and adheres to nothing.[33]

The eleventh condition of the fiery spirit is the rotundity of atoms or singles (*singulorum seu atomorum rotunditas*); or rather, their exact polyhedral atoms, because, says Warner, "roundness, whether spherical, or cylindrical or conic, or a mixture of these, is not found in nature."[34] He gives four lengthy arguments for the rotundity of fire atoms, which draw on both chymical and optical experimental evidence. His first argument, however, is an inference from the supposed essence of spirituality (*spiritualitatis essentia*): spiritual consistency requires an "omnimutual" reflection and reception between the singles or atoms of radiation, which necessarily supposes their rotundity.[35] Then, by this omnimutual passion and reaction, the singles – which are in a state of discontinuity or singularity – are protected from compression or from the impingement of surrounding bodies.[36] The third argument for the rotundity of fire atoms is derived from fire's penetrative or "transmentive" faculty. The atoms must be able to "enter into or penetrate the pores of continuous bodies, whether liquid or stable without impaction or adhesion." This necessarily

---

32 BL, Add. MS 4395, fols. 49v–50r.

33 Ibid., fol. 50r: "Facultas ista transmentiva tollit insulsu[m] illud de spiritu ignis salso, calcinatoru[m] pondus augmentante et varia huiusmodi empyreumatica […] spiritus eni[m] iste pyricis purus est nec quicqua[m] corpulenti aut ponderosi post se reliquit, omnia pervadit, nullibi haeret." "Empyreumatics" are things which smell or taste of charred organic matter (see *OED*, s.v. "empyreumatic").

34 Ibid., fol. 51v: "Vndecima spiritus ignei conditio est singuloru[m] seu atomoru[m] rotunditas seu […] polyhedria exacta eni[m] rotunditas siue sphaerica siue cylindrica siue conica siue ex his mixta in naturae non datur."

35 The *OED* has no entry for the word "omnimutual," which would appear to be Warner's own coinage. It seems to suggest a reciprocal action and passion of radiated atoms.

36 BL, Add. MS 4395, fol. 51v: "Nam ex omnimutua ista singuloru[m] inter se […] passione et reactione defenduntur in statu singularitatis seu discontinuitatis contra compressionem […] corporu[m] siue corporu[m] circumstantu[m]."

requires the atoms not only to be extremely subtle but also that they be round or at least polyhedral.[37] The reason for this is that the polyhedral atom is "more prone to roll and less apt to be checked or retarded when it meets or strikes the curved side of the pores, but either by repeated reflections or tortuous rolling finds its way to the exit."[38]

In the fourth argument for the rotundity of atoms – the incoaguability of 'ignible' matter – Warner draws on evidence from his chymical experiments:

> it has been observed in both common and chymical experience that oily, bituminous, fatty or unctuous liquids and similar things either by cold or by the evaporation of humidity become dense or even thicken, and a number of them coagulate or even crystallise (for this is what we understand by coagulation here) only rarely or with difficulty and if they are pure, not at all (sulphur, however, because of the heterogeneous parts which it has in it, yields many kinds of salt crystals by artificial manipulation). Not only are they found by experience not to crystallise by themselves, but also if they are mixed, either by art or by nature, with other things which are themselves coagulable, coagulation or crystallisation is found to be impeded.[39]

Warner also draws on his knowledge of optics to insist upon the rotundity of fire atoms. Fire is diaphanous, he argues, and all diaphanous bodies must be composed of round particles:

---

37 Ibid., fol. 52r: "Tertiu[m] est spiritus ignei facultas <penetratiua seu> transmentiua, qua scilicet poros corporu[m] continuorum tam liquidoru[m] quam stabiliu[m] absq[ue] haesione seu impactione [...] subintrat et penetrat seu transmeat. Ad quod perficiendu[m] necessario requiritur praeter singuloru[m] subtilitate[m] seu exilitatem si non exacta eoru[m] rotunditas saltem summa polyhedria."

38 Ibid.: "magis volubile est et minus aptu[m] vt sistatur aut omnino <retardetur> cu[m] in pororu[m] anfractus incidat seu impingat [...] sed vel iteratis reflexionibus vel provolutione tortuosa viam sibi ad exitu[m] invenit [...]."

39 Ibid., fol. 52r–v: "Quartu[m] rotunditatis argumentu[m] est marteriae [*sic* = materiae] ignibilis incoagulabilitas experientia eni[m] tu[m] communi tu[m] chymica observatu[m] est liquida oleosa, bituminosa, pingua seu vnctuosa et similia siue frigore siue humiditatis evaporatione densescere seu crassescere quidem aliquando eoru[m] nonnulla coagulari autem seu crystallescere (sic eni[m] coagulatione[m] hoc loco intelligi volumus) raro et difficulter et si mera sint non omnino (sulphur eni[m] propter heterogeneas partes quas in se habet no[n] paucas salini generis certo eius tractandi artificio crystallos reddit). Nec tantu[m] per se si mera sint presertim si salsulginis omnis sint expertia non crystallescunt, sed et aliorum quibus vel arte vel natura [...] admixta fuerint per se coagulabiliu[m] coagulationem seu crystallizationem impedire <experentia item> compertu[m] est."

> The diaphanous or transpicuous nature of bodies arises not only from their superficial equality, whether by nature or by artificial polishing, but also from the orderly and equable internal and formal composition of singles which is easily demonstrated from the elements of optics; since only round atoms or equally rotund polyhedra are capable of this kind of orderly and equable composition, it follows that the singles of diaphanous bodies are round.[40]

The truth of this argument, he claims, can be deduced from the law of refraction by means of optical principles.[41] Warner and his colleague Thomas Harriot had both engaged in a variety of optical experiments in the 1590s and early 1600s, including practical work on refraction.[42] Harriot, in fact, had established the Snell law of refraction twenty-four years before his Dutch contemporary Willebrord Snel van Royen (1580–1626), and Warner's tables relating to his own experiments on optical refraction were published posthumously by Marin Mersenne in 1644.[43] In some cryptic remarks in his correspondence with Johannes Kepler, Harriot had hinted that there was an atomistic explanation for the phenomenon of optical refraction, and although Harriot himself never explicitly articulated what this explanation was, these remarks in Warner's papers perhaps indicate the lines along which he was thinking.[44] Harriot and Warner's colleague Nathaniel Torporley had also been party to

40 Ibid., fol. 53r: "corporu[m] diaphanitas seu transpicuitas non tantu[m] ex aequabilitate superficiali siue natura id fiat siue artificiali politura sed etiam ex ordinata et aequabili singuloru[m] compositione interna et formali <necessario oriat[u]r quod ex elementis opticis facile est demonstratum>; sola autem atoma rotunda vel rotundis aequipollentia polyhedra huiusmodi ordinata et aequabilis compositionis capacia sunt; sequitur corporu[m] diaphanoru[m] singula rotunda esse." Cf. also ibid., fol. 71v on the atomic consistency of diaphanous bodies.

41 Ibid., fol. 53r–v: "Argumenta[m][…] efficacissimu[m] est, ipsa ratio refractionis ex principijs dioptricis deducta."

42 On Harriot's optics, see Lohne, "The Tycho Brahe of Optics" and Goulding, "Thomas Harriot's Optics."

43 On Harriot's anticipation of Snell's law, see Shirley, *Thomas Harriot*, 381 and idem, "An Early Experimental Determination of Snell's Law." For the publication of Warner's tables relating to optical refraction, see Mersenne, *Universae geometriae mixtaeque mathematicae synopsis*, 549–566. ("Problema ad tabulas refractionum. Ex obseruatis construendas, sequenti processu apodicitico soluendum"). Warner also wrote a treatise on optics, *De loco imaginis*, which was discussed by Charles Cavendish, Robert Payne and Thomas Hobbes in the 1630s; see letters between Warner and Cavendish in BL, Add. MS 4407, fol. 186, and Add. MS 4279, fols. 182 and 307.

44 On Harriot's correspondence with Kepler, see Shirley, *Thomas Harriot*, 385–388. On the relationship between Harriot's optics and his atomistic thinking, see Goulding, "*Chymicorum in more*."

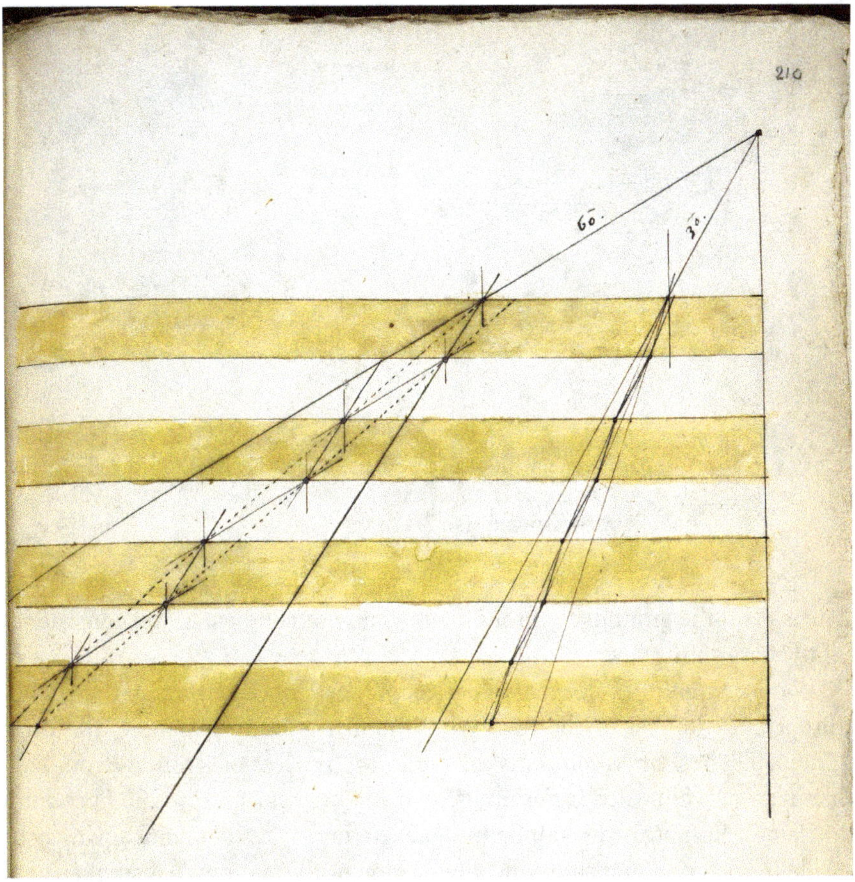

FIGURE 10.2 Refraction through a medium as a series of internal reflexions. Thomas Harriot, British Library, Add. MS 6789, f. 210r. CC0 1.0 Universal
COURTESY OF THE BRITISH LIBRARY

discussions of these atomistic theories, although he vehemently opposed them. In a short manuscript treatise entitled "A Synopsis of the controversie of Atoms," Torporley prepared counter-arguments against Harriot's atomistic theories, including his explanation of refraction as the result of the internal reflection of the rays of light off the spherical surfaces of the atoms in the medium (see Figs. 10.2 and 10.3).[45] Warner finally rejects these initial ideas about the rotundity of fire atoms and opts instead for an oxygonal or acute-angled

45 Torporley, "A Synopsis of the controversie of Atoms." BL, Add. MS 4458, fols. 6–8 (fol. 6r).

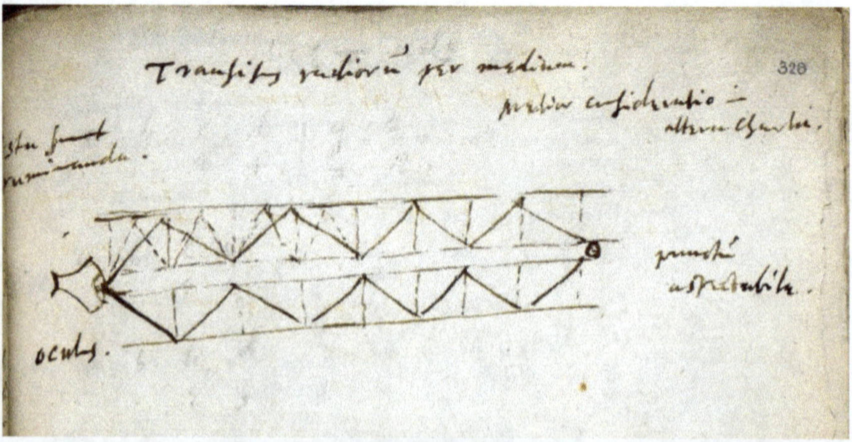

FIGURE 10.3 "The passage of rays through a medium." Thomas Harriot, British Library, Add. MS 6789, f. 328r. CC0 1.0 Universal
COURTESY OF THE BRITISH LIBRARY

particle, on the grounds that round atoms do not conduce to the consistency of the fiery spirit.[46]

Warner continues to make occasional use of experimental evidence throughout the rest of the text. When he discusses, for example, the three "graduall states" or "resolutions" of fire in the process of ignition (from ignible matter to fumosity, from fumosity to the fiery spirit, from the fiery spirit to flame), he notes the "reinflammable" nature of candle smoke, which he explains by a partial resolution, where some of the "minims" of smoke fail to become spiritualized and are expelled along with the "heterogeneities" (that is, non-atomic parts of the substance), and "do remaine commixed w[i]th them"

46 BL, Add. MS 4395, fols. 53v–54r: "Wherefore, notwithstanding those things which have been argued to the contrary above, it is manifest that the incoagulability of ignible matter cannot be explained by any other argument that that of the oxygoneity or non-rotundity of the ignible atoms with which it is overflowing. Whence it also follows that ignible atoms – the atoms themselves being unangular or round – are not apt to constitute the fiery spirit, but rather bits and pieces of atoms, acute-angled, pyramidal or wedge-shaped atoms, and for that reason we must posit not the rotundity, but the oxygoneity of singles as the eleventh condition of the spirit of fire." ("Quare non obstantibus ijs quae superius in contrariu[m] disputata sint ex his manifestu[m] est materiae ignibilis incoagulabilitate[m] alia ratione salvari non posse quam atomoru[m] ignibiliu[m] quibus abundat [...] oxygoneitate siue non rotunditate. Vnde etiam sequitur atoma ignibilia siue ad spiritu[m] igneu[m] constituendu[m] apta non esse, ipsa atoma deangulata <siue rotunda> sed atomoru[m] quisquilias seu deangulamenta oxygona pyramidalia vel cuneata atque <id circo> non rotunditatem singuloru[m] sed oxygoneitatem pro vndecima [...] spiritus ignei conditione ponendam esse.")

whilst still able to proceed to the next resolution.[47] He then goes on to posit a similar commixture in flammable liquids, where "ignible atoms" are combined with liquid heterogeneities.[48] "That the heterogeneities of the spirit of wine and such like are merely aqueous," says Warner,

> apperes by destilling the same from the salt of tartar but to [...] make the experiment more accommodate to this purpose some artifice may be devised for destilling it by incension or *per campana[m]* yf it be possible ether *per se* conducting <it> by files or by commixing it w[i]th some thing appropriate.[49]

What is particularly interesting here is that Warner is beginning to formulate a notion of proof by experiment. Although he begins with a known chymical experience or phenomenon – the "rectification" of "spirits of wine" (i.e., aqueous ethanol) by distillation with "salt of tartar" (i.e., potassium carbonate) – he shifts toward the possibility of contriving a particular experiment designed expressly for the proof of a particular natural philosophical proposition (an "experiment *more accommodate to this purpose*"). This marks a departure from Warner's reliance on an Aristotelian mode of reasoning and a movement towards a genuinely experimental mode of philosophising.

In another passage where Warner discusses the "evaporatio[n] or dissipation" of "aqueous humidity" as a necessary part of inflammation, he begins with a "common experience" – that of the "charcoling of wood," where water is dissipated whilst leaving behind the flammable "oleous or ignible humidity" – and then moves on to a chymical consideration of the same phenomenon:

> In the evaporation or dissipation of the aqueous humidity it is to be vnderstood that there flies away w[i]th it the volatile salt yf any there be ether covnited w[i]th it <and> inherent in it or adherent to the terreous [...] elements of the body, that there is in wood apperes by the making of salarmonack of the fuliginosities thereof [...].[50]

---

47 BL, Add. MS 4395, fol. 57r: "there may be [...] accounted three <graduall> states [...] w[hi]ch do proceed from three successiue and graduall resolutions, that first of fumosity from the resolution of the matter ignible *in minima speciei* [...] the second of igneity from the resolution of fumosity *in spiritu[m] igneu[m] hoc est in ipsu[m] ignem siue flamma[m]*." The other quotes are from ibid., fol. 57v.

48 For "ignible" (derived from the Neo-Latin adjective *ignibilis*), see *OED*, s.v.: "capable of ignition; ignitable. Obsolete. rare."

49 BL, Add. MS 4395, fols. 57v–58r.

50 Ibid., fol. 64v.

This chymical observation of the saline properties of potash (which was used to produce potassium permanganate) leads on to further philosophical conjecture about "what other volatilities there are or of what other <kindes> there can be besides these in combustible substances ether of vegetalls or other [things]."[51] The potential here for the reciprocal dependency of chymical experimentation and natural philosophical speculation is an early anticipation of the kinds of reflection prompted by fire analysis in the work of Robert Boyle.[52] Like Boyle, Warner's chymical reflections led him toward corpuscular explanations. He wonders, for instance, whether (given the example of sal ammoniac) "there be not allwais <ether> some aqueous vapour or some saline volatility commixt [...] <with> the oleous and ignible parts" acting as an "instrument of their spiritualization." This, in turn, leads him to consider "whether there be not some difference betwene the atomes of the fire incorporate and those of the flame ether *genere vel accidente*."[53]

Not only is Warner inclined toward atomistic explanations in his considerations of fire, but he also advances explicitly mechanical explanations for its violent motions. In the 1610s, Warner and Harriot had both worked on the mechanics of colliding bodies. Inspired by the example of the Swiss natural philosopher and mathematician Michel Varro, and his concept of *linea nutus*, both men developed elaborate mathematical theories of the collisions of round bodies.[54]

That these mechanical theories were not intended to be merely abstract mathematical exercises can be seen from the Warner papers here under discussion. When considering the "absolute propulsion or evacuation" of air in the process of ignition, Warner invokes a purely mechanical explanation based on the "materiall incussion and shok of the singles ... vpon those of the aire." Although the "firy singles" are said to be "far inferior to the airy in magnitude," he says, this inferiority is "recompensed [...] <by> the velocity, density and continuatio[n] [...] or continuall incidence of the firy singles vpon those of the aire."[55]

The "force or impetuosity" of the fire atoms proceeds, Warner says, "from the spherically <actiue or> extensiue force of spirituality."[56] This "sphere of actiuity" (or *sphaera actiuitatis*) is conceived by Warner in purely mechanical

---

51 Ibid.

52 See Eaton, *Boyle on Fire*; Holmes, "Analysis by Fire", and Debus, "Fire Analysis."

53 BL, Add. MS 4395, fol. 64v.

54 On the influence of Michel Varro on the mechanics of Warner and Harriot, see Clucas, "Thomas Harriot and Walter Warner on Collisions."

55 BL, Add. MS 4395, fol. 55v.

56 Ibid., fol. 58r.

terms. The "sphere of actiuity," he says, "doth necessarily consist of three graduall differences or states." The first of these is the original point of generation, where the ignible matter is "<first> resolved," conceived as the centre of a sphere (albeit "not punctuall but spacious"). The second state (which takes place in the "next or contiguous space" to the centre) is a state of spirituality or fluxion. This is produced by a "blow or nutus or impetus of the first [generation]" (*plagam seu nutum <seu impetus primi vel> principiantis*).[57] Warner then goes on to speculate on the location of the "firm and unmoved basis" which is "a thing of necessity in all principiated and violent motions."[58] This concept of a *nutus* by which a moving body can "impress or communicate its force" (*vim suam imprimere seu communicare possit*) is a direct product of Warner's mechanical reflections inspired by Varro. Despite confessing the "difficulty yf not impossibility of [...] sphericall indifferency of all circumstances" in nature, Warner still opts for the idea of the sphere's centre being the "base or firmament of the motion" and declares that the ideal state of fire is spherical and explains its departure from this form as "accidentall [...] deficiencies from the naturall" caused "by impediments and by material or locational anomalies" (*impedimentis et anomalijs materialibus et localibus*) – a position which was also adopted by Francis Bacon in one of the experiments recorded in his posthumously published *Sylva Sylvarum*.[59] Warner's reflections on heat and fire, then, are shaped by both his experience in chymistry and his engagement with late sixteenth-century mechanics, and both of these influences led him towards corpuscular interpretations of the phenomena.

2 Chymical Compounds and Atoms

Like Sennert, who sought to explain the chymical composition of substances by reference to different grades of compound atoms (or "molecules"), which are different in character from the simplest or "elemental" atoms, Warner also has recourse to theories of atomic composition when discussing chymical compounds.[60] In his discussion of the seventh condition of the fiery spirit ("continuous generation"), Warner raises the question of how fire is generated in the first instance, and identifies what he calls the "original subject of ignition"

57 Ibid., fol. 55v.
58 Ibid.
59 Ibid., fol. 56r. Cf. Bacon, *Sylva Sylvarum*, 10–11 (I. 31, "Experiment Solitary touching the Secret Nature of Flame").
60 See Michael, "Sennert's Sea Change," 351–354.

or "ignible matter" of substances. The ignible parts of matter are, he says, "oily matter, either fatty or bituminous in kind, although not sulphurous, properly speaking, since sulphur is a compound of some kind and may contain in itself besides bituminous parts, also aluminous salts or other more acrid salts."[61] This emphasis on the internal complexity of chymical bodies such as sulphur is explained by his distinction between "prime elements" which he calls "simples," and the "secondary elements" which he calls "minims."[62] Substances of a "simple elementation" are composed "of pure atoms" (*ex meris atomis*), but "all [...] continuate bodies in nature" are said to be of a "double elementation," that is, the "simples" are "combined with the atoms of some other kinde ether [aqueous], ayry, or saline or terrene."[63] Thus, for example, an "ignible liquor" is said to have a "double elementation," on account of the fact that if it "consisted of pure atoms without any commixture of heterogeneous parts" (*ex meris atomis ignibilibus absque commixtione heterogenorum*), it would ignite all at once, whereas, in fact, it ignites gradually.

Warner (in a classic piece of late-scholastic eclecticism) invokes Aristotle's dictum about the instantaneous nature of generation, but insists upon the unimaginability of generation without an interceding alteration, to argue for the necessity of a double resolution conceived in atomistic terms.[64] Thus, there is first a resolution of the liquid "into minimal species in the form of a fumous consistency," and then into the "spiritual consistency" of fire, "where the heterogeneall eleme[n]ts are quite separated" from the igneous singles. These atomic resolutions are meanwhile conceived within an Aristotelian teleological framework, for they are seen as alterations which are "nothing but a graduall approximation to the future forme the perfection whereof <being> acted

61 BL, Add. MS 4395, fol. 50v: "Subiectu[m] originale ignariu[m] seu materia ignibilis, est materia oleosa, seu pinguis seu bituminosa in genere, sulphurea non adeo proprie dici potest cum sulphur sit compositu[m] quiddam contineatq[ue] in se preter bitumineas partes etiam salsugineas alumineas nempe et alterius cuiusdam salis acrioris."

62 Ibid., fol. 63r. On this distinction, see Prins, *Walter Warner*, 22.

63 BL, Add. MS 4395, fol. 59v.

64 Ibid.: "For, even if we accept the old dictum that the generation of all things happens in an instant, that is granted for generation pure and simple, if, however, it is understood correctly it is neither possible nor imaginable that the generation of a determined thing from a determined thing, or the corruption of a determinate thing into a determinate thing can happen without an interceding alteration." ("Nam licet vetus dictu[m] sit, generatione[m] omne[m] in instante fieri quod quidem de generatione simpliciter accepta conceditur, si sano modo intelligatur, tamen generationem determinati ex determinato vel determinati corruptione[m] in determinatu[m] absque intercedente alteratione fieri nec possibile <est> nec imaginabile.") Cf. Aristotle, *Physics* VIII.8. On generation, in Aquinas and other thirteenth-century Aristotelian commentators, as an instantaneous change between contrary states, see Knuuttila "Change and Contradiction," 28.

in the very finall instant of the alteration," which is the point at which one substance can be said to have been destroyed and another generated. While it may be "rightly said," therefore, that generation happens in an instant, this is to be understood to refer to "the whole procese of the alteration or motion from the beginning to the end w[hi]ch [...] can not be conceved to be but *in tempore*, so as the difference is but only verball."[65] Warner thus equivocates, seeing generation either as a temporal series of atomic resolutions or as an "alteration" or "generation" in the Aristotelian sense.

The "commixtion" of simple and compound atomic forms is also used to explain why a candle flame burns with different colours. According to Warner, this is due to the difference between the behaviour of "igneous singles" in "mere flame" and "fire incorporate," where the igneous singles are "enclosed w[i]th the porosities and betweene the straits and commissures of the grosser corporeall elements." The "igneous singles" or "specificall minims" of which fire consists, are "violently evacuated" from their spiritual consistence by the "continuall penetration of the solid substance" being consumed, which "cut[s] and divide[s] the continuity" of the spirit. These "singularized" minims are distributed "into severall spaces or porosities" of the substance being burned, causing the "intermixtion or interpolation" of the igneous singles "w[i]th the prime elements or simples or w[i]th the secondary elements or minims of the body continent," and it is this "intermixtion" that causes the incorporated fire to vary in colour. The whiter the "elements" of the burning body, the "yellower or brighter" will be the flame; the less white they are, the redder will be the flame.[66]

Warner places great emphasis on *contact* as a principle of the process of inflammation, and explains, for example, the "tardity of resolution" – i.e., the gradual inflammation of a piece of ignible material – as the product of atomic "interpolation" – the "distance between the fired part and the vnfired being edged vp between the grosse atoms of the body or by adhering vnto them fixed [...] and thereby the prompt disposition thereof to resolution hindred."[67] Working backwards from the phenomenon of the "calcination or incineration of [...] combustible substances," he deduces that atomic commixture is also a principle of consolidation, at least in substances which are ignible. "By this effect of calcination <or incineration>," says Warner,

65 BL, Add. MS 4395, fol. 59v–60r.

66 Ibid., fol. 63r.

67 Ibid., fol. 63r–v.

> it doth appere that the consolidation or continuity and coherence of these combustible substances doth arise from and depend on the composition or commixtion of the volatile atoms or elements w[i]th the terreous and fixed.[68]

This leads into a lengthy discussion of consolidation, a subject on which Warner made notes when reading chapter XI of Sennert's *De chymicorum consensu et dissensu*, where both inflammability and concretion and coagulation are discussed as "principles of the chymists" ("De principiis chymicorum").[69] Consolidation, in fact, requires the existence of composite particles (that is "the compound fragments of bodies compounded"), because "the prime elements or simple atoms are to be vnderstood [to be] of that degree of parvity or subtility as [...] is much vnder the prime mesure of consolidability, that is that the spaces or intervalls of their accumulation are exclusive of the aire."[70] The incoherence of pulverous bodies which are subject to consolidation (like cement or plaster of Paris) suggests that air is interposed "in the spaces and intervalls that are betwene the corporeall particles" which, Warner argues, are "not to be vnderstood" to be "mere atoms or prime elements," but rather as "elementated, composite, specious minima, or greater than minima."[71] If they were "prime elements" and thus "reduced to their absolute atomicall simplicity," he explains, they would be so closely packed that all air would be excluded, and they would already be a perfect solid.[72] He goes on to suggest that the exclusion of air is itself a universal cause of the coherence and incoherence of bodies: "the [...] vertue seperatiue universall being the interposition of aire: and the cause of consolidation besides the prime vertue compressive or confirmative universall being the exclusion of aire."[73] Warner also suggests that different kinds of solutions in water can be explained by differing degrees of atomization – those which are transparent suggest an "atomicall" resolution, while those which are opaque and non-transparent are said to be the result of "no perfect atomization" (or what modern chemists would refer to as 'colloidal' suspension).[74]

68 Ibid., fol. 65r.

69 See Isham-Lamport, IL 3422, notebook VIII, fol. 13r, and see Sennert, *De chymicorum*, 297–298 (on inflammation) and 301–307 (on concretion).

70 BL, Add. MS 4395, fol. 68r.

71 Ibid., fol. 65v: "elementata seu composita siue minima specialia siue plus qua[m] minima."

72 Ibid.

73 Ibid., fol. 66r.

74 Ibid., fol. 68r.

In these papers, then, Warner relies on a number of observations drawn from his chymical experience to advance a corpuscularian account of fire and the consolidation of certain substances, which is, however, coloured by certain late-scholastic natural philosophical themes.[75]

3 Medical Spirits, Mechanism, and Intellection

These kinds of atomistic considerations can be found not only in those papers where Warner explicitly treats the atomic constitution of matter and the role played by "vis radiatiua" (seen as a kind of cohesive force acting on the inert parts of matter), but also (and more surprisingly) in Warner's papers on intellection, volition, and physiological processes. In the first forty-nine folios of Add. MS 4395, there is a group of notes, for example, concerning the "loco-motiue faculty," where Warner discusses the interaction between different faculties of the soul in mechanical terms – as the physical interaction of spirits of different kinds.[76] Warner treats spirits as "a substance fluid" which can exert force and exhibits the physical properties of a fluid. Thus, he describes the "nerveous parts" of the "main lyms or organs" of the body as being "animated w[i]th spirits motory." This movement of the nerves is preceded by "a change and alteration" of the motory spirits "by way of extension or dilation [...] or by influxion and augmentation." These alterations in the motory spirits are said, in turn, to be caused by "some previous alteration of the spirits cognoscitiue on w[hi]ch they depend and to w[hi]ch they are continuate." The cogniscitive spirits are altered "by way of extension or ampliation or dilatation" either by "compression or constriction" or "some kind of ince[n]sion or inflam[m]ation."[77] This last detail suggests that Warner conceived of the process of cognition as analogous to certain kinds of chymical process, as well as subject to the physical laws of a fluid mechanics.[78] That Warner sees cognition as obeying mechanical laws can be seen from his statement that the cognoscitive spirits have "power [...] or force" which is "communicable vnto the spirits motory."[79] Intellection, too – as

75 On Warner's eclectic Aristotelianism, see Clucas, "The 1595 *Advice*," lvii–lviii and idem, "The Infinite Variety of Formes and Magnitudes," 258–265.

76 On the development of the concept of medical spirits in the later seventeenth century, see Clericuzio, "The Internal Laboratory."

77 BL, Add. MS 4395, fol. 15r–v.

78 In his physiological writings, Warner explicitly rejects explanations which are "against physical or chymical principles" (*contra principia physica seu chimica*); see BL, Add. MS 4395, fol. 177v.

79 Ibid., fol. 15v.

distinct from cognition – is seen as involving spirits, the only difference being that "spirits intellectiue are more subtile and alterable then the sensitiue."[80] In a metaphor which is distinctly Lucretian, Warner compares the fluid substance of the intellective spirits to "water in a vessel," and like water that refuses to retain the impressions of physical objects but "returneth instantly to its former state retayning no shew or print of that impression." As such, it is "vnapte to retayn any impression" from perceptions, and yet it is capable of other "alterations formall" by means of "dilatation or extension or rarefaction and contraction or condensation or inspissation."[81] Warner's atomism and mechanism, then, was extended by him beyond the realm of matter and the elements into physiology. He considered a wide range of phenomena – such as magnetism, the transmission of infectious diseases, and the movement of odours through the air – as explicable by means of spherically-active material spirits emanating from bodies:

> there is some consideration to be had of [...] odorable and contagious <(and also magneticall)> spirits w[hi]ch though they <be> originally exhaled from some fixed and corporeall subiect yet they appere to retayne their actiuity being vagous and solute <w[i]thin their sphere> though according to their distance from their originall subiect <w[hi]ch is the center thereof> they be degraded by their confusion w[i]th the como[n] aire.[82]

There is evidence in some of Warner's papers on mechanics that he was a planning what was almost a complete mechanical philosophy, which sought to explain a wide range of physical phenomena through theories of impact. In a fragmentary work headed *De motu et quiete*, which survives in the Isham-Lamport papers, there is a tabular summation of his scheme, which was to present: "a doctrine of motion and rest universally and scientifically explaining both the causes and effects and the passions or accidents of motion and rest, whose differences arise from diverse conditions."[83]

While his work made use of a variety of late-scholastic natural philosophical concepts, it is clear from the totality of Warner's papers that, in the 1620s,

80 Ibid., fol. 22v.

81 Ibid., fol. 28v; see Lucretius, *De rerum natura*, 3.434–439.

82 BL, Add. MS 4394, fol. 183v. On Warner's use of the medieval concept of the sphere of activity (*sphaera actiuitatis*), see Clucas, "Mediaeval Concepts of Force," 3067–3068.

83 *De motu et quiete*, IL 3422, notebook IV, fol. 9r: "Doctrina de Motu et Quiete vniversaliter et scientificè explicat tu[m] causas tu[m] effectus et passiones seu accidentia motus et quietis. Quorum differentiae oriuntur ex diversis conditionibus."

he was beginning to piece together mechanical and corpuscular explanations for a wide variety of natural phenomena.[84] In his notes on natural philosophy, Aristotelian notions of alteration and generation co-exist uneasily with mechanical processes involving collisions, the exclusion of air, and the interactions of corpuscular particles of different sizes. The example of Sennert's chymical atomism was obviously an inspiration for the address made by Warner's atomistic thinking to chymical phenomena and his making use of chymical examples.

Warner has not, as yet, figured significantly in accounts of the rise of the mechanical philosophy in England, but – as I hope this chapter has shown – there are many reasons why he ought to be included in these accounts. While his extensive notes on natural philosophy were not published, and thus had no significant influence on later English atomists, the ideas themselves, I would argue, have an inherent interest in relation to the "shared knowledge" of natural philosophers experimenting with corpuscular explanations of natural phenomena in the late sixteenth and early seventeenth century.[85]

84 On Warner's blend of late-scholastic and corpuscular themes, see Clucas, "The Infinite Variety of Formes and Magnitudes," 258–263.

85 On the value of the "shared knowledge" approach in historical epistemology, see Schemmel, *The English Galileo*, vol. 1, 4–5, 232. It should, perhaps, be noted that while Warner's manuscript writings were not published, there is evidence that he circulated some of them among his contemporaries, including Robert Payne, Thomas Hobbes, John Pell and Charles Cavendish.

CHAPTER 11

Isaac Newton's Atomist Sources: The Case of Bernhard Varenius

William R. Newman

When one thinks of the sources for Isaac Newton's early matter theory, the famous names associated with the mechanical philosophy of the seventeenth century immediately spring to mind. Robert Boyle, Robert Hooke, and René Descartes all feature prominently in Newton's pre-*Principia* writings on the structure of matter and light, before he had advanced from the mechanical philosophy to his own rational mechanics. Boyle and Descartes, in particular, were instrumental in Newton's framing of his fundamental, early optical theory, as reflected in the student notebook *Certain Philosophical Questions* and in Newton's important experimental notebook Cambridge University Additional MS. 3975.[1] Newton was also immediately captivated by Hooke's 1665 *Micrographia*, as demonstrated both in his reading notes on that text and in his work on the optical phenomenon known to us today as Newton's rings. Thanks to historians of Newtonian optics and to the scholarly edition of Newton's student notebook *Certain Philosophical Questions*, we now have a solid understanding of the Cambridge savant's use of the mechanical philosophy in the 1660s and 1670s.

Newton was also mining the works of other well-known authors with atomistic leanings in his student years: Walter Charleton, Thomas Hobbes, Kenelm Digby, and Henry More all provided the budding natural philosopher with food for thought about the nature of material composition.[2] Things become far less clear, however, when we pass from optics and natural philosophy in general to the realm of chymistry. The welter of exotic names stretching from

1 McGuire and Tamny, *Certain Philosophical Questions*, 127–194 and *passim*. For Boyle's general influence on Newton's optics, see Shapiro, *The Optical Papers of Isaac Newton*, vol. 1, 4–7, and idem, *Fits, Passions, and Paroxysms*, 99–102, 120; see also Newman, *Newton the Alchemist*, ch. 6, where other sources are cited. The reader can consult Newton's experimental notebook Cambridge University Additional MS. 3975 on the *Chymistry of Isaac Newton* project website at <www.chymistry.org>.

2 This point is made by Dobbs, "Newton's Alchemy and His Theory of Matter," 512. See also McGuire and Tamny's introduction and notes to their edition of *Certain Philosophical Questions*.

 | DOI:10.1163/9789004528925_012

the *Turba philosophorum*'s "Acsubofen" and "Astanus" up to Newton's contemporary Johann de Monte Snyders presents a sort of *thesaurus obscurorum virorum* whose dicta Newton painstakingly recorded in the multiple drafts of his extensive *Index Chemicus* and elsewhere.[3] Yet even if we exclude the writings of chymical authors who devoted most of their time to chrysopoeia and if we focus rather on those who concerned themselves primarily with the nature and structure of matter, we at once encounter several names that find little if any place in the standard histories of the Scientific Revolution. One of these authors is David von der Becke, whose 1672 *Epistola ad praecellentissimum virum Joelem Langelottum* provided Newton with a corpuscular theory of the formation of ammonium chloride that would influence the English scientist's chymical laboratory practice over the course of decades. A committed follower of Johann Baptista Van Helmont, von der Becke developed the Belgian chymist's ideas about the formation of volatile and fixed alkalis to explain the composition of sal ammoniac and the formation of "volatile salt of tartar." His largely correct view that sal ammoniac consisted of an acid spirit in combination with a volatile, alkaline salt would provide Newton with a corpuscular model of composition based on elective affinities between particles, which could be transferred to a variety of compounds.[4] Very little is presently known about von der Becke's life and career, however, making him a less than ideal subject for the present chapter. I will examine instead another figure at the intersection of chymistry and natural philosophy who exercised a demonstrable influence on Newton. I will also consider the traditions upon which this author drew, in order to show that they extend back to chymistry over the *longue durée*.

1 Newton and Bernhard Varenius

The little-appreciated Newtonian source whom I have in mind is Bernhard Varenius (1622–1650), a German scholar whose career endured for just the single decade of the 1640s, and who expressed ideas that he himself termed as "atomist." Varenius is known to historians as a geographer rather than a matter-theorist or chymist. Indeed, Newton himself edited Varenius' influential

3 For Acsubofen and Astanus in the *Turba philosophorum*, see Lacaze, *Turba philosophorum*, 597. For references to these names by Newton, see National Library of Israel MS. Var. 259 in the edition of the *Chymistry of Isaac Newton* project, <http://webapp1.dlib.indiana.edu/newton/mss/norm/ALCH00063>, accessed 13 January 2022.

4 See my treatment of von der Becke and his influence on Newton, in *Newton the Alchemist*, ch. 15.

Geographia generalis in 1672 and then re-edited it in 1681, possibly for the use of his students.[5] Newton seems to have left no records that explicitly discuss his interest in Varenius' work, and historians of Newton's non-chymical pursuits have failed to pinpoint any serious interest that he may have had in the *Geographia generalis*. His major editorial contribution is generally acknowledged to have been constituted by his creation of thirty-one "schemes" or figures to accompany the Latin text.[6] While most scholars have focused on the topographic and mathematical features of Newton's editions, however, substantial evidence has recently emerged to indicate that the celebrated physicist had other reasons to be interested in Varenius' work. Traces of Varenius' influence can be found in an early manuscript by Newton kept at the Smithsonian Institution, Dibner MS. 1031B. The document consists of two short texts, both denominated by their incipits, *Humores minerales* and *Of Natures obvious laws & processes in vegetation*. One of the many striking features of these short treatises lies in the extensive reference that they make to fixed and volatile subterranean materials. Alchemical writers had long held that the principles of metals, mercury and sulfur, could exist in either fixed or volatile forms. Metals that fused only at high temperatures and refused to evaporate, such as gold, were thought to consist of the principles in their fixed form, whereas metals that liquefied at relatively low temperatures and tended to pass off when molten as fumes, such as tin, were believed to consist of unfixed, volatile principles.[7] Numerous alchemical texts, especially those influenced by the pseudonymous high-medieval *Summa perfectionis* ascribed to Geber, provided a corpuscular explanation for volatility and fixity, linking them respectively to the small or large particles respectively making up the principles. Newton is clearly aware of this longstanding tradition of "chymical atomism," though his own corpuscular theory depends on other sources as well.

As I have argued elsewhere, Newton's own theory of subterranean generation owes a significant debt to Varenius. In *Humores minerales*, Newton describes a circulatory theory according to which primordial fumes consisting of vaporized sulfur and mercury, making up a sort of *prima materia*, pervade our planet and "wander over the earth and bestow life on animals and

5 According to a later editor of Varenius, Newton edited the text for his Lucasian lectures, which he began to deliver in 1670 (see Newton, *Corr.*, vol. 2, 264 n. 1). This claim is reiterated by Warntz in his "Newton, the Newtonians, and the *Geographia Generalis Varenii*," 177.

6 There is some minor disagreement as to the extent of Newton's textual interventions in Varenius' text. See Mayhew, "From Hackwork to Classic," esp. 244–245, who criticizes Warntz for overstating the degree of Newton's involvement in the text.

7 See Newman, "Mercury and Sulphur Among the High Medieval Alchemists."

vegetables. And they make stones, salts, and so forth."[8] The ultimate sources for this theory lie primarily in the *Novum lumen chemicum* (1604) of the influential Polish courtier and alchemist Michael Sendivogius and in the work of Johann Grasseus, a German lawyer and alchemist. But Newton has fleshed out their work by putting a much greater emphasis on the volatility or fixity of the underground constituents of metals, an area in which the influence of Varenius is particularly clear.[9] Newton begins by pointing out that subterranean metals are constantly undergoing corrosion by contact with acidic waters present in the ground, and dripping downward. He then argues that there must be a way by which metals are regenerated and brought back to the surface of the earth; otherwise the upper layer of the world would soon be depleted of all metallic material. But most acidic solutions of metals, as every chymist of Newton's era knew, cannot be distilled as such. The acid will vaporize, and the metal will remain below in the form of a fixed salt. How then can the dissolved metals return to the surface of the earth, if they are not simply driven back up by subterranean heat? Newton responds that the dissolved metals, as they approach the center of the earth, encounter rising fumes of mercury and sulfur which break them apart and cause them to undergo a process of vegetation. The dissolved metals then disengage from their acidic solvent in decompounded form and rise up again to the surface, where they are literally regenerated as new metals. Hence the fixity of the dissolved metals is first overcome by the volatility of the subterranean principles and then restored upon the regeneration of the metals.

The second treatise in Dibner MS. 1031B is a fragmentary work that probably grew out of a commonplace book rather like *Certain Philosophical Questions*, but which represents a slightly later phase in the development of Newton's thought. As the title (or rather, the incipit) implies, *Of Natures obvious laws & processes in vegetation* treats the "vegetation" alluded to in *Humores minerales* as a central issue for discussion. In accordance with its Latin etymology (from *vegetatio*), Newton uses "vegetation" primarily to mean "growth" and "animation." In seventeenth-century English, the term did not have the strict association with the plant world that it typically does today. In short, the term "vegetation" covered a range of phenomena belonging to growth and development that presented serious problems for the billiard-ball world of the

8 Newton, *Humores minerales*, in Dibner MS. 1031B, fol. 6r, from the *Chymistry of Isaac Newton* project, <http://webapp1.dlib.indiana.edu/newton/mss/norm/ALCH00081>, accessed 13 January 2022.

9 Rafał Prinke has written numerous valuable articles on Sendivogius, a number of which are cited in my *Newton the Alchemist*, 66–67. For Grasseus, see Thomas Lederer, "Leben, Werk und Wirkung," and idem, "Der Kölner Kurfürst Herzog Ernst von Bayern (1554–1612)."

mechanical philosophy. Newton's primary objective in writing *Of Natures obvious laws* lay in the integration or reconciliation of mechanism with the deeper agencies that he believed to be at work in vegetation. Surprisingly, he found material in Varenius' *Geographia* that would help him toward that goal.

One of the more unusual theories propounded by Newton in *Of Natures obvious laws* is that niter, or saltpeter, and common salt represent essentially the same material though in different degrees of corpuscular "packing." Both niter and sea salt, according to Newton, are products of active, subterranean metallic fumes interacting with water, but niter has a looser, more subtle structure. When the fumes emerging from subterranean passages meet with "subtile invisible" water vapor, niter is produced, whereas sea salt originates from the combination of the volatilized metals with liquid water or mist. A preponderance of water causes the fumes to be "overwhelmed & drowned," which kills their fermentative or vegetative activity and results in the immediate formation of sea salt. Furthermore, Newton claims that seawater can never be entirely freed of its salt, as it still contains some of the volatile material left over from its formation. Let us quote Newton on this interesting point:

> Because the sea is perpetually replenished wth fresh vapours it cannot bee freed from a salin tast by destillation, that salt arising wth ye water wch is not yet ~~indurated~~ concreted to a grosser body. [...] hence in destillation there is not only a gross salt left behind but also a subtil spirit not yet coagulated ariseth with the ▽.[10]

In the work of Varenius, as we will see, Newton found both the claim that salts can compound with water while in a vaporous or "spirituous" form and the belief that their combination of volatile and fixed components makes it impossible to entirely separate the salt from seawater. Let us therefore turn to the German geographer and consider his education and background.

2 Varenius' Educational Background

The author of the *Geographia generalis*, who died quite prematurely at the age of twenty-eight, received initial academic training at Uelzen and Helmstedt in northern Germany, but then moved to Hamburg, where he matriculated at

10 Isaac Newton, *Of Natures obvious laws and processes in vegetation*, Dibner MS. 1031B, 1v, available at the *Chymistry of Isaac Newton* edition, <http://webapp1.dlib.indiana.edu/newton/mss/norm/ALCH00081>, accessed 13 January 2022.

the academic gymnasium Johanneum in 1640. In Hamburg he studied under the well-known scholar Joachim Jungius until 1643, and the two established an enduring relationship. Varenius left for Königsberg in 1643, and in 1645 we find him enrolled at the University of Leiden.[11] A number of interesting letters from Varenius to Jungius have survived, and these provide important clues about the former's matter theory. In the midst of his descriptions of contemporary Cartesian disputes at Leiden, Varenius reports, for example, that a professor of medicine named "Walleus," a man of unusual ingenuity and experiments, "relates natural science for the most part in the Syndiacritical fashion, but not in the way of Descartes, for he retains qualities."[12] The editors of Jungius' correspondence identify this "Walleus" as Jan de Wale, the famous Leiden medical professor who became an ardent supporter of Harvey's model of blood circulation.[13] But what is of most interest to us is Varenius' mentioning of the "syndiacritical" approach, an explicit reference to Jungius' signature brand of atomism, according to which semi-permanent "hypostatical parts" separate from one another and recombine in processes of analysis and synthesis (*diacrisis* and *synkrisis*), hence providing Jungius with the term *syndiacriticus*. The terms *diacrisis* and *syncrisis* had been used by the ancient atomists to describe the separation and combination of atoms, and they had been made famous by Aristotle's refutations of atomism.[14] It is worth noting that Varenius himself employs the qualitative syndiacritical approach that he attributed to "Walleus" in his *Geographia*. There we encounter *atomi* (atoms) of water, gold, common salt, vitriol, sulfur, and other substances: clearly these are not examples of pure, unqualified Cartesian *res extensa*.[15]

To changes brought about by *syndiacrisis*, Varenius' teacher Jungius contrasted the *mutatio actupotentialis* of scholastic Aristotelianism. The latter relied on the imposition and removal of form to explain fundamental

11 Margret Schuchard, "Varenius and his Family," 19. See also Lehmann, "Der Bildungsweg des jungen Varenius."

12 Bernhard Varenius to Joachim Jungius, from Leiden, 10/20 March 1646: "Dominus Cartesius paucos habet asseclas, habet tamen, sed publice eius sententia non proponitur. Geometrica ipsius opera in Latinum sermonem translata edenda sunt, nondum factum est initium. Physicam syndiacritica hypothesi magna ex parte explicat quidam medicinae doctor, Walleus, sed non modo Cartesiano, retinet enim qualitates. Est hic vir experientiae perquam deditus et ingenii non vulgaris," in Elsner and Rothkegel, *Der Briefwechsel des Joachim Jungius*, 613.

13 Schouten, "Johannes Walaeus."

14 The terms *diacrisis* and *syncrisis* acquired particular prominence in the early seventeenth-century works of Daniel Sennert. For a discussion of this, see Newman, *Atoms and Alchemy*.

15 Varenius, *Geographia generalis* [Newton, ed.]. 171, 190–193, and *passim*.

processes of change such as transmutation.[16] Jungius himself was heavily influenced by the atomism of Daniel Sennert, and even went so far as to publish an *Auctarium epitomes physicae … Danielis Sennerti* (1635), which was in large part an abridgement of Sennert's *De chymicorum cum Aristotelicis et Galenicis consensu ac dissensu* (1619 and 1629).[17] The *Auctarium* contains the very passages in which Sennert announces his allegiance to Democritean atomism, albeit an atomism to which Sennert attached Aristotelian forms and qualities, forgoing the ontological austerity of the Abderite.[18] According to Jungius' modern editor Christoph Meinel, the Hamburg pedagogue prepared the *Auctarium* specifically for the use of his students: hence it is very likely that Varenius was exposed to the syndiacritical hypothesis not only in the form in which his master conveyed it, but also in the more hylomorphic version propounded by Sennert. Varenius' positive comment about "Walleus'" use of the syndiacritical method therefore owes a manifest debt to the chymical atomism of his German predecessors.

The atomistic features of the *Geographia generalis* also received support from Descartes, however, whose work clearly impressed the young German during his stay in the Netherlands. There is a small literature devoted to the determination of Cartesian features in Varenius' work, but most of the focus up to now has been on more metaphysical than material debts.[19] Once again, Varenius' letters to Jungius help us understand his position towards Descartes; in a letter sent from Amsterdam in 1647, Varenius first states that Cartesianism has found many adherents in Leiden and Utrecht, and that they express their views more freely than before in open theses and lectures (*publicis thesibus et praelectionibus*). Nonetheless, the theologians oppose them, attacking

16 For the term *mutatio actupotentialis*, see Meinel, *Joachim Jungius, Praelectiones Physicae*, 73, 81–82, and *passim*.

17 See Meinel, *Joachim Jungius, Praelectiones Physicae*, 12 n. 18 for evidence that the anonymous *Auctarium* was written by Jungius. Sennert's influence notwithstanding, Jungius vigorously attacked his doctrine according to which forms inhere in atoms, as already noted by Emil Wohlwill, "Joachim Jungius und die Erneuerung atomistischer Lehren," 63 and *passim*. It is worth noting that the 1650 edition of the *Auctarium* (I have not seen the 1635 edition) reveals that Jungius initially relied only on the 1619 *editio princeps* of Sennert's *De chymicorum*, but that the editor came across a copy of the heavily revised 1629 edition in the late stages of publication, and added in a number of Sennert's own emendations and expansions.

18 [Joachim Jungius,] *Auctarium epitomes physicae Danielis Sennerti* (1650), 79.

19 See Richter, "Varenius – Ein Geograph zwischen allen Stühlen?" and Staffhorst, "Miszellen zur *Geographia generalis*." For a more general account of Varenius' response to Descartes, see Günther, *Varenius, passim*. Günther suggests that Varenius was planning to write a monograph on Cartesian physics; see ibid., n. 470.

both Descartes' "*Meditationes Metaphysicae* and the atom-hypothesis, since Descartes avoids the term 'atom' to no avail."[20] From this one could conclude that Varenius regarded Descartes as a crypto-atomist, and that the French philosopher sought to deflect his critics merely by substituting other terms for what were actually atoms. In a certain sense, this opinion is supported by a consultation of the *Geographia generalis* itself, which I quote here from the English translation of 1682:

> There are divers opinions of *Philosophers* concerning this matter. The *Peripateticks* number four *Elements* of the *Earth*, and the whole *sublunary World*, being now sufficiently known to the very Vulgar, *Fire*, *Air*, *Water*, and *Earth*. Many of the Ancients, as *Democritus* and *Leucippus*, determined that the whole World consisted of very little solid pieces, which differ only in their various *figures*, *shapes* and *magnitude*: and them many of the later *Philosophers* do follow; and of late *Cartesius* endeavoured by such an *hypothesis* to declare all natural appearances.[21]

Again, Varenius associates Descartes with the atomists, but his comments are revealing in another respect. It appears that for Varenius, the most salient aspect of atomism lay in its reduction of sensible phenomena to the structural characteristics of "very little solid pieces" (*ex minimis particulis solidis*) or microparticles, not in such features as the indivisibility of atoms or the existence of void between atoms. In effect, Varenius assimilates Descartes to the atomists by imposing on the sources his own selective reading of the essential features both of Cartesianism and of atomism. Nor is Varenius committed to Cartesianism as the *explanans fundamentale* of the natural world, as the continuation to the above passage shows:

> *Chymists* make three Principles, *Sal*, *Sulphur*, and *Mercury*, to whom some do rightly add *Caput mortuum* or the *Dead head*, when as the [*the*

20 Varenius to Jungius, from Amsterdam, December 14/24, 1647: "De conditione in academiis vicinis si quaeras, scito non magis Aristotelicam quam Cartesianam excoli, majoremque esse jam libertatem hanc profitentibus, quam antehac fuerat, et tam in Leydensi, quam Ultrajectina Academia publicis thesibus et praelectionibus defendi multosque adstipulatores nactam esse, etsi Theologici utriusque loci eam impugnent, imprimis Meditationes Metaphysicas et atomorum hypothesin, frustra Cartesio vocabulum atomi in scriptis suis vitante," *Briefwechsel des Joachim Jungius*, 686–687.

21 Varenius [and Sanson], *Cosmography and Geography*, 30. The Latin of this passage may be found on p. 42 of Newton's 1672 edition. Although Mayhew, "From Hackwork to Classic," views Blome's translation as inferior, it is in some places closer to Varenius' original Latin text than the later Dugdale-Shaw translation. In the present paper, I have used both translations, quoting in each case from the one that better reflects the meaning of the Latin.

> *translation has* "they" *for* "the"] three are fruitful. But to me, doubtful terms and words being laid aside, and the things themselves well considered, there seem to be five *simple Bodies* the first Principles of all things, to wit, *Water, Oyl* or *Sulphur, Salt, Earth,* and a certain Spirit which the *Chymists* call *Mercury.* For indeed all *Bodies* and the parts of the *Earth* are resolved into those five *Elementary substances.* Notwithstanding I deny not that those differ not so much in essence, as in the singular variety of their *shapes* and *magnitudes.* ¶Therefore the whole *Earth* consisteth of these *simple Bodies,* which are divers ways commixed, from whence ariseth so great variety of *Bodies,* which do appear different from one another, and similar or *Bodies* of like parts.[22]

Here Varenius affirms the existence of "five simple bodies," namely the chymical principles in the fivefold version that they had been given by the textbook tradition extending back at least to Jean Beguin and his followers in the first decades of the seventeenth century.[23] These consisted of the Paracelsian *tria prima* to which the chymical textbook tradition added two additional principles, a watery phlegm and "a sandy earth" also called *terra damnata* or *caput mortuum.* It is striking that Varenius explicitly states these five chymical principles to be the immediate "bodies" out of which "the whole *Earth* consisteth," while at the same time refusing to deny the Cartesian possibility that "those differ not so much in essence, as in the singular variety of their *shapes* and *magnitudes.*" Although this might at first seem to be the product of an uncritical eclecticism, I believe instead that it reflects the empiricist approach of Varenius' teacher Jungius, whose syndiacritical method made considerable appeal to chymical analysis as a tool for revealing the nature of matter. Even though Jungius was skeptical of the fundamental character that chymists ascribed to the three principles – mercury, sulfur, and salt – he still argued that the labors of the chymists had "brought the syndiacritical hypothesis back from oblivion to light in recent times after its being virtually buried."[24] This helps to explain Varenius' acceptance of the chymical principles, since their reality was affirmed by spagyrical analysis in the laboratory. Yet the very same

22 Varenius [and Sanson], *Cosmography and Geography*, 30. For Varenius' Latin, see Newton, 1672, 42.

23 For Beguin's introduction of phlegm and *terra damnata* or *caput mortuum* to the three Paracelsian principles – mercury, sulfur, and salt – see Beguin, *Tyrocinium chymicum.* 31–32. For the remarkable influence of Beguin, see the classic study by Patterson, "Jean Beguin."

24 Jungius, *Disputationes hamburgenses*, 396, #46: "Qui chymiae exercitijs incumbunt et Spagirici atque Hermetici philosophi cognominari avent, syndiacriticam hypothesin oblivione ferè sepultam industriâ suâ in lucem hisce seculis revocarunt."

emphasis on the experiential basis of knowledge made it difficult or impossible to determine whether the Cartesian denial of essence or the Aristotelian affirmation of multiple substances was correct; hence we encounter Varenius' careful words, which appear in his Latin original as "*Non tamen negaverim, illas ipsas non tam essentia interna, quam singulari figurarum & magnitudinum varietate differre.*" At a deeper level than can be attained by chymical analysis, matter may well be differentiated "not so much" (*non tam*) by essential variety as by difference in structure, though Varenius will neither deny nor affirm this.

3 Varenius and Thurneisser zum Thurm

Varenius' acceptance of the chymical principles as the constituent parts of our world opens up the possibility that the *Geographia generalis* may contain other points of intersection with chymistry, and an examination of the text shows this indeed to be the case. Although Varenius is parsimonious in referring to his sources, we can identify at least one chymical writer upon whom he definitely relied. The *Geographia* deals at length with mineral waters of underground origin, and also with salts found in the oceans. For information on mineral waters, in the course of these discussions Varenius frequently draws on a chymical author to whom he refers simply as "Thurnhuserus".[25] This is surely the sixteenth-century German author Leonhart Thurneisser zum Thurm, whose multifarious publishing activities ranged from the production of almanacs and herbals to works of chymical medicine.[26] It appears that Varenius had read Thurneisser's 1572 work *Pison: Das erst Theil. Von kalten/warmen minerischen und metallischen Wassern*, although the geographer never mentions the title of the book.

Thurneisser's *Pison* relies on ideas drawn both from medieval alchemy and Paracelsian *spagyria* to explain the activity of mineral waters, particularly invoking the fixed and unfixed principles that I introduced in section 1 of this chapter. He applies the division between fixed and unfixed not only to sulfur and mercury, but to the Paracelsian principle salt, and to metals and minerals more generally. In the case of mineral salt, alum, and vitriol, for example, Thurneisser distinguishes between their "spirit, soul, and subtlety" (*Geist/ Seel und subtilitet*) on the one hand, and their "corporeal salt" (*Corporalisches Salz*)

25 I have identified eight citations of "Thurnhuserus" in Newton's 1672 edition of the *Geographia generalis*. They occur on pp. 45, 159, 192 (twice), 193 (three times), and 197.

26 For some recent work on Thurneisser, see Moran, "Preserving the Cutting Edge"; see also idem, "Art and Artisanship."

on the other.[27] What is Thurneisser's evidence that common salt contains both a spiritual and a corporeal part? He appeals to the well-known fact that "spirit of salt" (our hydrochloric acid in aqueous solution) can be "extracted" from common salt by a process involving destructive distillation (usually performed with the addition of vitriol, though *Pison* is silent on this point). As Thurneisser puts it,

> Now if you wish to extract the spirits of salt (*Spiritus Salis*) from salt, you will have to use one hundred pounds of salt for 5 *loten* of spirits. Why? Because there is much salt, but little spirit therein.[28]

Thus, common salt contains both a volatile spirit and a fixed, non-volatile body. Interestingly, Thurneisser employs the same distinction between unfixed, volatile saline spirit and a fixed corporeal salt when discussing the mere boiling of saltwater (as opposed to the destructive distillation that would yield hydrochloric acid). He argues that boiling the water away drives off almost all the *subtilitet*, or spirit, of the salt and leaves behind an earthy part "as the dregs and tartar" (*als die* Feces *und Drüsen*). The spirit that is driven off by the mere boiling of saltwater is obviously not the same as the corrosive, acid spirit produced in the manufacture of hydrogen chloride gas, however. It is instead a hypothetical product, and one that enters into Thurneisser's speculations in various ways. He claims, for example, that the method of separating salt from sweet water by means of a wax container, mentioned by Aristotle in *Meteorology* II 358b34 ff., cannot separate the spirit of salt from the water that has percolated through the pores of the container, even if the product tastes sweet.[29] As we will see, Thurneisser's comments about a putative, unfixed salt contained in seawater would find a resonance also in Varenius' discussion of salt.

Thurneisser's speculations about the fixed and volatile components of minerals such as common salt extend to the main subject of *Pison*, metallic and mineral waters. In order to understand his theory of their formation, we must first consider his explanation of metallogenesis more generally. Echoing traditional alchemical theory, he argues that the metals are made from sulfur and mercury, in the following fashion. The two principles are first vaporized within the earth by subterranean heat. Thanks to repeated heating and cooling

27 Thurneisser, *Pison*, iiii.

28 Ibid., vi: "So du nu die Spiritus Salis von Salz ausziehen wilſt/ wirſtu hundert pfunde ſaltz zu loten der Spirituum brauchen müſſen/ Warumb: Es iſt vil saltz/ aber wenig Spiritus darinnen."

29 Ibid., ix.

within the earth, a thick steam containing the vaporized principles gradually collects on the stone found there. This is cooked gradually to a "hard, thick juice," which collects in seams "like smoke in the flue (*Camin*)."[30] At this point, Thurneisser reveals a knowledge of mining literature, or possibly of earlier authors who had combined the knowledge of miners with the beliefs of alchemists, such as the renowned Lutheran preacher Johann Mathesius and his less famous but equally pastoral coreligionist, Nicolaus Solea.[31] He refers to the condensed subterranean fumes as *witternus* or *brent*, and argues that when the miners encounter these deposits, they know that ores are nearby. The metallic juice gradually enters the pores within stones and hardens into something like *Gummi*. This sounds a great deal like the immature metallic material that Mathesius called *Gur* or *Guhr*, popularized in his famous 1562 collection of sermons to the miners of Joachimsthal, *Sarepta oder Bergpostill*.[32] Thurneisser's metallic juice in time hardens further, burns up the stone upon which it has been deposited, and remains as a vein of metal within it.[33]

Thurneisser also uses his theory of metallic generation to explain how mineral waters are formed beneath the earth. As he points out, ordinary water does not extract a metallic color or virtue from fully formed metals; otherwise metal cooking vessels and water pitchers left full overnight would gradually lose their *metallische krafft* and "fall to ashes."[34] The formation of a metallic water can occur, however, when the metal is still in its immature, uncoagulated form, because the metallic juice "has something in common with water on account of its own moistness."[35] The water therefore extracts the immature metal's *subtilitet* and *krafft*, and becomes a mineral or metallic water. According to Thurneisser, this mode of formation accounts for the fact that spiritual mineral waters, as he calls them, are clear and do not settle when allowed to sit. It is true, he admits, that some mineral waters contain metallic matter in corporeal form as fully formed metals, but this is only because the rushing subterranean rivers have broken small bits of the metal off as they pass by; otherwise the metal in mineral waters is purely *geistlich*.

Let us now return to Varenius in order to see how the German geographer makes use of his predecessor's description of subterranean geological

30 Ibid., xiiii–xv.

31 For Mathesius and Solea, see Newman, *Newton the Alchemist*, 73–80, where other literature is cited.

32 Ibid., 76, 156, 212, 217–220, 419, 422.

33 Thurneisser, *Pison*, xiiii–xv.

34 Ibid., iiii: "[…] letstlich zu aschen fallen."

35 Ibid., xx: "[…] so hat es noch von wegen seiner feuchtigkeit/ etwass gemeinschaft mit dem wasser."

processes. In essence, Varenius adopts Thurneisser's distinction between corporeal and spiritual mineral waters, but subdivides the former category into two further subdivisions. Being an explicit atomist, Varenius also provides a corpuscular explanation of these subcategories such as is largely absent from *Pison*:

> *Mineral Waters are of three kinds.* SOME are corporeal, (we want a better Word for it) others spirituous, and the rest both corporeal and spirituous. Those we call corporeal mineral Waters do contain fixed and solid Particles of Minerals and Fossils, which can be separated from the Water, and seen with the naked Eye and such as these are of two sorts: Some carry large Particles of Minerals and Fossils, which may be perceived with little or no trouble in the Water itself [...]. BUT corporeal mineral Waters are more properly such as indeed contain solid Particles of Fossils, but so small and minute that they are entirely mixed, and cannot quickly be distinguished by the Sight, unless they are made to subside by Art, or a long space of Time, or by Concretion are brought to a visible Mass, such as Salt and sulphureous Fountains, &c. and chymical Waters in which Metals are dissolved. SPIRITUOUS Waters are those that contain only a volatile Spirit, such as is found in Minerals; but have no fixed Particles in them; and therefore their Composition can never be made visible.[36]

As one can see, Varenius argues that metallic waters exist in three forms, and in the first two cases, at least, these forms are explicitly determined by the size of the metal particles. In the case of the grossest form of corporeal mineral waters, visible flakes or nuggets are carried in the water, the chunks are large and therefore remain unmixed: in our terms, they do not go into solution. The second type of corporeal mineral waters consists of smaller particles in solution, as in the case of "chymical Waters," by which Varenius means mainly acids in which metals have been dissolved. In the case of these chymical waters, the dissolved metal can be made to "fall down" by the addition of a precipitating agent (such as an alkali, if the solution was made with an acid). Such chymical waters also exist within the earth:

> as *Aqua Fortis* dissolves Metals into Atoms, and easily unites them with itself, so that they do not subside at the Bottom, unless they be separated by Art: in like manner when such Water runs thro' a metallic Earth, it

36 Varenius, *A Compleat System*, 361–362. For the corresponding Latin, see Varenius, *Geographia* (1672), 189.

> may dissolve the metalline Particles and unite them with itself; and thus are the corporeal mineral Waters of the second Sort, accounted for and explained.[37]

By this point, then, Varenius has nuanced Thurneisser's category of corporeal waters by creating a separate category for those that consist of metals in an acidic solution. But his third category, "spirituous" waters, comes directly out of his German predecessor's *Pison*, and in order to explain it, Varenius even recapitulates Thurneisser's theory of metallic generation:

> BEFORE Metals are formed in the Bowels of the Earth, Steams and Vapours are condensed about the extant Corners of the Rocks, to which they stick fast; being at first but of a soft Substance, though they are afterwards hardened by degrees; if therefore the Water should run or gleet thro' the Places where such Vapours are in Commotion, it is impregnated with them; and thus spirituous mineral, and metalline, Waters are produced. Imperfect Minerals also make mineral Waters of their own Nature, after another Method; viz. when, being heated by a subterraneous, or their own proper, heat, they send forth Spirits and Vapours, as Sulphur, Vitriol, Salt, Coal, &c. And such Fumes and Exhalations are always stirred up where there are such Minerals; among which the permeating Water is impregnated with that Spirit.[38]

Here Varenius faithfully reproduces Thurneisser's combination of themes drawn from alchemical and mining sources. It is not the fully formed metals that contribute to Varenius' spirituous waters, but rather the metallic material before it has hardened into a metal per se. It seems, however, that Varenius has placed greater emphasis than Thurneisser on the metallic vapors before their condensation. For him, it is the vapors themselves "in commotion," rather than the semi-solidified immature metals, that settle in the flowing subterranean water. As it says in Newton's Latin edition, "ubi tales vapores sunt & excitantur, impregnantur illis, & ita fiunt aquae minerales spirituosae metallicae."[39] For Varenius, then, the volatile metallic fumes must be in a state of excitation when they encounter the water, in order to impregnate it. As a result of their impregnation by spirits, and in contradistinction to the "corporeal mineral waters" formed by acid dissolution, the "spirituous mineral waters" are

---

37 Varenius, *A Compleat System*, 363. For the Latin, see Varenius, *Geographia* (1672), 190.

38 Varenius, *A Compleat System*, 363. For the Latin, see Varenius, *Geographia* (1672), 190.

39 Idem, *Geographia generalis* (1672), 190.

fully volatile – they do not leave behind a fixed residue when evaporated. In sum, Varenius allows for two modes by which genuine metallic waters can be generated – either by solution of the fully formed metals in subterranean acidic solutions or by direct solution of the immature metallic fumes in water. In the former case, metallic waters of a fixed nature arise, while in the latter a totally volatile solution is formed.

4 Conclusion: Newton's Use of Varenius

From the earlier synopsis of *Humores minerales* it is not difficult to see how Newton has employed Varenius' theory of corporeal and spirituous waters consisting respectively of fixed and unfixed ingredients. Like Varenius, Newton distinguishes between fully volatile metallic fumes and fixed metallic salts that are the product of dissolution in subterranean acids. Yet Newton has combined Varenius' discussion of fixed and volatile principles with the circulatory theory of metallic generation famously articulated by Sendivogius. The acidic, metallic solutions of *Humores minerales* with their fixed salts cannot regain, and thus replenish with ores, the surface of the earth unless they are broken down and rendered volatile by the immature fumes of the metals themselves. Newton, of course, sees all of this in atomistic terms, at least in the sense of the chymical atomism employed by Varenius.

If we now turn from *Humores minerales* to the other text found in Dibner MS. 1031B, *Of Natures obvious laws*, the influence of Varenius also emerges. It is highly likely that Newton's introductory lines about saline generation are loosely inspired by Varenius' discussion of sea-salt, for both Newton and the author of the *Geographia generalis* make the seemingly odd claim that while the sea is saltier in the tropics thanks to the higher volume of fresh water evaporated off by the tropical sun, seawater cannot be rid of all its salt by means of distillation.[40] Indeed, Newton's words betray the direct influence of Varenius' assertion that seawater contains both a fixed salt that is left behind in distillation and a volatile salt that evaporates with the water: "Because the sea is perpetually replenished wth fresh vapours it cannot bee freed from a salin tast by destillation, that salt arising wth ye water wch is not yet ~~indurated~~ concreted to a grosser body."[41] Newton's words surely recapitulate the salient part of the following passage from Varenius:

40 For these two claims in Varenius, see *Geographia generalis* (1672), 109 and 112.

41 Dibner MS. 1031B, 1v, available at the *Chymistry of Isaac Newton* edition, <http://webapp1.dlib.indiana.edu/newton/mss/norm/ALCH00081>.

> Those that have diligently searched into the Secrets of Nature, I mean the learned Chymists, (not those ignorant Pretenders to Chymistry) have hitherto laboured in vain to find out a Method of distilling or extracting fresh Water from the Sea-Water, which would be of great use and advantage to Navigation. And tho' both by Decoction and Distillation, which are in Effect the same, there is Salt left in the Bottom of the Vessel, yet the Water thus separated is still salt, and not fit for drinking, which seems strange to those that are ignorant of the Cause of it. This is taught by Chymistry (which is the truest Philosophy), by the help of which there are found two kinds of Salt in all Bodies, which tho' they perfectly agree in Taste, yet they exceedingly differ in other Qualities: Artists call the one fixed Salt, the other volatile. The fixed Salt, because of it's Gravity, is not evaporated by Distillation, but remains in the Bottom of the Vessel: but the volatile Salt is spirituous, and indeed nothing but a most subtile Spirit, which is easily raised with a very gentle Fire; and therefore in Distillation ascends with the sweet Water, and is well mixed with it by Reason of the subtilty of it's Particles [the Latin has *propter atomorum subtilitatem*]. This fixed and volatile Salt is found, by Chymists, to be not only in Sea-Water, but almost in all Bodies, tho' more in some than in others; in Herbs that taste sharp there is more, but in oily and insipid Things less. The Difficulty therefore lies in separating the volatile Salt, or the salt Spirit from the Water; for it is this which hath rendered all the Efforts hitherto fruitless.[42]

Varenius is employing much the same line of reasoning that he used in his discussion of volatile "spirituous metallic waters" and "corporeal metallic waters." In nature, tiny atoms of light weight are found mingled with larger, heavier ones; distillation merely separates the two types of particles by raising the smaller and leaving behind the bigger. Hence it is possible for the smaller atoms of the volatile salt to ascend while the larger, fixed ones remain behind, just as the spiritual metallic waters could be completely distilled while the corporeal waters left behind a residue upon their distillation. The same ideas linking subtlety to volatility, and grossness to fixity, pervade Newton's reasoning as well, and indeed are standard features of medieval and early modern alchemy.

It would be an easy matter to argue for a direct and evolutionary progression from the qualitative treatment of metallic generation and the formation

42 Varenius, *A Compleat System*. 216–218. The earlier English translation of this passage in Varenius [and Sanson], *Cosmography and Geography*, 79, is less satisfactory at this point; for the Latin of this passage, see Varenius, *Geographia* (1672), 111–112.

of mineral waters in Thurnessier zum Thurn, to a corpuscularism of Cartesian inspiration in Varenius, followed by the fully atomistic worldview of Newton. The attentive reader will understand, however, just how misleading such an account would be. First, Varenius was no simple acolyte of Descartes. His approving reference to the syndiacritical hypothesis in reference to "Walleus," his claim that Descartes was a crypto-atomist, his refusal to commit to Cartesian as opposed to chymical principles in the *Geographia generalis*, and his perennial use of the term "atomus" to mean any small, semi-permanent corpuscle endowed with what we would call secondary qualities, reveal the complexity of Varenius' atomism. As for the sources of his matter theory, they descend as much from Jungius' syndiacritical hypothesis and Sennert's chymical atomism – which conditioned Varenius' reworking of Thurneisser – as they do from Descartes. The case is even more complex for Newton, of course, whose encounter with atomism probably began with his *Certain Philosophical Questions* of the mid-1660s and continued to evolve through the *Principia* and the later editions of the *Opticks* until his death in 1727. Perhaps the best way to approach Newton's debt to the atomism of the chymists is to think in terms of very specific material phenomena as opposed to top-down theories. Just as modern biology has its model organisms, so early modern atomists like Newton picked specific phenomena and reactions from which they could generalize. Newton's use of Von der Becke's corpuscular explanation of sal ammoniac is one example, and his reworking of Varenius' ideas about volatility and fixity in the realm of metals, salts, and saline waters is another. Varenius' work taught Newton how a circulatory theory of metallic generation could actually be made to account for mineral depletion, and provided him with the basis for his explanation of the formation of niter and salt. If one considers that *Humores minerales* and *Of Natures obvious laws* served as the theoretical bases for over three decades of alchemical experimentation by Newton, as I have argued elsewhere, this was no mean accomplishment.[43]

43 See my *Newton the Alchemist*, chs. 7 and 8.

Bibliography

Manuscript Sources

Cremonini, Cesare: *Expositio in quatuor libros Meteorologicorum.* Padova Biblioteca Universitaria, MS 120 (1616–1617).

Harriot, Thomas: British Library, Additional MS 6788, 6789.

Severinus, Petrus: "A Mappe of Medicyne or Philosophicall Path," London, British Library, Sloane MS 11.

Torporley, Nathaniel: British Library, Additional MS 4458.

Warner, Walter: "Generall Principells or conclusions," MS Sion L40.2/E.10. London, Lambeth Palace Library.

Warner, Walter: British Library, Additional MS 4394, 4395.

Warner, Walter: Northamptonshire Record Office, Isham-Lamport Papers [IL] MS 3422.

Primary Sources

Adam, Melchior: *Vitae Germanorum medicorum, qui seculo superiori, et quod excurrit, claruerunt* (Heidelberg, 1620).

Albertus Magnus: *Meteora*, ed. Paul Hossfeld, in *Opera omnia*, ed. Bernhard Geyer and Wilhelm Kübel (Münster, 2003) 6/1.

Andrea, Valerius, ed.: *Fasti academici studii generalis Lovaniensis* (Louvain, 1635).

Anon.: *Rosarium philosophorum* (Frankfurt, 1550).

Aristoteles Latinus: *Meteorologica liber quartus: Translatio Henrici Aristippi*, ed. Elisa Rubino (Brussels, 2010), 10/1.

Aristoteles Latinus: *Meteorologica: Translatio Guillelmi de Morbeka*, ed. Gudrun Vuillemin-Diem (Turnhout, 2008), 10/2.2.

Aristotle: *Metaphysics*, transl. David Ross, in *The Internet Classics Archive*, <http://classics.mit.edu/Aristotle/metaphysics.7.vii.html>, last accessed 6 May 2022.

Aristotle: *Meteorologica*, transl. Henry Desmond Pritchard Lee (Loeb Classical Library 397) (Cambridge, 1952).

Aristotle: *Physics*, transl. R.P. Hardie and R.K. Gaye, in *The Internet Classics Archive*, <http://classics.mit.edu/Aristotle/physics.4.iv.html>, last accessed 6 May 2022.

Aubrey, John: *Brief Lives*, ed. Oliver Lawson Dick (London, 1949; Boston, MA, 1999).

Averroes: *Averrois Cordubensis Aristotelis de Physico audito*, in *Aristotelis Opera cum Averrois Commentariis*, vol. IV (Venice, 1562–1574; reprint, Frankfurt am Main, 1962).

Averroes: *In libros metaphysicorum*, in *Opera omnia* (Venice, 1562) 8.

Avicenna: *Avicenne perhypatetici philosophi … opera* (Venice, 1508).

Avicenna: *Canon medicinae* (Venice, 1595), 2 vols.

Avicenna: *Liber canonis* (Venice, 1507).

Avicenna: *Liber de philosophia prima sive Scientia diuina*, 9.5, ed. Simone van Riet and Gerard Verbeke (Louvain-la-Neuve/Leiden, 1977).

Avicenna: *Liber tertius naturalium: de generatione et corruptione*, ed. Simone van Riet and Gerard Verbeke (Louvain-la-Neuve/Leiden, 1987).

Avicenna: *The Physics of the Healing: Book I and II*, transl. Jon McGinnis (Provo, 2009), 2 vols.

Bacon, Francis: *Instauratio magna* [= *Novum organum*] (London, 1620).

Bacon, Francis: *Sylva Sylvarum: or A Naturall Historie. In Ten Centuries* (London, 1626).

Basson, Sébastien: *Philosophiae naturalis adversus Aristotelem libri XII, in quibus abstrusa veterum physiologia restauratur, et Aristotelis errores solidis rationibus refelluntur* (Geneva, 1621).

Becanus, Martin: *Tractatus de Deo et attributis divinis* (Mainz, 1611).

Bedenckingen over de beroepinghe D.D. Conradi Vorstii (s.l., 1611).

Beguin, Jean: *Tyrocinium chymicum recognitum et auctum* (Paris, 1613).

Bernardus Trevisanus: *De chymico miraculo, quod lapidem philosophiae appellatio* (Basel, 1583).

Biesius, Nicolaus, ed.: *Cicero: De T. Annio Milone oratio*, with annotations by Biesius (Valencia, 1546).

Biesius, Nicolaus: *De republica libri IV, quibus universa de moribus philosophia continetur* (Antwerp, 1556; expanded 2nd ed. 1564).

Biesius, Nicolaus: *De universitate libri tres, quibus universa de natura philosophia continetur* (Antwerp, 1556).

Biesius, Nicolaus: *Theoreticae medicinae libri VI, quibus omnia huius artis praecepta explicantur* (Antwerp, 1558).

Biesius, Nicolaus: *In artem medicam Galeni commentarii* (Antwerp, 1560).

Biesius, Nicolaus: *De methodo medicinae* (Antwerp, 1564).

Biesius, Nicolaus: *De varietate opinionum liber I* (Louvain, 1567).

Biesius, Nicolaus: *De arte dicendi libri II* (Antwerp, 1573).

Biesius, Nicolaus: *De natura libri V* (Antwerp, 1573; 2nd ed. 1613).

Boccaccio, Giovanni: *Decameron*, vol. 2, ed. Vittore Branca (Turin, 1992).

Boccadiferro, Lodovico: *Lectiones in quartum Meteororum Aristotelis librum* (Venice, 1563).

Bodin, Jean: *Universae naturae theatrum* (Lyon, 1596).

Boyle, Robert: *Some Specimens of an Attempt to Make Chymical Experiments Useful to Illustrate the Notions of the Corpuscular Philosophy* (London, 1661).

Boyle, Robert: *The Works of Robert Boyle*, ed. Michael Hunter and Edward B. Davis, 14 vols. (London, 1999–2000).

Bruno, Giordano: *De innumerabilibus, immenso et infigurabili*, in *Poemi filosofici latini. Ristampa anastatica delle cinquecentine*, ed. Eugenio Canone (La Spezia, 2000), 147–655.

Bruno, Giordano: *De triplici minimo et mensura* (Frankfurt, 1591).

Bruno, Giordano: *De umbris idearum*, ed. Rita Sturlese (Florence, 1991).

Bruno, Giordano: *Dialoghi filosofici italiani*, ed. Michele Ciliberto (Milan, 2000).

Bruno, Giordano: *Opere italiane*, ed. Giovanni Aquilecchia, Nuccio Ordine et al., 2 vols. (Turin, 2013).

Bruno, Giordano: *Opera latine conscripta*, ed. Felice Tocco et al., 3 vols. in 8 tomes (Naples – Florence, 1879–1891).

Bruno, Giordano: *Poemi filosofici latini. Ristampa anastatica delle cinquecentine*, ed. Eugenio Canone (La Spezia, 2000).

Burton, Robert: *The Anatomy of Melancholy* (Oxford, 1990).

Cabeo, Niccolò: *In libros Meteorologicorum Aristotelis commentaria, et quaestiones* (Rome, 1646).

Campanella, Tommaso: *Apologia pro Galileo* (Frankfurt, 1622).

Campanella, Tommaso: *Apologia pro Galileo*, ed. Michel-Pierre Lerner, trans. Germana Ernst (Pisa, 2006).

Cardano, Girolamo: *De subtilitate libri XXI* (Basel, 1554).

Carpenter, Nathanael: *Philosophia libera* (Frankfurt, 1621).

Charleton, Walter: *Physiologia Epicuro-Gassendo-Charelton: or a Fabric of Science Natural, upon the Hypothesis of Atoms* (London, 1654).

Charleton, Walter: *The Darknes of Atheism Dispelled by the Light of Nature. A Physico-Theologicall Treatise* (London, 1652).

Chiaramonti, Scipione: *In quartum Metheorum commentaria* (Cesena, 1654).

Cicero, Marcus Tullius: *On the Nature of the Gods*, transl. H. Rackham (Loeb Classical Library) (London, Cambridge, MA, 1940).

Cudworth, Ralph: *The True Intellectual System of the World* (London, 1678).

Cusanus, Nicolaus: *Dialogus de ludo globi*, ed. Hans Gerhard Senger (Hamburg, 1998).

Da Monte, Giambattista: *In primam fen libri primi Canonis Avicennae explanatio* (Venice, 1554).

Daneau, Lambert: *Physica christiana sive de rerum creatarum cognitione et usu* (Lyon, 1576).

Daneau, Lambert: *Physices christianae pars altera, sive de rerum creatarum natura* (Geneva, 1580).

Dante: *The Divine Comedy*, transl. Allen Mandelbaum, in *Digital Dante*, Columbia University, <https://digitaldante.columbia.edu/dante/divine-comedy/inferno/inferno-10/> [last accessed 6 May 2022].

Descartes, René: *Principia philosophiae* (Amsterdam, 1644).

Die Vorsokratiker, Auswahl der Fragmente, Übersetzung und Erläuterungen von Jaap Mansfeld, 2 vols. (Stuttgart, 1983–1986).

Diogenes Laertius: *Lives of Eminent Philosophers*, transl. R.D. Hicks, 2 vols. (Cambridge, MA – London, 1979–1980).

DuBartas, Guillaume de Saluste: *La semaine, ou Création du monde* (Paris, 1578).

DuClo, Gaston [Gasto Claveus]: *Apologia chrysopoeiæ et argyropoeiæ aduersus Thomam Erastum* [...] *De triplici praeparatione auri et argenti* (Oberursel, 1602).

Epicurus: *Sulla Natura, libro II*, ed., transl. and comm. Giuliana Leone (Naples, 2012).

Erastus, Thomas: *Disputationum de nova Philippi Paracelsi medicina pars tertia* (Basel, 1572).

Erastus, Thomas: "Disputatio de saporibus," in *Varia opuscula medica* (Frankfurt, 1590).

Estienne, Robertus & Henricus: *Thesaurus Graecae linguae* (Paris, 1572).

Facciolati, Jacobo: *Fasti gymnasii patavini* (Padua, 1757).

Fernel, Jean: *Universa Medicina* (Paris, 1567).

Fernel, Jean: "The *Physiologia* of Jean Fernel (1567)," transl. John M. Forrester, *Transactions of the American Philosophical Society*, 93 (2003), i–636.

Fernel, Jean: *On the Hidden Causes of Things: Forms, Souls and Occult Diseases in Renaissance Medicine*, transl. John M. Forrester (Leiden-Boston, MA, 2005).

Ficino Marsilio: *Compendium in Timæum* (Florence, 1496).

Ficino Marsilio: *Diuini Platonis Opera omnia Marsilio Ficino interprete* (Lyon, 1557).

Ficino Marsilio: *Platonic Theology*, vol. 1, Books I–IV, transl. Michael J.B. Allen with John Warden; Latin text ed. James Hankins with William Bowen (Cambridge, 2001–2006).

Fracastoro, Girolamo: *De sympathia, et antipathia rerum*, ed. Concetta Pennuto (Rome, 2008).

Frachetta, Girolamo: *Spositione sopra la canzone di Guido Cavalcanti Donna mi prega etc.* (Venice, 1586).

Frachetta, Girolamo: *Breve spositione di tutta l'opera di Lucretio. Nella quale si disamina la dottrina di Epicuro, & si mostra in che sia conforme col vero, & con gl'insegnamenti d'Aristotile, & in che differente* (Venice, 1589).

Francus, Raphael: *Raphaelis franci florentini In Lucretium paraphrasis cum appendice de animi immortalitate* (Bologna, 1504).

Fromondus, Libertus: *Labyrinthus; sive, de compositione continui liber unus* (Antwerp, 1631).

Galen: *Opera omnia*, ed. Karl G. Kühn (Leipzig, 1821–33), 20 vols.

Galen: *On the Elements According to Hippocrates*, transl. Philip de Lacy (Berlin, 1996).

Galen: *De elementis ex Hippocrate sententia*, transl. Johann Winter von Andernach (Paris, 1528).

Galen: *De elementis libri duo*, transl. Niccolò Leoniceno, in Galen, *Prima classis humani corporis* (Venice, 1541).

Galen: *De elementis libri duo*, transl. Vittore Trincavelli (Lyon, 1548).

Galilei, Galileo: *Il saggiatore* (Rome, 1623).

Gassendi, Pierre: *Animadversiones in decimum librum Diogenis Laertii, qui est de vita, moribus placitisque Epicuri* (Lyon, 1649).

Gilbert, William: *De magnete, magneticisque corporibus, et de magno magnete tellure; Physiologia nova, plurimis et argumentis, et experimentis demonstrata* (London, 1600).

Gorlæus, David: *Exercitationes philosophicae quibus universa fere discutitur philosophia theoretica* (Leiden, 1620).

Healey, John: *St. Augustine of the Citie of God: with the learned Commentarie of Jo. Lod. Vives. Englished by J.H.* (London, 1610).

Highmore, Nathanael: *The History of Generation, Examining the several Opinions of divers Authors, especially that of Sir Kenelm Dibgy, in his Discourse of Bodies* (London, 1651).

Hill, Nicholas: *De philosophia epicuraea democritiana theophrastica proposita simpliciter, non edocta* (Paris, 1601; Geneva, 1619).

Hill, Nicholas: *Philosophia Epicuraea Democritiana Theophrastica*, ed. Sandra Plastina (Pisa-Rome, 2007).

Hooke, Robert: *Micrographia, or some Physiological Descriptions of Minute Bodies Made by Magnifying Glasses* (London, 1665).

Isaac Israeli: *Omnia opera Ysaac* (Lyon, 1515).

Jacobo da Forlì: *Expositio et quaestiones in primum Canonem Avicennae* (Venice, 1547).

Jonson, Ben: *Works*, ed. William Gifford (London, 1875).

Jungius, Joachim: *Disputationes hamburgenses: kritische Edition*, ed. Clemens Müller-Glauser (Göttingen, 1988).

[Jungius, Joachim]: *Auctarium epitomes physicae clarissimi atque experientissimi viri Danielis Sennerti* (Frankfurt, 1651).

Kepler, Johannes: *Strena seu de nive sexangula* (Frankfurt, 1611).

Libavius, Andreas: *Medicinae hermeticae ... Expositio fidelis* (Frankfurt, 1599).

Libavius, Andreas: *Novus de medicina veterum tam Hippocratica quam Hermetica tractatus* (Frankfurt, 1599).

Libavius, Andreas: *Alchymia triumphans* (Frankfurt, 1607).

Libavius, Andreas: *Syntagma selectorum undiquaeque et perspicue traditorum alchymiae arcanorum* (Frankfurt, 1611).

Lucretius: *On the Nature of the Things*, transl. John Selby Watson (London, 1880).

Magnen, Jean-Chrysostôme: *Democritus reviviscens, sive, De atomis* (Pavia, 1646).

Mersenne, Marin: *Universae geometriae, mixtaque mathematicae synopsis, et bini refractionum demonstratarum tractatus* (Paris, 1644).

Michelius, Josephus: *Apologia chymica adversus invectivas calumnias A. Libavii* (Middelburg, 1597).

Newton, Isaac: *The Correspondence of Isaac Newton*, ed. H.W. Turnbull (Cambridge, 1960), vol. 2.

Newton, Isaac: *Chymistry of Isaac Newton* project, <www.chymistry.org>.

Nifo, Agostino: *Commentaria in libro De mixtis qui a veteribus quartus Meteororum liber inscribitur* (Venice, 1560).

Paracelsus: *Bücher und Schrifften*, ed. Johannes Huser (Basel, 1589–1591), 10 vols.

Paracelsus: *Das Buch Paramirum darinn die ware ursachen der Kranckheyten und vollkomne Cur in Kürtze erkleret wird allen Artzten nützlich unnd notwendig* (Mülhausen, 1562).

Paracelsus: De caducis liber secundus, nemlich de caduco matricis das ist, vom hinfallenden siechtagen der mutter, so allein den frauen anhangt, in *Theophrast von Hohenheim genannt Paracelsus sämtliche Werke*, ed. Karl Sudhoff, Abt. 1, v. 8 (Munich, 1924), 319–368.

Paracelsus: *Liber secundus de causa et origine morborum. Das ist: Von ursachen und herkommen der kranckheiten* (Cologne, 1565).

Paracelsus: *Paracelsus, Operum Medico-Chimicorum sive Paradoxorum tomus genuinus primus-duodecimus* (Frankfurt, 1603–05).

Patrizi, Francesco: *Discussionum peripateticarum libri XIII* (Venice, 1571).

Patrizi, Francesco: *De paralleli militari nei quali si fa paragone delle Milizie antiche in tutte le parti loro colle moderne* (Rome, 1594).

Patrizi, Francesco: *Discussiones peripateticae quibus aristotelicae philosophiae historia atque dogmata cum veterum placitis collata eleganter et erudite declarantur*, 4 vols. (Basel, 1581).

Patrizi, Francesco: *La militia romana di Polibio, di Tito Livio e di Dionigi di Alicarnasso* (Ferrara, 1584).

Patrizi, Francesco: *Lettere ed opuscoli inediti*, ed. Danilo Aguzzi Barbagli (Florence, 1975).

Patrizi, Francesco: *Nova de universis philosophia libris quinquaginta comprehensa* (Venice, 1593 [Ferrara, 1591]).

Pereira, Benedictus: *De communibus omnium rerum naturalium principiis et affectionibus libri quindecim* (Lyon, 1588).

Piccolomini, Francesco: *"De mixtione,"* in Piccolomini, *Librorum ad scientiam de natura attinentium partes quinque*, vol. 2 (Venice, 1600).

Pius, Johannes Baptista: *Praelectio in Titum Lucretium et Svetonium Tranquillum* (Bologna, 1501).

Pius, Johannes Baptista: *In Carum Lucretium poetam Commentarij a Ioanne Baptista Pio editi, codice Lucretiano diligenter emendato …* (Bologna, 1511).

Plato: *Platonis Opera*, ed. John Burnet (Oxford, 1903).

Pomponazzi, Pietro: *De incantationibus*, ed. Vittoria Perrone Compagni (Florence, 2011).

Pomponazzi, Pietro: *Dubitationes in quartum Meteorologicorum* (Venice, 1563).

Pomponazzi, Pietro: *Tractatus de immortalitate animae*, ed. Gianfranco Morra (Bologna, 1954).

Power, Henry: *Experimental Philosophy, in three books: containing new experiments microscopical, mercurial, magnetical: with some deductions, and probable hypotheses, raised from them, in avouchment and illustration of the now famous atomical hypothesis* (London, 1664).

Pseudo-Democritus: *De arte magna sive de rebus naturalibus*, transl. Domenico Pizzimenti (Padua, 1572).

Quercetanus, Josephus: *De priscorum philosophorum verae medicinae Materia* (Saint Gervais, 1603).

Regius, Henricus: *Disputatio medica tertia, de illustribus aliquot quaestionibus physiologicis* (Utrecht, 1641).

Rhodius, Ambrosius: *Dialogus de transmigratione animarum Pythagorica* (Copenhagen, 1638).

Rhodius, Ambrosius: *Disputationes supra Ideam medicinæ philosphicæ Petri Severini Dani philosophi & Friderici II Daniæ & Septentrionalis Regis archiatri olim felicissimi, quibus loca illius libri obscura & difficilia illustrantur, adversarij refutantur, & multi discursus ex intimis naturæ adytis deprompti moventur* (Copenhagen, 1643).

Ricardus Anglicus: *Correctorium alchemiae*, in Geber, *De alchemia* (Nuremberg, 1541).

Riolan, Jean: *Opera omnia* (Paris, 1610).

de Rupescissa, Johannes: *De consideratione quintae essentiae rerum* (Basel, 1561).

Santorio, Santorio: *Commentaria in Artem medicinalem Galeni* (Lyon, 1632).

Santorio, Santorio: *Commentaria in primam fen primi libri Canonis Avicennae* (Venice, 1625).

Santorio, Santorio: *Methodi vitandorum errorum omnium, qui in arte medica contingunt libri* (Venice, 1603).

Scaliger, Julius Caesar: *In libros duos … de plantis, Aristotele autore, libri duo* (Paris, 1556).

Scaliger, Julius Caesar: *Exotericarum exercitationum liber quintus decimus de subtilitate adversus Cardanum* (Paris, 1557).

Sennert, Daniel: *Institutionum medicinae libri V* (Wittenberg, 1620).

Sennert, Daniel: *De chymicorum cum Aristotelicis et Galenicis consensu ac dissensu liber* (Wittenberg, 1619, rev. ed., 1629).

Severino, Marco Aurelio: *Zootomia Democritea, idest anatome generalis totius animantium opficii* (Nuremberg, 1645).

Severinus, Petrus: *Idea medicinæ philosophicæ fundamenta continens totius doctrinæ Paracelsicæ, Hippocraticæ et Galenicæ* (Basel, 1571).

Shakespeare, William: *Comedy of Errors*, ed. T.S. Dorsch (Cambridge, 1988).

Sorel, Charles: *De la perfection de l'homme* (Paris, 1655).

Sorel, Charles: *La science universelle* (Paris, 1668).

Sozzini, Fausto: *Christianae religionis brevissima institutio* (Raków, 1618).

Sozzini, Fausto: *Opera omnia*, 2 vols. (Irenopolis [Amsterdam], 1656).

Taurellus, Nicolaus: *Kosmologia, hoc est physicarum et metaphysicarum discussionum de mundo libri II* (Amberg, 1603).

Taurellus, Nicolaus: *Philosophiæ triumphus, hoc est, metaphysica philosophandi methodus* (Basel, 1573).

Taurellus, Nicolaus: *Philosophiae triumphus, hoc est, metaphysica philosophandi methodus*, ed. and transl. Henrik Wels (Stuttgart, 2012).

Telesio, Bernardino: *De rerum natura iuxta propria principia libri IX* (Naples, 1586).

Terrence: *The Comedies of Terence*, transl. Frederick W. Clayton (Exeter, 2006).

Thomas Aquinas: *In Aristotelis libros De caelo et mundo, De generatione et corruptione, Meteorologicorum expositio*, ed. Raimondo M. Spiazzi (Rome, 1952).

Thomas Aquinas: *De unitate intellectus contra Averroistas*, in *Corpus Thomisticum*, <www.corpusthomisticum.org/oca.html> [last accessed 6 May 2022].

Thurneisser, Leonhart: *Pison, Das erst Theil* (Frankfurt an der Oder, 1572).

Turretin, Francis: *Institutes of Elenctic Theology*, transl. George Musgrave Giger, 3 vols. (Phillipsburg, NJ, 1992–1997).

Ugo Benzi: *In primi Canonis Auicenne fen primam luculentissima expositio* (Venice, 1523).

Vallés, Francisco: *Commentarius in quartum Aristotelis Meteorologicorum* (Padua, 1591).

Vallés, Francisco: *Controversiarum medicarum et philosophicarum libri decem* (Frankfurt, 1582).

Varenius, Bernhard: *A Compleat System of General Geography*, ed. Peter Shaw (London, 1734), 361–362.

Varenius, Bernhard: *Geographia generalis*, ed. Isaac Newton (Cambridge, 1672).

Varenius, Bernhard: *Geographia generalis* [ed. Isaac Newton] (Cambridge, 1681).

Varenius, Bernhard [and Nicolas Sanson]: *Cosmography and Geography in Two Parts*, transl. Richard Blome (London, 1682).

Vimercati, Francesco: *In Aristotelis quartum librum Meteorologicorum commentarii* (Paris, 1556).

Voetius, Gisbertus: *Testimonium academiae Ultraiectinae, et narratio historica qua defensae, qua exterminatae novae philosophiae* (Utrecht, 1643).

Voetius, Gisbertus: *Selectarum disputationum pars quinta* (Utrecht, 1669).

Vorstius, Conrad: *Tractatus theologicus de Deo, sive de natura et attributis Dei* (Steinfurt, 1610).

Vorstius, Conrad: *Oratio apologetica habita in pleno consessu illustrium ac praepotentium Hollandiae et Westfrisiae Ordinum* (Leiden, 1612).

Wood, Anthony: *Athenae Oxonienses*, ed. Philip Bliss, 4 vols (London, 1812–1820).

Zanchi, Girolamo: *De natura Dei, seu de divinis attributis* (Heidelberg, 1577).

Secondary Sources

Alback, Gustav and F.J. Billeskov Jansen, eds.: *Dansk Litteratur Historie*, vol. 1 (Copenhagen, 1967).

Algra, Keimpe: *Concepts of Space in Greek Thought* (Leiden, 1994), 72–118.

Amato, Barbara: "Atomo," *Bruniana & Campanelliana*, 18 (2012), 579–586.

Atanasijević, Ksenjia: *La doctrine métaphysique et géométrique de Bruno exposée dans son ouvrage 'De triplici minimo'* (Belgrade, 1923).

Bakelants, Louis: "Le *De Universitate* de Nicolas Biesius," in *Hommage à Leon Herrmann* (Brussels, 1960), 138–145.

Baldini, Artemio Enzo: "Frachetta, Girolamo," in *Dizionario biografico degli italiani*, XLIX (1997), 567–573.

Baldini, Artemio Enzo: "Girolamo Frachetta: vicissitudini e percorsi culturali di un pensatore politico nell'Italia della Controriforma," in *Archivio della ragion di Stato online*, <www.ragionidistato.it/2021/02/07/girolamo-frachetta-un-pensatore-politico-nell italia-della-controriforma/>, last accessed 7 May 2022.

Banchetti-Robino, Marina Paola: *The Chemical Philosophy of Robert Boyle: Mechanism, Chymical Atoms and Emergence* (Oxford, 2020).

Batho, Gordon R.: "The Library of the 'Wizard' Earl: Henry Percy Ninth Earl of Northumberland (1564–1632)", *The Library*, 15 (1960), 246–261.

Bayle Pierre: *Dictionary Historical and Critical of Mr Peter Bayle*, vol. 5 (London, 1738).

Beck, Andreas J.: *Gisbertus Voetius (1589–1676): sein Theologieverständnis und seine Gotteslehre* (Göttingen, 2007).

Beddard, Robert A.: "The Source of Anthony Wood's Life of Nicholas Hill," *Archives*, 29 (2004), 1–11.

Beretta, Marco: "Gli scienziati e l'edizione del *De rerum natura*," in *Lucrezio: La natura e la scienza*, ed. Marco Beretta and Francesco Citti (Florence, 2008), 177–224.

Beretta, Marco: *La rivoluzione culturale di Lucrezio. Filosofia e scienza nell'antica Roma* (Rome, 2015).

Berman, Brad: "Aristotle on Like-Partedness and the Like-Parted Bodies," *Early Science and Medicine*, 20 (2015), 27–47.

Berthelot, Marcellin: *Les Origines de l'Alchemie* (Paris, 1885).

Bigotti, Fabrizio: "A Previously Unknown Path to Corpuscularism in the Seventeenth Century: Santorio's Marginalia to the *Commentaria in Primam Fen Primi Libri Canonis Avicennae* (1625)," *Ambix*, 64 (2017), 29–42.

Billeskov Jansen, F.J.: *Ludwig Holberg* (New York, 1974).

Bizer, Ernst: *Frühorthodoxie und Rationalismus* (Zurich, 1963).

Blair, Ann: "Mosaic Physics and the Search for a Pious Natural Philosophy in the Late Renaissance," *Isis*, 91 (2000), 32–58.

Blair, Ann: *The Theater of Nature: Jean Bodin and Renaissance Science* (Princeton, NJ, 1997).

Blank, Andreas: "Jean Fernel on Divine Immanence and the Origin of Simple Forms," in *Nature et surnaturel : philosophies de la nature et métaphysique aux XVI*ᵉ*–XVIII*ᵉ *siècles*, ed. Vlad Alexandrescu and Robert Theis (Hildesheim – Zürich – New York, 2010), 9–21.

Blank, Andreas: "Nicolaus Taurellus on Forms and Elements," *Science in Context*, 27 (2014), 659–682.

Blank, Andreas: "Sixteenth-Century Pharmacology and the Controversy between Reductionism and Emergentism," *Perspectives on Science*, 26 (2018), 157–184.

Blyth, Dougal: *Aristotle's Ever-Turning World in* Physics *8: Analysis and Commentary* (Leiden, 2015).

Boas Hall, Marie: "The Establishment of the Mechanical Philosophy," *Osiris*, 10 (1952), 412–442.

Bono, James J.: "Medical Spirits and the Medieval Language of Life," *Traditio*, 40 (1984), 91–130.

Borsic, Luka: "Patrizi and the History of Philosophy," *Francesco Patrizi. Philosopher of the Renaissance*, ed. Tomas Nejeschleba and Paul Richard Blum (Olomouc, 2014), 104–117.

Boudon-Millot, Véronique: "La notion de mélange dans la pensée médicale de Galien: mixis ou crasis?," *Revue des études grecques*, 124 (2011), 261–279.

Brockett, Oscar G.: *History of the Theatre*, 6th ed. (Boston, MA, 1991).

Brown, Alison: *The Return of Lucretius to Renaissance Florence* (Cambridge, MA – London, 2010).

Butterfield, David: *The Early Textual History of Lucretius'* De rerum natura (Cambridge, 2013).

Bylebyl, Jerome J.: "The Medical Meaning of *Physica*," *Osiris*, 6 (1990), 16–41.

Bylebyl, Jerome J.: "The School of Padua: Humanistic Medicine in the Sixteenth Century," in *Health, Medicine and Mortality in the Sixteenth Century*, ed. Charles Webster (Cambridge, 1979), 335–370.

Byrne, Christopher: "Prime Matter and Actuality," *Journal of the History of Philosophy*, 33 (1995), 197–224.

Calvet, Antoine: "Étude d'un texte latin du XIVᵉ siècle : le *Rosarius philosophorum* attribué au médecin Arnaud de Villeneuve (ob. 1311)," *Early Science and Medicine*, 11 (2006), 162–206.

Canone, Eugenio: "Fenomenologie dell'anima," in Giordano Bruno, *Poemi filosofici latini. Ristampa anastatica delle cinquecentine*, ed. Eugenio Canone (La Spezia, 2000), XIII–XXXIX.

Carlson, Marvin: *Theories of the Theatre: A Historical and Critical Survey, from the Greeks to the Present* (Ithaca, NY – London, 1984).

Cassirer, Ernst: *Das Erkenntnisproblem in der Philosophie und Wissenschaft der neueren Zeit* (Darmstadt, 1995).

Caullery, Maurice: *La science française depuis le XVIIe siècle* (Paris, 1933).

Ceccarelli, Andrea: "Reading Lucretius in Padua: Gian Vincenzo Pinelli and the Sixteenth-Century Recovery of Ancient Atomism," in *Lucretius Poet and Philosopher*, ed. Philip R. Hardie, Valentina Prosperi and Diego Zucca (Berlin – Boston, MA, 2020), 219–234.

Chandelier, Joel: *Avicenne et la médecine en Italie : le Canon dans les Universités, 1200–1350* (Paris, 2017).

Chandelier, Joel, and Aurélien Robert: "Nature humaine et complexion du corps chez les médecins italiens de la fin du Moyen Âge," *Revue de Synthèse*, 134 (2013), 473–510.

Charlton, William: "Prime Matter: A Rejoinder," *Phronesis*, 28 (1983), 197–211.

Clericuzio, Antonio: "A Redefinition of Boyle's Chemistry and Corpuscular Philosophy," *Annals of Science*, 47 (1990), 561–589.

Clericuzio, Antonio: *Elements, Principles and Corpuscles: A Study of Atomism and Chemistry in the Seventeenth Century* (Dordrecht, 2000).

Clericuzio, Antonio: "L'atomisme de Gassendi et la philosophie corpusculaire de Boyle," in *Gassendi et l'Europe (1592–1792): Actes du Colloque International de Paris "Gassendi et sa postérité,"* ed. Sylvia Murr (Paris, 1997), 227–235.

Clucas, Stephen: "'All the mistery of infinites': Mathematics and the Atomism of Thomas Harriot," in *Mathématiques et connaissance du monde réel avant Galilée, Histoire des Savoirs*, ed. Sabine Rommevaux (Montreuil, 2010), 113–154.

Clucas, Stephen: "Corpuscular Theories in the Northumberland Circle," in *Late Mediaeval and Early Modern Corpuscular Matter Theories*, ed. Christoph H. Lüthy, John E. Murdoch and William R. Newman (Leiden, 2001), 181–207.

Clucas, Stephen: "Hill, Nicholas" in *The Dictionary of Seventeenth-Century British Philosophers*, ed. Andrew Pyle, 2 vols. (Bristol, 2000), 424–426.

Clucas, Stephen: "Mediaeval Concepts of Force in the Atomism of the Northumberland Circle," in *Science and Cultural Diversity, Proceedings of the XXIst International Congress of History of Science*, ed. Juan José Saldaña, CD-ROM (Mexico City, 2005), 3060–3073.

Clucas, Stephen: "*Scientia* and *Inductio Scientifica* in the *Logica Hamburgensis* of Joachim Jungius," in *Scientia in Early Modern Philosophy*, ed. John G. A. Rogers, Tom Sorrel, and Jill Kraye (Dordrecht, 2010), 53–70.

Clucas, Stephen: "The Atomism of the Cavendish Circle: A Reappraisal," *Seventeenth Century*, 9 (1994), 247–273.

Clucas, Stephen: "The 1595 Advice and the Philosophical Milieu of the Northumberland Circle," in *The Wizard Earl's Advices to his Son. A Facsimile and Transcript from the Manuscripts of Henry Percy, Ninth Earl of Northumberland at Petworth House*, ed. Gordon R. Batho and Stephen Clucas (Otley, 2002), xli–lxi.

Clucas, Stephen: "'The infinite variety of formes and magnitudes': Sixteenth- and Seventeenth-century English Corpuscular Philosophy and Aristotelian Theories of Matter and Form," *Early Science and Medicine*, 2 (1997), 251–271.

Clucas, Stephen: "Thomas Harriot and Walter Warner on Collisions," *British Journal for the History of Science* [forthcoming].

Cohen, H. Floris: *The Scientific Revolution: A Historiographical Inquiry* (Chicago, IL, 1994).

Coleman, James K.: "Translating Impiety: Girolamo Frachetta and the First Vernacular Commentary on Lucretius," *Quaderni d'Italianistica*, 35 (2014), 55–71.

Conti, Daniele: "Pio, Giovanni Battista," in *Dizionario biografico degli italiani*, 84 (2015), 87–91.

Copenhaver, Brian P.: *Magic in Western Culture: From Antiquity to the Enlightenment* (Cambridge, 2015).

Cordonier, Valérie: "Matière, qualités, mélange : la physique élémentaire d'Aristote chez Galien et Alexandre d'Aphrodise," *Quaestio*, 7 (2007), 79–103.

Crisciani, Chiara: "Il corpo nella tradizione alchemica: Teorie, similitudini, immagini," *Micrologus*, 1 (1993), 189–233.

Cunningham, Andrew: "The Pen and the Sword: Recovering the Disciplinary Identity of Physiology and Anatomy before 1800. Part 1: Old Physiology: The Pen," *Studies in History and Philosophy of Science*, 33 (2002), 631–665.

Daniel, Dane: "Paracelsus on Baptism and the Acquiring of the Eternal Body," in *Paracelsus. The Man and His Reputation, His Ideas and Their Transformation*, ed. Ole Grell (Leiden, 1998), 117–134.

Daugirda, Kęstutis: *Die Anfänge des Sozinianismus: Genese und Eindringen des historisch-ethischen Religionsmodells in den universitären Diskurs der Evangelischen in Europa* (Göttingen, 2016).

Davidson, Herbert A.: *Alfarabi, Avicenna and Averroes, On Intellect: Their Cosmologies, Theories of Active Intellect, and Theories of Human Intellect* (New York, 1992), 74–126.

Davidson, William: *Commentariorum in sublimis philosophi & incomparibilis viri Petri Severini Dani Ideam Medicinæ Philosophicæ ... Prodromus* (The Hague, 1660).

Davidson, William: *Commentaria in Idœam Medicinæ Philosophicæ Petri Dani, Medici incomparabilis & Philosophi sublimis* (The Hague, 1663).

Debru, Armelle: *Le corps respirant : la pensée physiologique chez Galien* (Leiden – New York, 1996).

Debus, Allen G.: "Chemistry and the Quest for a Material Spirit of Life in the Seventeenth Century," in *Spiritus. IV. Colloquio internazionale del Lessico intelletuale europeo*, ed. Marta Fattori and Massimo Luigi Bianchi (Rome, 1984), 245–263.

Debus, Allen G.: "Fire Analysis and the Elements in the Sixteenth and Seventeenth Centuries," *Annals of Science*, 23 (1967), 127–147.

Debus, Allen G.: *Science, Medicine and Society in the Renaissance* (New York, 1972), 151–165.

Deer Richardson, Linda: "The Generation of Disease: Occult Causes and Diseases of the Total Substance," in *The Medical Renaissance of the Sixteenth Century*, ed. Andrew Wear, Roger K. French and Ian M. Lonie (Cambridge, 1985), 175–194.

De Haas, Frans A.J. de: "Mixture in Philoponus: An Encounter with a Third Kind of Potentiality," in *The Commentary Tradition on Aristotle's* De generatione et corruptione, ed. J.M.M.H. Thijssen and H.A.G. Braakhuis (Turnhout, 1999), 21–46.

Deitz, Luc: "*Falsissima Est Ergo Haec De Triplici Substantia Aristotelis Doctrina*, A Sixteenth-Century Critic of Aristotle: Francesco Patrizi Da Cherso on Privation, Form, and Matter," *Early Science and Medicine*, 2 (1997), 227–250.

Del Nero, Valerio: "Note sulla vita di Giovan Battista Pio (con alcune lettere inedite)," *Rinascimento*, 21 (1981), 247–263.

Del Prete, Antonella: *Universo infinito e pluralità dei mondi. Teorie cosmologiche in età moderna* (Naples, 1998).

Del Soldato, Eva: "Francesco Vimercato's *De placitis naturalibus Platonis et Aristotelis, ac inter eos de illis consensione et dissensione*," *Rinascimento*, 52, second series (2012), 117–177.

Denifle, Heinrich and Emile Chatelain, ed.: *Chartularium Universitatis Parisiensis*, vol. 1 (Paris, 1889).

DeVun, Leah: *Prophecy, Alchemy, and the End of Time: John of Rupecissa in the Late Middle Ages* (New York, 2014).

Dhanani, Alnoor: *The Physical Theory of Kalām: Atoms, Space, and Void in Basrian Mu'tazilī Cosmology* (Leiden, 1994).

Dick, Steven J.: *Plurality of Worlds: The Origins of the Extraterrestrial Life Debate from Democritus to Kant* (Cambridge, 1982).

Diels, Hermann and Walther Kranz: *Die Fragmente der Vorsokratiker*, 6th ed. (Berlin 1951).

Dijksterhuis, E.J.: *The Mechanization of the World Picture*, transl. C. Dikshoorn (Oxford, 1961).

Di Liscia, Daniel A.: "*Operosum negotium*: Jungius' doxoskopische Betrachtung des Aristotelismus von Zabarella," in *La presenza dell'aristotelismo padovano nella filosofia della prima modernità*, ed. Gregorio Piaia (Padua, 2002), 215–255.

Dobbs, Betty Jo Teeter: "Newton's Alchemy and His Theory of Matter," *Isis*, 73 (1982), 511–528.

Donahue, William H.: *The Dissolution of the Celestial Spheres* (New York, 1981).

Dondaine, Antoine and Louis-Jacques Bataillon: "Le commentaire de saint Thomas sur les *Météores*," *Archivum Fratrum Praedicatorum*, 36 (1966), 81–152.

Doni, Manuela: "Il 'De incantationibus' di Pietro Pomponazzi e l'edizione di Guglielmo Grataroli," *Rinascimento*, 15, 2nd series (1975), 183–230.

Dorandi, Tiziano: "Introduction," in Diogenes Laërtius, *Lives of Eminent Philosophers, ed. Dorandi* (Cambridge, 2013).

Dox, Donalee: *The Idea of the Theater in Latin Christian Thought: Augustine to the Fourteenth Century* (Ann Arbor, MI, 2004).

Duhem, Pierre: *Le mixte et la combinaison chimique: essai sur l'évolution d'une idée* (Paris, 1902).

Düring, Ingemar: *Aristotle's Chemical Treatise Meteorologica, book IV* (Göteborg, 1944).

Eichholz, David E.: "Aristotle's Theory of the Formation of Metals and Minerals," *Classical Quarterly*, 43 (1943), 141–146.

Elford, Daniel: "William of Conches" in *A History of Twelfth-Century Western Philosophy*, ed. Peter Dronke (Cambridge, 1992), 308–327.

Eloy, Nicolas François Joseph: *Dictionnaire historique de la médecine contentant son origine, ses progrès, ses révolutions, ses sectes & son etat chez différens peuples* [...], 2 vols. (Liège – Frankfurt, 1755).

Elsner, Bernd and Martin Rothkegel: *Der Briefwechsel des Joachim Jungius* (Göttingen, 2005).

Emerton, Norma E.: *The Scientific Reinterpretation of Form* (Ithaca, NY, 1984).

Fabbri, Natacha: "Looking at an Earth-like Moon and Living on a Moon-like Earth in Renaissance and Early Modern Thought," *Early Modern Philosophers and the Renaissance Legacy*, ed. Cecilia Muratori and Gianni Paganini (Cham, 2016), 135–151.

Fatio, Olivier: *Méthode et théologie: Lambert Daneau et les débuts de la scolastique réformée* (Geneva, 1976).

Favaro, Antonio, ed.: *Atti della nazione germanica artista nello Studio di Padova* (Venice, 1912).

Fechner, Gustav Theodor: *Über die physikalische und philosophische Atomenlehre*, 2nd rev. ed. (Leipzig, 1864).

Ferrari, Giovanna: "Public Anatomy Lessons and the Carnival: The Anatomy Theatre of Bologna," *Past and Present*, 117 (1987), 50–117.

Fleischmann, Wolfgang B.: "Lucretius Carus, Titus," *Catalogus translationum et commentariorum*, 11 (1971), 349–365.

"Focus: Getting Back to 'The Death of Nature': Rereading Carolyn Merchant," *Isis*, 97 (2006), 485–533.

Foppens, Johannes Franciscus: *Bibliotheca belgica, sive Virorum in Belgio vita, scriptisque illustrium catalogus, librorumque nomenclatura* (Brussels, 1739).

French, Roger: "Astrology in Medical Practice," in *Practical Medicine from Salerno to the Black Death*, ed. Luis Garcia-Ballester, et al. (Cambridge, 1994), 30–59.

Freudenthal, Gad: *Aristotle's Theory of Material Substance: Heat and Pneuma, Form and Soul* (Oxford, 1995).

Furley, David: "The Mechanics of *Meteorologica IV*: A Prolegomenon to Biology," in *Zweifelhaftes im Corpus aristotelicum*, ed. Paul Moraux and Jürgen Wiesner (Berlin, 1983), 73–93.

Gambino Longo, Susanna: "La *spositione* de Lucrèce par Girolamo Frachetta et les théories poétiques de la fin du XVIe siècle en Italie," in *Renaissance de Lucrèce, Cahiers du Centre V.L. Saulnier*, 27 (2010), 185–200.

Garin, Eugenio: *La filosofia dal Rinascimento al Risorgimento* (Milan, 1947).

Gatti, Hilary: "Giordano Bruno: The Texts in the Library of the Ninth Earl of Northumberland," *Journal of the Warburg and Courtauld Institutes*, 46 (1983), 63–77.

Gatti, Hilary: "Giordano Bruno's Soul-Powered Atoms: From Ancient Sources towards Modern Science," in *Late Medieval and Early Modern Corpuscular Matter Theories*, ed. Christoph Lüthy, John E. Murdoch and William R. Newman (Leiden, 2001), 163–180.

Gatti, Hilary: *Essays on Giordano Bruno* (Princeton, NJ, 2011).

Gemelli, Benedino: *Aspetti dell'atomismo classico nella filosofia di Francis Bacon e nel Seicento* (Florence, 1996).

Gerhard, Johann: *Theological Commonplaces: On the Nature of God and on the Most Holy Mystery of the Trinity*, transl. Richard J. Dinda (St. Louis, MO, 2007).

Giard, Luce: "Histoire de l'université et histoire du savoir; Padoue (XIVe–XVIe siècles)," *Revue de synthèse*, 104 (1983), 139–169.

Gilbert, Neal W.: "Francesco Vimercato of Milan: A Bio-Bibliography," *Studies in the Renaissance*, 12 (1965), 188–217.

Gilbert, Otto: *Die meteorologischen Theorien des griechischen Altertums* (Leipzig, 1907).

Gill, Mary Louise: "The Limits of Teleology in Aristotle's *Meteorology* IV.12," *HOPOS: The Journal of the International Society for the History of Philosophy of Science*, 4 (2014), 335–350.

Gillespie, Stuart, and Philip R. Hardie: *The Cambridge Companion to Lucretius* (Cambridge, 2007).

Gilly, Carlos: "The 'Fifth Column' within Hermetism: Andreas Libavius," in *Magic, Alchemy and Science, 15th–18th Centuries: The Influence of Hermes Trismegistus*, ed. Carlos Gilly and Cis van Heertum (Florence, 2002), vol. 1, 409–415.

Gilly, Carlos: "'Theophrasta Sancta' – Paracelsianism as a Religion, in Conflict with the Established Churches," in *Paracelsus: The Man and His Reputation, His Ideas and Their Transformation*, ed. Ole Grell (Leiden, 1998), 151–185.

Gomes, Alan W.: "The Rapture of the Christ: The 'Pre-Ascension Ascension' of Jesus in the Theology of Faustus Socinus (1539–1604)," *Harvard Theological Review*, 102 (2009), 75–99.

Gordon, Cosmo Alexander: *A Bibliography of Lucretius* (London, 1962).

Gottschalk, Hans B.: "The Authorship of *Meteorologica*, Book IV," *Classical Quarterly*, 11 (1961), 67–79.

Goulding, Robert: "*Chymicorum in morem*. Refraction, Matter Theory and Secrecy in the Harriot-Kepler Correspondence," in *Thomas Harriot and His World: Mathematics,*

Exploration, and Natural Philosophy in Early Modern England, ed. Robert Fox (Farnham, 2012; reprint, London – New York, 2016), 27–52.

Goulding, Robert: "Thomas Harriot's Optics, between Experiment and Imagination: the Case of Mr Bulkeley's Glass," *Archive for History of Exact Sciences*, 68 (2014), 137–178.

Goudriaan, Aza: *Philosophische Gotteserkenntnis bei Suárez und Descartes: im Zusammenhang mit der niederländischen reformierten Theologie und Philosophie des 17. Jahrhunderts* (Leiden, 1999).

Goudriaan, Aza: *Reformed Orthodoxy and Philosophy, 1625–1750: Gisbertus Voetius, Petrus van Mastricht, and Anthonius Driessen* (Leiden, 2006).

Goudriaan, Aza: "Gorlaeus en de Arminiaanse kwestie," *It Beaken*, 76 (2014), 229–246.

Graham, Daniel W.: "The Paradox of Prime Matter," *Journal of the History of Philosophy*, 25 (1987), 475–490.

Graiff, Franco: "I prodigi e l'astrologia nei commenti di Pietro Pomponazzi al *De caelo*, alla *Meteora* e al *De generatione*," *Medioevo*, 2 (1976), 331–361.

Granada, Miguel Ángel & Dario Tessicini: "Introduzione," in *Giordano Bruno, "De Immenso". Letture critiche*, ed. Granada & Tessicini (Pisa, 2020), 9–28.

Grant, Edward: "The Medieval Doctrine of Place: Some Fundamental Problems and Solutions," in *Studi sul XIV secolo in memoria di Anneliese Maier*, ed. Alfonso Maieru and Agostino Paravicini Bagliani (Rome, 1981), 57–79.

Grant, Edward: *Much Ado about Nothing: Theories of Space and Vacuum from the Middle Ages to the Scientific Revolution* (Cambridge, 1981).

Grant, Edward: *Planets, Stars, and Orbs: The Medieval Cosmos, 1200–1687* (Cambridge, 1994).

Greenblatt, Stephen: *The Swerve. How the World Became Modern* (New York, 2012).

Grellard, Christophe: "Nicholas of Autrecourt's Atomistic Physics," in *Atomism in Late Medieval Philosophy and Theology*, ed. Aurélien Robert and Christophe Grellard (Leiden, 2009), 107–126.

Grendler, Paul F.: *The Universities of the Italian Renaissance* (Baltimore, MD, 2004).

Gründler, Otto: *Die Gotteslehre Girolami Zanchis und ihre Bedeutung für seine Lehre von der Prädestination* (Neukirchen, 1965).

Gulizia, Stefano: "Francesco Patrizi da Cherso and the anti-Aristotelian Tradition: Interpreting the *Discussiones peripateticae* (1581)," *Intellectual History Review*, 29 (2019), 561–573.

Günther, Siegmund: *Varenius* (Leipzig, 1905).

Guyomarc'h, Gweltaz and Stéphane Marchand, eds.: "Studies on Galen's De elementis," *Aitia*, 7.2 (2017). doi:10.4000/aitia.1804.

Haak, Hans L.: "Blood, Clotting and the Four Humours," in *Blood, Sweat and Tears – The Changing Concepts of Physiology from Antiquity to Early Modern Europe*, ed. Manfred Horstmanshoff, Helen Kind and Claus Zittel (Leiden, 2012), 295–306.

Hall, Thomas S.: "Origini scientifiche del problema del protoplasma," in *Le radici del pensiero scientifico*, ed. Philip P. Wiener and Aaron Noland, transl. Margherita Cerletti Novelletto (Milan, 1977), 37–56.

Halleux, Robert: "Les ouvrages alchimiques de Jean de Rupescissa," in *Histoire littéraire de la France*, ed. Académie des Inscriptions et des Belles Lettres (Paris, 1981), vol. 41, 242–277.

Halleux, Robert: "Atomisme," in *La science classique, XVI*e*–XVIII*e *siècle: dictionnaire critique*, ed. Michel Blay and Robert Halleux (Paris, 1998) 440–448.

Hammer-Jensen, Ingeborg: "Das sogenannte IV. Buch der Meteorologie des Aristotles," *Hermes*, 50 (1915), 118–136.

Hannaway, Owen: *The Chemists and the Word: The Didactic Origins of Chemistry* (Baltimore, MD, 1975).

Hannequin, Arthur: *Essai critique sur l'hypothèse des atomes dans la science contemporaine* (Paris, 1899).

Hardie, Philip R., Valentina Prosperi and Diego Zucca, eds.: *Lucretius Poet and Philosopher. Background and Fortunes of* De rerum natura (Berlin – Boston, MA, 2020).

Hardison, O.B.: *Christian Rite and Christian Drama in the Middle Ages: Essays in the Origin and Early History of Modern Drama* (Baltimore, MD, 1965).

Harris, John Wesley: *Medieval Theatre in Context: An Introduction* (London, 1992).

Hasse, Dag Nikolaus: "Avicenna's 'Giver of Forms' in Latin Philosophy, Especially in the Works of Albertus Magnus," in *The Arabic, Hebrew and Latin Reception of Avicenna's Metaphysics*, ed. Dag G. Hasse and Amos Bertolacci (Berlin – Boston, MA, 2012), 227–249.

Hasse, Dag Nikolaus: *Success and Suppression: Arabic Sciences and Philosophy in the Renaissance* (Cambridge, MA, 2016).

Hattab, Helen: *Descartes on Forms and Mechanisms* (Cambridge, 2009).

Hattab, Helen: "Aristotelianism and Atomism Combined: Gorlaeus on Knowledge of Universals," *Perspectives on Science*, 24 (2016), 285–304.

Henry, John: "Francesco Patricius da Cherso's Concept of Space and its Later Influence," *Annals of Science*, 36 (1979), 549–573.

Henry, John: "Void Space, Mathematical Realism and Francesco Patricius da Cherso's Use of Atomistic Arguments," in *Late Medieval and Early Modern Corpuscular Matter Theories*, ed. Christoph Lüthy, John E. Murdoch and William R. Newman (Leiden, 2001), 133–161.

Henry, John: "Jean Fernel and the Importance of the *De abditis rerum causis*," in Fernel, Jean: *On the Hidden Causes of Things: Forms, Souls and Occult Diseases in Renaissance Medicine*, transl. John M. Forrester (Leiden – Boston, MA, 2005), 3–66.

Henry, John, and John M. Forrester: "Introduction. Tradition and Reform: Jean Fernel's *Physiologia* (1567)," in Jean Fernel, *The Physiologia of Jean Fernel*, transl. John Forrester (Philadelphia, PA, 2003), 1–13.

Hershbell, Jackson P.: "Democritus and the Beginnings of Greek Alchemy," *Ambix*, 24 (1987), 5–20.

Hirai, Hiro: *Le concept de semence dans les théories de la matière à la Renaissance: de Marsile Ficin à Pierre Gassendi* (Turnhout, 2005).

Hirai, Hiro: "Concepts of Seeds and Nature in the Work of Marsilio Ficino," in *Marsilio Ficino: His Theology, His Philosophy, His Legacy*, ed. Michael J.B. Allen and Valery Rees (Leiden, 2002), 257–284.

Hirai, Hiro: "Humanisme, néoplatonisme et *prisca theologia* dans le concept de semence de Jean Fernel," in *Corpus, revue de philosophie*, 41 (2002), 43–70.

Hirai, Hiro: *Medical Humanism and Natural Philosophy: Renaissance Debates on Matter, Life, and the Soul* (Leiden, 2011).

Holmes, Frederic L.: "Analysis by Fire and Solvent Extractions: The Metamorphosis of a Tradition," *Isis*, 62 (1971), 128–148.

Hooykaas, R.: *Religion and the Rise of Modern Science* (1972; reprint, Grand Rapids, MI, 1978).

Howlett, D.R., ed.,: *Dictionary of Medieval Latin from British Sources*, fascicle VI (Oxford, 2002).

Ingegno, Alfonso: *Saggio sulla filosofia di Cardano* (Florence, 1980).

Jacquart, Danielle: "Minima in Twelfth-Century Medical Texts from Salerno," in *Late Medieval and Early Modern Corpuscular Matter Theories*, ed. Christoph Lüthy, John E. Murdoch and William R. Newman (Leiden, 2001), 39–56.

Jacquart, Danielle: "De *crasis* à *complexio* : note sur le vocabulaire du tempérament en latin medieval," in *Mémoires V: textes médicaux latins antiques* (Saint-Etienne, 1985), 71–76.

Jacquot, Jean: "Harriot, Hill, Warner and the New Philosophy," in *Thomas Harriot: Renaissance Scientist*, ed. John W. Shirley (Oxford, 1974), 107–128.

Jacquot, Jean: "Les idées de Francesco Patrizi sur l'histoire et le rôle d'Acontius dans leur diffusion en Angleterre," *Revue de littérature comparée*, 26 (1952), 333–354.

Jaeger, Werner: *Aristotle, Fundamentals of the History of his Development* (Oxford, 1934).

Jones, Richard F.: *Ancients and Moderns. A Study of the Rise of the Scientific Movement in Seventeenth Century England* (Berkeley – Los Angeles, CA, 1961).

Joy, Lynn Sumida: *Gassendi the Atomist: Advocate of History in a Age of Science* (Cambridge, 1987).

Kahn, Didier: "Recherche sur le livre attribué au prétendu Bernard le Trévisan (fin du XVe siècle)," in *Alchimia e medicina nel Medioevo*, ed. Chiara Crisciani and Agostino Paravicini Bagliani (Florence, 2003), 265–336.

Kahn, Didier: *Alchimie et paracelsisme en France à la fin de la Renaissance (1567–1625)* (Geneva, 2007), 354–357.

Kany-Turpin, José, ed.: "Jean Fernel," *Corpus, Revue de Philosophie*, 41 (2002), 5–197.

Kargon, Robert Hugh: *Atomism in England from Hariot to Newton* (Oxford, 1966).

Kessler, Eckhard: "Alexander of Aphrodisias and His Doctrine of the Soul: 1400 Years of Lasting Significance," *Early Science and Medicine*, 16 (2011), 1–93.

Klemm, Matthew: "A Medical Perspective on the Soul as Substantial Form of the Body: Peter of Abano and the Reconciliation of Aristotle and Galen," in *Psychology and the Other Disciplines: A Case of Cross-Disciplinary Interaction (1250–1750)*, ed. Paul J.J.M. Bakker, Sander W. de Boer and Cees Leijenhorst (Leiden, 2012), 275–295.

Knox, Dilwyn: "Wonder and the Philosopher's Perfection: Giordano Bruno," in *Poetry, Philosophy and Theology in Conversation. Thresholds of Wonder*, ed. Francesca Bugliani Knox & Jennifer Reek (London, 2020), 29–49.

Knuuttila, Simo: "Change and Contradiction in Henry of Ghent," in *The Instant of Change in Medieval Philosophy and Beyond*, ed. Frédéric Goubier and Magali Roques (Leiden – Boston, MA, 2015), 22–35.

Køning, Mauritz: *Dissertatio de rerum principiis et mechanica seminum liturgia* (Copenhagen, 1663).

Kraye, Jill: "Daniel Heinsius and the Author of *De mundo*," in *The Uses of Greek and Latin: Historical Essays*, ed. A.C. Dionisotti, Anthony Grafton, and Jill Kraye (London, 1988), 171–197.

Kraye, Jill: "Aristotle's God and the Authenticity of *De mundo*: An Early Modern Controversy," *Journal of the History of Philosophy*, 28 (1990), 339–358.

Kraye, Jill: "Disputes over the Authorship of *De mundo* between Humanism and *Altertumswissenschaft*," in *Cosmic Order and Divine Power: Pseudo-Aristotle,* On the Cosmos, ed. Johan C. Thom (Tübingen, 2014), 181–197.

Kubbinga, H.H. (Henk): "Les premières théories 'moléculaires'; Isaac Beeckman et Sébastien Basson: Le concept d' 'individu substantiel' et d' 'espèce substantielle,'" *Revue d'Histoire des Sciences*, 37 (1984), 215–233.

Lacaze, Grégoire: *Turba philosophorum, Congrès pythagoricien sur l'art d'Hermès* (Leiden, 2018).

Lasswitz, Kurd: *Geschichte der Atomistik vom Mittelalter bis Newton*, 2 vols. (Hamburg-Leipzig, 1890).

Latham, R.E., ed.: *Revised Medieval Latin Word-List from British and Irish Sources* (London, 1965).

Lederer, Thomas: "Der Kölner Kurfürst Herzog Ernst von Bayern (1554–1612) und sein Rat Johann Grasse (um 1560–1618) als Alchemiker der Frühen Neuzeit: Ein Beitrag zur Geschichte des Paracelsismus" (Inaugural diss., Ruprecht-Karls-Universitat Heidelberg, 1992).

Lederer, Thomas: "Leben, Werk und Wirkung des Stralsunder Fachschriftstellers Johann Grasse (nach 1560–1618)," in *Pommern in der Frühen Neuzeit*, ed. Wilhelm Kühlmann and Horst Langer (Tübingen, 1994), 227–237.

Lehmann, Klaus: "Der Bildungsweg des jungen Bernhard Varenius," in *Bernhard Varenius (1622–1650)*, ed. Margret Schuchard (Leiden, 2007), 59–90.

Leijenhorst, Cees, and Christoph Lüthy: "The Erosion of Aristotelianism: Confessional Physics in Early Modern Germany and the Dutch Republic," in *The Dynamics of Aristotelian Natural Philosophy from Antiquity to the Seventeenth Century*, ed. Cees Leijenhorst, Christoph Lüthy, and Johannes M.M.H. Thijssen (Leiden, 2002), 375–411.

Leinkauf, Thomas: *Il neoplatonismo di Francesco Patrizi da Cherso come presupposto della sua critica ad Aristotele* (Florence, 1990).

Leinsle, Ulrich Gottfried: *Das Ding und die Methode. Methodische Konstitution und Gegenstand der frühen protestantischen Metaphysik*, 2 vols. (Augsburg, 1985).

Lewis, Charlton T. and Charles Short: *A Latin Dictionary* (1879; Oxford, 1989).

Lewis, C.S.: *The Discarded Image: An Introduction to Medieval and Renaissance Literature* (Cambridge, 1964).

Liddell, Henry and Robert Scott: *An Intermediate Greek-English Lexicon* (1889; Oxford, 1997).

Lines, David A.: "Science and Universities of Early Modern Europe: Teaching, Specialization, Professionalization," *Early Science and Medicine*, 6 (2001), 267–323.

Lloyd, G.E.R.: *Aristotelian Explorations* (Cambridge, 1996).

Lloyd, G.E.R.: *Magic, Reason and Experience: Studies in the Origins and Development of Greek Science* (Cambridge, 1979).

Lohne, Johannes: "Thomas Harriott (1560–1621): The Tycho Brahe of Optics. Preliminary Notice," *Centaurus*, 6 (1959), 113–121.

Lohr, Charles H.: *Renaissance Authors. Latin Aristotle Commentaries*, vol. 2 (Florence, 1988).

Lüthy, Christoph: "An Aristotelian Watchdog as Avant-Garde Physicist: Julius Caesar Scaliger," *The Monist*, 84 (2001), 542–561.

Lüthy, Christoph: "Atomism in the Renaissance," in *Encyclopedia of Renaissance Philosophy*, ed. Marco Sgarbi (Cham, 2015). doi:10.1007/978-3-319-02848-4_252-1.

Lüthy, Christoph: "Bruno's *Area Democriti* and the Origins of Atomistic Imagery," *Bruniana et Campanelliana*, 4 (1998), 59–92.

Lüthy, Christoph: "Centre, Circle, Circumference: Giordano Bruno's Astronomical Woodcuts," *Journal for the History of Astronomy*, 41 (2010), 311–327.

Lüthy, Christoph: *David Gorlaeus (1591–1612): An Enigmatic Figure in the History of Philosophy and Science* (Amsterdam, 2012).

Lüthy, Christoph: "David Gorlaeus' Atomism, or: The Marriage of Protestant Metaphysics with Italian Natural Philosophy," in *Late Medieval and Early Modern Corpuscular Matter Theories*, ed. Christoph Lüthy, John E. Murdoch, and William R. Newman (Leiden, 2001), 245–290.

Lüthy, Christoph: "Ontology and the Place of Metaphysics: Taurellus, Gorlaeus, Regius and Descartes," in *Physics and Metaphysics in Descartes and in His Reception*, ed. Delphine Antoine-Mahut and Sophie Roux (New York, 2018), 85–116.

Lüthy, Christoph: "The Fourfold Democritus on the Stage of Early Modern Science," *Isis*, 91 (2000), 443–479.

Lüthy, Christoph: "The Invention of Atomist Iconography," in *The Power of Images in Early Modern Science*, ed. Wolfgang Lefèvre, Jürgen Renn and Urs Schoepflin (Basel, 2003), 117–138.

Lüthy, Christoph: "Thoughts and Circumstances of Sébastien Basson. Analysis, Micro-history, Questions," *Early Science and Medicine*, 2 (1997), 2–73.

Lüthy, Christoph: John E. Murdoch, and Newman, William R., eds: *Late Medieval and Early Modern Corpuscular Matter Theories* (Leiden – Boston, MA – Cologne, 2001).

Lüthy, Christoph: and William R. Newman, eds.: "The Fate of Hylomorphism: 'Matter' and 'Form' in Early Modern Science," *Early Science and Medicine*, 2 (1997), 215–359.

Lüthy, Christoph: and Leen Spruit: "The Doctrine, Life, and Roman Trial of the Frisian Philosopher Henricus de Veno (1574?–1613)," *Renaissance Quarterly*, 56 (2003), 1112–1151.

Mabilleau, Léopold: *Histoire de la philosophie atomistique* (Paris, 1895).

Maier, Anneliese: *An der Grenze von Scholatik und Naturwissenschaft. Studien zur Naturphilosophie des 14. Jahrhunderts* (Essen, 1943; Rome, 1952).

Maier, Anneliese: "Die Struktur der materiellen Substanz," in eadem, *An der Grenze von Scholastik und Naturwissenschaft. Studien zur Naturphilosophie des 14. Jahrhunderts* (Essen, 1943), 7–140.

Maier, Anneliese: *On the Threshold of Exact Science: Selected Writings of Anneliese Maier on Late Medieval Natural Philosophy*, ed. and transl. Steven D. Sargent (Philadelphia, PA, 1982).

Malagola, Carlo, ed.: *Statuti delle università e dei college dello studio bolognese* (Bologna, 1888).

Mallet, Charles E.: *A History of the University of Oxford*, 2 vols. (London, 1924).

Manning, Gideon: "Three Biased Reminders about Hylomorphism in Early Modern Science and Philosophy," in *Matter and Form in Early Modern Science and Philosophy*, ed. Manning (Leiden, 2012), 1–33.

Manning, Gideon: "The History of 'Hylomorphism'," *Journal of the History of Ideas*, 74 (2013), 173–18.

Manning, Gideon, ed.: *Matter and Form in Early Modern Philosophy* (Leiden – Boston, MA, 2012).

Manzo, Silvia: "Francis Bacon and Atomism: a Reappraisal," in *Late Medieval and Early Modern Corpuscular Matter Theories*, Christoph Lüthy, John E. Murdoch and William R. Newman (Leiden, 2001), 209–243.

Martin, Craig: "Alchemy and the Renaissance Commentary Tradition on *Meteorologica* IV," *Ambix: The Journal of the Society for the History of Alchemy and Chemistry*, 51 (2004), 245–262.

Martin, Craig: "Francisco Vallés and the Renaissance Reinterpretation of Aristotle's *Meteorologica* IV as a Medical Text," *Early Science and Medicine*, 7 (2002), 1–30.

Martin, Craig: *Renaissance Meteorology: Pomponazzi to Descartes* (Baltimore, MD, 2011).

Martin, Craig: "With Aristotelians Like These, Who Needs Anti-Aristotelians? Chymical Corpuscular Matter Theory in Niccolò Cabeo's Meteorology," *Early Science and Medicine*, 11 (2006), 135–161.

Matteoli, Marco: "Materia, minimo e misura: la genesi dell'atomismo 'geometrico' in Giordano Bruno," *Rinascimento*, 50 (2010), 425–449.

Mayhew, Robert J.: "From Hackwork to Classic: The English Editing of the *Geographia Generalis*," in *Bernhard Varenius (1622–1650)*, ed. Margret Schuchard (Leiden, 2007), 239–257.

McGinnis, Jon: *Avicenna* (Oxford – New York, 2010).

McGuire, J.E. and Martin Tamny: *Certain Philosophical Questions: Newton's Trinity Notebook* (Cambridge, 1983).

McPherson, David: "Ben Jonson's Library," *Studies in Philology*, 21, supplement (1974), 1–106.

Meinel, Christoph: *Die Bibliothek des Joachim Jungius: Ein Beitrag zur Historia litteraria der frühen Neuzeit* (Hamburg, 1992).

Meinel, Christoph: "Early Seventeenth-Century Atomism: Theory, Epistemology, and the Insufficiency of Experiment," *Isis*, 79 (1988), 68–103.

Meinel, Christoph: *Joachim Jungius, Praelectiones Physicae: Historisch-kritische Edition* (Göttingen, 1982).

Merchant, Carolyn: *The Death of Nature: Women, Ecology, and The Scientific Revolution* (San Francisco, CA, 1980).

Merchant, Carolyn: "The Scientific Revolution and The Death of Nature," *Isis*, 97 (2006), 485–533.

Michael, Emily: "Averroes and the Plurality of Forms," *Franciscan Studies*, 52 (1992), 155–182.

Michael, Emily: "Daniel Sennert on Matter and Form: At the Juncture of the Old and the New," *Early Science and Medicine*, 2 (1997), 272–299.

Michael, Emily: "Sennert's Sea Change: Atoms and Causes," in *Late Medieval and Early Modern Corpuscular Matter Theories*, ed. Christoph H. Lüthy, John E. Murdoch and William R. Newman (Leiden, 2001), 331–362.

Michel, Paul: "L'atomisme de Giordano Bruno," in *La science au seizième siècle. Colloque international de Royaumont, 1–4 juillet 1957* (Paris 1960), 249–264.

Milanese Guido: "Italian Commentaries on Lucretius," in *Classical Commentaries: Explorations in a Scholarly Genre*, ed. Christina Shuttleworth Kraus and Christopher Stray (Oxford, 2016), 195–215.

Monti, Carlo: "Incidenza e significato della tradizione materialistica antica dei poemi latini di Girodano Bruno: la mediazione di Lucrezio," *Nouvelles de la République des Lettres*, 2 (1994), 75–87.

Monti, Carlo: "Introduzione," in Giordano Bruno, *Opere latine*, transl. Monti (Turin, 1980), 9–62.

Moran, Bruce T.: *Andreas Libavius and the Transformation of Alchemy: Separating Chemical Cultures with Polemical Fire* (Sagamore Beach, MA, 2007).

Moran, Bruce T.: "Art and Artisanship in Early Modern Alchemy," *Getty Research Journal*, 5 (2013), 1–14.

Moran, Bruce T.: "Preserving the Cutting Edge: Traveling Woodblocks, Material Networks, and Visualizing Plants in Early Modern Europe," in *The Structures of Practical Knowledge*, ed. Matteo Valleriani (Cham, 2017), 393–419.

Moreau, Elisabeth: "Elements, Mixture and Temperament: The Body's Composition in Renaissance Physiology," in *Oeconomia corporis: The Body's Normal and Pathological Constitution at the Intersection of Philosophy and Medicine*, ed. Chiara Beneduce and Denise Vincenti (Pisa, 2018), 51–58.

Moreau, Elisabeth: "Reforming the *Prisca Medicina*: Libavius' Axioms of Elements and Mixture," in *Natural Knowledge and Aristotelianism at Early Modern Protestant Universities*, ed. Pietro D. Omodeo and Volkhard Wels (Wiesbaden, 2019), 255–270.

Moreau, Elisabeth: "Atoms, Mixture, and Temperament in Early Modern Medicine: The Alchemical and Mechanical Views of Sennert and Beeckman," in *Santorio Santori and the Emergence of Quantified Medicine, 1614–1790: Corpuscularianism, Technology and Experimentation*, ed. Jonathan Barry and Fabrizio Bigotti (Cham, 2022), 137–164.

Mortimer, Sarah: "Early Modern Socinianism and Unitarianism," in *The Oxford Handbook of Early Modern Theology, 1600–1800*, ed. Ulrich L. Lehner, Richard A. Muller, and A. G. Roeber (Oxford, 2016), 361–372.

Mucillo, Maria: "La concezione dello spazio di Francesco Patrizi (1529–1597) e la sua fortuna nell'ambito della reazione anticartesiana inglese," *Dal cartesianismo all'Illuminismo radicale*, ed. Carlo Borghero and Claudio Buccolini (Florence, 2010), 49–71.

Mucillo, Maria: "Marsilio Ficino e Francesco Patrizi," in *Marsilio Ficino e il ritorno di Platone*, ed. Gian Carlo Garfagnini (Florence, 1986), 615–679.

Mucillo, Maria: *Platonismo, ermetismo e "prisca theologia." Ricerche di storiografia filosofica rinascimentale* (Florence, 1996).

Mulsow, Martin, and Jan Rohls, eds.: *Socinianism and Arminianism: Antitrinitarians, Calvinists and Cultural Exchange in Seventeenth-Century Europe* (Leiden, 2005).

Multhauf, Robert P.: "John of Rupescissa and the Origin of Medical Chemistry," *Isis*, 45 (1954), 359–367.

Murdoch, John E.: "Mathesis in philosophiam scolasticam introducta: the Rise and Development of the Application of Mathematics in Fourteenth Century Philosophy and Theology," in *Arts libéraux et philosophie au moyen âge* (*Actes du IVe Congrès International de Philosophie Médiévale*) (Montreal/Paris, 1969), 215–254.

Murdoch, John E.: "Naissance et développement de l'atomisme au bas Moyen Âge latin," in *La science de la nature: théories et pratiques*, ed. Guy H. Allard et al. (Montreal, 1974), 11–32.

Murdoch, John E.: "The Medieval and Renaissance Tradition of *Minima Naturalia*," in *Medieval and Early Modern Corpuscular Matter Theories*, ed. Christoph Lüthy, John E. Murdoch and William R. Newman (Leiden, 2001), 91–131.

Nail, Thomas: *Lucretius I: An Ontology of Motion* (Edinburgh, 2018).

Nardi, Bruno: *Studi su Pomponazzi* (Florence, 1965).

Navarro-Brotons, Victor: "Matter and Form in Sixteenth-Century Spain: Some Case Studies," in *The Mechanization of Natural Philosophy*, ed. Daniel Garber and Sophie Roux (Dordrecht, 2012), 99–116.

Neumann, Hanns-Peter: *Monaden im Diskurs. Monas, Monaden, Monadologien* (*1600 bis 1770*) (Stuttgart, 2013).

Newman, William R.: *Atoms and Alchemy: Chymistry and the Experimental Origins of the Scientific Revolution* (Chicago, IL, 2006).

Newman, William R.: *The Summa Perfectionis of pseudo-Geber* (Leiden, 1991).

Newman, William R.: "Boyle's Debt to Corpuscular Alchemy," in *Robert Boyle Reconsidered*, ed. Michael Hunter (Cambridge, 1994), 107–118.

Newman, William R.: "Corpuscular Alchemy and the Tradition of Aristotle's *Meteorology*, with Special Reference to Daniel Sennert," *International Studies in the Philosophy of Science*, 15 (2001), 145–153.

Newman, William R.: "Experimental Corpuscular Theory in Aristotelian Alchemy: From Geber to Sennert," in *Late Medieval and Early Modern Corpuscular Matter Theories*, ed. Christoph Lüthy, John E. Murdoch, and William R. Newman (Leiden, 2001), 291–329.

Newman, William R.: "Mercury and Sulphur Among the High Medieval Alchemists: From Rāzī and Avicenna to Albertus Magnus and pseudo-Roger Bacon," *Ambix*, 61 (2014), 327–344.

Newman, William R.: *Newton the Alchemist* (Princeton, NJ, 2019).

Newman, William R.: *Promethean Ambitions: Alchemy and the Quest to Perfect Nature* (Chicago, IL, 2004).

Newman, William R.: "The Alchemical Sources of Robert Boyle's Corpuscular Philosophy," *Annals of Science*, 53 (1996), 567–585.

Newman, William R.: "The Significance of 'Chymical Atomism'," *Early Science and Medicine*, 14 (2009), 248–264.

Newman, William R., and Principe, Lawrence M.: "Alchemy vs. Chemistry: The Etymological Origins of a Historiographic Mistake," *Early Science and Medicine*, 3 (1998), 32–65.

Nicoli, Elena: "The Earliest Renaissance Commentaries on Lucretius and the Issue of Atomism" (PhD thesis, Radboud University Nijmegen, 2017).

Nicoli, Elena: "Ficino, Lucretius and Atomism," *Early Science and Medicine*, 23 (2018), 330–361.

Nicoli, Elena: "Atoms, Elements, Seeds: A Renaissance Interpreter of Lucretius' Atomism," in *Lucretius Poet and Philosopher*, ed. Philip R. Hardie, Valentina Prosperi and Diego Zucca (Berlin – Boston, MA, 2020), 235–250.

Norbrook, David, Stephen Harrison and Philip Hardie: *Lucretius and the Early Modern* (Oxford, 2016), 45–68.

Novoa, James W. Nelson: "Leonardo Marso d'Avezzano's Oratio on the Death of Giovanni Battista Pio," *Bruniana & Campanelliana*, 16 (2010), 247–253.

Nutton, Vivian: "Physiologia from Galen to Jacob Bording," in *Blood, Sweat, and Tears: The Changing Concepts of Physiology from Antiquity into Early Modern Europe*, ed. Manfred Horstmanshoff, Helen King and Claus Zittel (Leiden – Boston, MA, 2012), 27–41.

Nye, Mary Jo: *Molecular Reality. A Perspective on the Scientific Work of Jean Perrin* (London, 1972).

Obrist, Barbara: "Les rapports d'analogie entre philosophie et alchimie médiévales," in *Alchimie et philosophie à la Renaissance*, ed. Jean-Claude Margolin and Sylvain Matton (Paris, 1993), 43–64.

Oliva, Cesare: "Note sull'insegnamento di Pietro Pomponazzi," *Giornale critico della filosofia italiana*, 7 (1926), 83–103, 179–190, 254–275.

Olson, Glendening: "The Medieval Fortunes of 'Theatrica'," *Traditio*, 42 (1986), 265–86.

O'Malley, Charles Donald: *Andreas Vesalius of Brussels 1514–1564* (Berkeley, CA, 1964).

Pabst, Bernhard: *Atomentheorien des lateinischen Mittelalters* (Darmstadt, 1994).

Pagel, Walter: *Paracelsus: An Introduction to Philosophical Medicine in the Era of the Renaissance*. 2nd ed. (Basel, 1982).

Pagnoni, Maria Rita: "Prime note sulla tradizione medievale ed umanistica di Epicuro," *Annali della Scuola normale superiore di Pisa*, 4 (1974), 1443–1477.

Palmer, Ada: *Reading Lucretius in the Renaissance* (Cambridge, MA – London, 2014).

Palmer, Ada: "The Persecution of Renaissance Lucretius Readers Revisited," in *Lucretius Poet and Philosopher*, ed. Philip R. Hardie, Valentina Prosperi and Diego Zucca (Berlin – Boston, MA, 2020), 167–197.

Palmerino, Carla Rita: "Libertus Fromondus' Escape from the Labyrinth of the Continuum (1631)," *Lias*, 42 (2015), 3–36.

Palmerino, Carla Rita: "The Isomorphism of Space, Time and Matter in Seventeenth-century Natural Philosophy," *Early Science and Medicine*, 16 (2011), 296–330.

Pasnau, Robert: *Metaphysical Themes, 1274–1671* (Oxford, 2011).

Patterson, T.S.: "Jean Beguin and his Tyrocinium Chymicum," *Annals of Science*, 2 (1937), 243–298.

Pereira, Michela: "Heavens on Earth: From the *Tabula Smaragdina* to the Alchemical Fifth Essence," *Early Science and Medicine*, 5 (2000), 131–144.

Perfetti, Stefano: *Aristotle's Zoology and its Renaissance Commentators, 1521–1601* (Leuven, 2000).

Perfetti, Stefano: "Docebo vos dubitare. Il commento inedito di Pietro Pomponazzi al *De partibus animalium* (Bologna, 1521–24)," *Documenti e studi sulla tradizione filosofica medievale*, 10 (1999), 439–466.

Petersen, Peter: *Geschichte der aristotelischen Philosophie im protestantischen Deutschland* (Leipzig, 1921).

Petrescu, Lucian: "John Duns Scotus and the Ontology of Mixture," *Res philosophica*, 91 (2014), 315–337.

Pioli, Giovanni: *Fausto Socino: vita, opere, fortuna* (Modena, 1952).

Pizzani, Ubaldo: "Dimensione cristiana dell'umanesimo e messaggio lucreziano: la *Paraphrasis in Lucretium* di Raphael Francus," in *Validità perenne dell'Umanesimo*, ed. Giovannangiola Tarugi (Florence, 1986), 313–333.

Plastina, Sandra: "Nicholas Hill and Giordano Bruno: The New Cosmology in the *Philosophia Epicurea*," *Physis*, 38 (2000), 415–432.

Plastina, Sandra: "Nicholas Hill: *The English Campanella?*," *Bruniana&Campanelliana*, 4 (1998), 207–212.

Plastina, Sandra: "Nicholas Hill: un medico-filosofo 'epicureo' della tarda età elisabettiana," in Nicholas Hill, *Philosophia Epicuraea Democritiana Theophrastica*, ed. Plastina (Pisa-Rome, 2007), 11–75.

Plastina, Sandra: "*Philosophia lucis proles verissima est.* Nicholas Hill lettore di Francesco Patrizi," *Bruniana&Campanelliana*, 10 (2004), 175–182.

Plastina, Sandra and Tiziana Provvidera: "Il 'De infinitate et aeternitate mundi' attribuito a N. Hill (Wood, M.S. Bodleian Library 42, ff. 174–175)," *Bollettino Filosofico*, 16 (2000), 50–74.

Podestà, Bartolomeo: "Di alcuni documenti inediti risguardanti Pietro Pomponazzi lettore nello studio bognese cavati dall'antico archivio del reggimento in oggi della prefettura," *Atti e memorie della regia deputazione di storia patria per le provincie di Romagna*, 6 ([1867] 1868), 133–180.

Popa, Tiberiu: "Scientific Method in *Meteorology* IV," *HOPOS: The Journal of the International Society for the History of Philosophy of Science*, 4 (2014), 306–334.

Popkin, Richard H.: *The History of Scepticism from Erasmus to Spinoza* (Berkeley, CA, 1979).

Principe, Lawrence M.: "Diversity in Alchemy: The Case of Gaston 'Claveus' DuClo, a Scholastic Mercurialist Chrysopoeian," in *Reading the Book of Nature: The Other Side of the Scientific Revolution*, ed. Allen G. Debus and Michael Walton (Kirksville, MO, 1998), 181–200.

Prins, Johannes L.M.: "Walter Warner (ca. 1557–1643) and His Notes on Animal Organisms" (PhD thesis, Utrecht University, 1992).

Prosperi, Valentina: *Di soavi licor gli orli del vaso. La fortuna di Lucrezio dall'Umanesimo alla Controriforma* (Turin, 2004).

Prosperi, Valentina: "Lucretius in the Italian Renaissance," in *The Cambridge Companion to Lucretius*, ed. Stuart Gillespie and Philip R. Hardie (Cambridge, 2007), 214–226.

Prosperi, Valentina: "Lost in Translation: The Sixteenth Century Vernacular Lucretius," in *Lucretius Poet and Philosopher*, ed. Philip R. Hardie, Valentina Prosperi and Diego Zucca (Berlin – Boston, MA, 2020), 145–166.

Prosperi, Valentina: *Il fantasma di Lucrezio: La perduta traduzione del De rerum natura di Giovan Francesco Muscettola* (Pisa, 2022).

Puliafito Bleuel, Anna Laura: "Per uno studio della *Nova de universis philosophia* di Francesco Patrizi da Cherso: note alla Panaugia," *Atti e Memorie dell'Accademia toscana di Scienze e Lettere "La Colombaria,"* 52 (1987), 160–199.

Pullman, Bernard: *The Atom in the History of Human Thought*, transl. Axel Reisinger (New York, 1998).

Purnell, Fred: "Francesco Patrizi," *The Stanford Encyclopedia of Philosophy* (Spring 2017 Edition), Edward N. Zalta (ed.), https://plato.stanford.edu/archives.

Pyle, Andrew: *Atomism and Its Critics: From Democritus to Newton* (Bristol, 1998).

Ragland, Evan R.: "Chymistry and Taste in the Seventeenth Century: Franciscus Dele Boë Sylvius as a Chymical Physician between Galenism and Cartesianism," *Ambix*, 59 (2012), 1–21.

Redondi, Pietro: *Galileo Heretic*, transl. Raymond Rosenthal (Princeton, NJ, 1987).

Rees, Graham: "Atomism and Subtlety in Francis Bacon's Philosophy," *Annals of Science*, 37 (1980), 549–571.

Reeve, Michael: "Lucretius in the Middle Ages," in *The Cambridge Companion to Lucretius*, ed. Stuart Gillespie and Philip R. Hardie (Cambridge, 2007), 205–213.

Rhodes, Dennis E.: "Raphael Franciscus Equals Raphael Francus, florentinus," *Gutemberg-Jahrbuch*, 55 (1980), 79–81.

Richter, Frank: "Varenius – ein Geograph zwischen allen Stühlen?" in *Bernhard Varenius (1622–1650)*, ed. Margret Schuchard (Leiden, 2007), 191–213.

Robert, Aurélien: "Atomism," in *Encyclopedia of Medieval Philosophy*, ed. Hendrik Lagerlund (Dordrecht, 2011), 1: 122–125.

Robert, Aurélien: "Epicure et les épicuriens au Moyen Âge," *Micrologus*, 21 (2013), 3–46.

Robert, Aurélien: *Epicure aux Enfers: Hérésie, athéisme et hédonisme au Moyen Âge* (Paris, 2021).

Robin, Léon: *Aristote* (Paris, 1944).

Robinson, Howard M.: "Prime Matter in Aristotle," *Phronesis*, 19 (1974), 168–188.

Robson, John Adam: *Wyclif and the Oxford Schools* (Cambridge, 1961).

Ross, William David: *Aristotle* (London, 1923).

Rudolph, Hartmut: "Hohenheim's Anthropology in the Light of His Writings on the Eucharist," in *Paracelsus. The Man and His Reputation, His Ideas and Their Transformation*, ed. Ole Grell (Leiden, 1998), 187–201.

Sailor, Danton B.: "Moses and Atomism," *Journal for the History of Ideas*, 25 (1964), 3–16.

Saint-Genois, Jules de: "Bièse, Nicolas," in *Biographie nationale de Belgique*, 28 vols. (Brussels, 1866–1944), vol. 2, s.v.

Sakamoto, Kuni: *Julius Caesar Scaliger, Renaissance Reformer of Aristotelianism: A Study of His* Exotericae Exercitationes (Leiden, 2016).

Salatowsky, Sascha: "God in Time and Space: Socinian Physics and Its Opponents," in *Natural Knowledge and Aristotelianism at Early Modern Protestant Universities*, ed. Pietro Daniel Omodeo and Volkhard Wel (Wiesbaden, 2019), 47–82.

Schemmel, Matthias: *The English Galileo. Thomas Harriot's Work on Motion as an Example of Preclassical Mechanics*, 2 vols. (Dordrecht, 2008).

Schouten, Johannes: "Johannes Walaeus (1604–1649) and His Experiments on the Circulation of the Blood," *Journal of the History of Medicine and Allied Sciences*, 29 (1974), 259–279.

Schuchard, Margret: "Varenius and his Family: A Dynasty Dedicated to Scholarship and Rooted in Christian Philosophy," in *Bernhard Varenius (1622–1650)*, ed. Schuchard (Leiden, 2007), 11–26.

Schweizer, Alexander: "Conrad Vorstius. Vermittlung der reformirten Centraldogmen mit den socinianischen Einwendungen. Erster Artikel," *Theologische Jahrbücher*, 15 (1856), 435–486.

Scribano, Maria Emanuela: *Da Descartes a Spinoza: percorsi della teologia razionale nel Seicento* (Milan, 1988).

Sedley, David: *Lucretius and the Transformation of Greek Wisdom* (Cambridge, 1998).

Sedley, David: "Lucretius' Use and Avoidance of Greek," *Proceedings of the British Academy*, 93 (1999), 227–246.

Shackelford, Jole: *A Philosophical Path for Paracelsian Medicine: The Ideas, Intellectual Context, and Influence of Petrus Severinus (1540/2–1602)* (Copenhagen, 2004).

Shackelford, Jole: "La transplantation comme transformation chez Paracelse: de la structure à la function," in *Transplanter: une approche transdisciplinaire*, ed. François Delaporte, Bernard Devauchelle, and Emmanuel Fournier (Paris, 2015), 35–40.

Shackelford, Jole: "Paracelsian Uroscopy and German Chemiatric Medicine in the *Medicina Pensylvania* of George de Benneville," in *Medical Theory and Therapeutic Practice in the Eighteenth Century: A Transatlantic Perspective*, ed. Jürgen Helm and Renate Wilson (Stuttgart, 2008), 13–35.

Shackelford, Jole: "Providence, Power, and Cosmic Causality in Early Modern Astronomy: The Case of Tycho Brahe and Petrus Severinus," in *Tycho Brahe and Prague: Crossroads of European Science*, ed. J.R. Christianson, Alena Hadravova, Petr Hadrava, and Martin Šolc (Frankfurt am Main, 2002), 46–69.

Shackelford, Jole: "Rosicrucianism, Lutheran Orthodoxy, and the Rejection of Paracelsianism in Early Seventeenth-Century Denmark," *Bulletin of the History of Medicine*, 70 (1996), 181–204.

Shackelford, Jole: "Seeds with a Mechanical Purpose: Severinus' Semina and Seventeenth-Century Matter Theory," in *Reading the Book of Nature: The Other Side of the Scientific Revolution*, ed. Allen G. Debus and Michael T. Walton (Kirksville, MO, 1998), 15–44.

Shackelford, Jole: "Transplantation and Corpuscular Identity in Paracelsian Vital Philosophy," in *Early Modern Medicine and Natural Philosophy*, ed. Peter Distelzweig, Benny Goldberg, and Evan Ragland (New York, 2016), 229–253.

Shackelford, Jole: "Tycho Brahe, Laboratory Design, and the Aim of Science: Reading Plans in Context," *Isis*, 84 (1993), 211–230.

Shackelford, Jole: "Western Esotericism and the History of European Science and Medicine in the Early Modern Period," in *Western Esotericism*, ed. Tore Ahlbäck. Scripta Instituti Donneriani Aboensis xx (Åbo, 2008), 173–203.

Shapiro, Alan: *Fits, Passions, and Paroxysms* (Cambridge, 1993).

Shapiro, Alan: *The Optical Papers of Isaac Newton*, vol. 1: *The Optical Lectures, 1670–1672* (Cambridge, 1984).

Shirley, John W.: *Thomas Harriot: A Biography* (Oxford, 1983).

Silverstein, Theodor: "*Elementatum*: Its Appearance Among the Twelfth-Century Cosmogonists," *Mediaeval Studies*, 16 (1954), 156–162.

Silverstein, Theodor: "Guillaume de Conches and the Elements: Homoiomeria und Organica," *Mediaeval Studies*, 26 (1964), 363–367.

Singer, D. Waley: *Bruno, Giordano: His Life and Thought; with Annotated Translation of his Work,* On the Infinite Universe and Worlds (New York, 1950).

Siraisi, Nancy G.: *Avicenna in Renaissance Italy: The* Canon *and Medical Teaching in Italian Universities after 1500* (Princeton, NJ, 1987).

Skautrup, Peter: *Det Danske Sprogs Historie* (Copenhagen, 1944).

Solmsen, Friedrich: *Aristotle's System of the Physical World* (Ithaca, NY, 1960).

Solmsen, Friedrich: "The Vital Heat, the Inborn Pneuma and the Aether," *The Journal of Hellenic Studies*, 73 (1957), 119–123.

Sorabji, Richard: *The Philosophy of the Commentators 200–600 AD. A Sourcebook*. Vol. 2: *Physics* (London, 2004).

Spijker, Willem van 't: "Heidelberger Gutachten in Sachen Vorstius," in *Späthumanismus und reformierte Konfession. Theologie, Jurisprudenz und Philosophie in Heidelberg an der Wende zum 17. Jahrhundert*, ed. Christoph Strohm, Joseph S. Freedman, and Herman J. Selderhuis (Tübingen, 2006), 207–225.

Spruit, Leen: "Giordano Bruno's Use of Experience in Context," *La filosofia di Giordano Bruno. Problemi ermeneutici storiografici*, Convegno internazionale, Roma 23–24 ottobre 1998, ed. Eugenio Canone (Florence, 2003), 145–164.

Spruit, Leen: "The Pomponazzi Affair: The Controversy over the Immortality of the Soul," in *The Routledge Companion to Sixteenth Century Philosophy*, ed. Henrik Lagerlund and Benjamin Hill (London, 2017), 225–246.

Staffhorst, Ulrich: "Miszellen zur *Geographia generalis des Bernhard Varenius*," in *Berhard Varenius (1622–1650)*, ed. Margret Schuchard (Leiden, 2007), 215–225.

Stolberg, Michael: "Particles of the Soul: The Medical and Lutheran Context of Daniel Sennert's Atomism," *Medicina nei secoli*, 15 (2003), 177–203.

Stone, Abraham: "Avicenna's Theory of Primary Mixture," *Arabic Sciences and Philosophy*, 18 (2008), 99–119.

Stückelberger, Alfred: "Lucretius reviviscens: Von der antiken zur neuzeitlichen Atomphysik," *Archiv für Kulturgeschichte*, 54 (1972), 1–25.

Subow, Walter: "Zur Geschichte des Kampfes zwischen Atomismus und dem Aristotelismus im 17. Jahrhundert (Minima naturalia und mixtio)," in Nikolai Alexandrovich Figurowski et al., *Sowjetische Beiträge zur Geschichte der Wissenschaft* (Berlin, 1960), 161–191.

Sytsma, David S.: "Calvin, Daneau, and 'Physica Mosaica': Neglected Continuities at the Origins of an Early Modern Tradition," *Church History and Religious Culture*, 95 (2015), 457–476.

Szulakowska, Ursula: *The Sacrificial Body and the Day of Doom: Alchemy and Apocalyptic Discourse in the Protestant Reformation* (Leiden, 2006).

Tartarkiewitz, W.: "Theatrica, the Science of Entertainment from the XIIth to the XVIIth Century," *Journal of the History of Ideas*, 26 (1965), 263–272.

Taub, Liba: *Ancient Meteorology* (London, 2003).

Taub, Liba: "Physiological Analogies and Metaphors in Explanations of the Earth and the Cosmos," in *Blood, Sweat and Tears*, ed. Manfred Horstmanshoff et al. (Leiden, 2012), 41–64.

Taylor, C.C.W.: *The Atomists: Leucippus and Democritus: Fragments. A Text and Translation with Commentary* (Toronto 1999).

Telle, Joachim: "Remarques sur le *Rosarium philosophorum* (1550), avec une liste sélective d'ouvrages sur l'alchimie médiévale," *Chrysopoeia*, 5 (1992–1996), 265–319.

Telle, Joachim: "Ricardus Anglicus," in *Die deutsche Literatur des Mittelalters: Verfasserlexicon*, ed. Karl Ruh (Berlin, 1992), vol. 8, 38–41.

Thom, Johan C.: "The Cosmotheology of *De mundo*," in *Cosmic Order and Divine Power: Pseudo-Aristotle*, On the Cosmos, ed. Thom (Tübingen, 2014), 107–120.

Tilmans, Karin: "From *institutio* to *educatio*: The Origin of Political Education in the Habsburg Netherlands," in *The Education of a Christian Society: Humanism and the Reformation in Britain and the Netherlands*, ed. N. Scott Amos, Andrew Pettegree and Henk van Nierop (London, 1999), 41–61.

Trevor-Roper, Hugh: *Catholics, Anglicans and Puritans: Seventeenth-Century Essays* (London, 1987).

Trevor-Roper, Hugh: "Nicholas Hill," in *Oxford Dictionary of National Biography* (Oxford, 2004–2006), art. 13287.

Tydeman, William: *The Theatre in the Middle Ages: Western European Stage Conditions, c. 800–1576* (Cambridge, 1978).

Van Melsen, Andrew G.: *From Atomos to Atom: The History of the Concept Atom*, transl. Henry J. Koren (Pittsburgh and Oxford, 1952).

Vanpaemel, Geert: "Libert Froidmont et l'atomisme," in Anne-Catherine Bernès, ed., *Libert Foridmont et les résistances aux révolutions scientifiques. Actes du Colloque Château d'Oupeye, 26 et 27 septembre 1987* (Haccourt, 1988).

van Ruler, Han: *The Crisis of Causality: Voetius and Descartes on God, Nature and Change* (Leiden, 1995).

van Ruler, Han: "Substantiële entiteiten zonder vorm: Gorlaeus' worsteling met Aristoteles," *It Beaken*, 76 (2014), 197–228.

Vasoli, Cesare: "Il 'platonico machiavellico': gli scritti militari," in *Francesco Patrizi da Cherso*, ed. Vasoli (Rome, 1980), 229–259.

Védrine, Hélène: "L'Obstacle réaliste en mathématique chez deux philosophes du XVI siècle: Bruno et Patrizzi," in *Platon et Aristote à la Renaissance*, ed. Jean-Claude Margolin and Maurice de Gandillac (Paris, 1976), 239–248.

Verbeek, Theo: *René Descartes et Martin Schoock: La querelle d'Utrecht* (Paris, 1988).

Verde, Armando F.: *Lo studio fiorentino, 1473-1503, Ricerche e documenti.* Vol. 2: *Docenti, dottorati* (Florence, 1973).

Verde, Armando F.: "Il secondo periodo dello studio fiorentino (1504–1528)," in *L'Università e la sua storia*, ed. Paolo Renzi (Siena, 1998), 105–131.

Viano, Cristina: *La matière des choses: Le livre IV des* Météorologiques *d'Aristote et son interprétation par Olympiodore* (Paris, 2006).

Walker, Daniel Pickering: "The Astral Body in Renaissance Medicine," *Journal of the Warburg and Courtauld Institutes*, 21 (1958), 119–133.

Walker, Daniel Pickering: *The Ancient Theology: Studies in Christian Platonism from the Fifteenth to the Eighteenth Century* (London, 1972).

Walker, Daniel Pickering: "Medical 'Spirits' and God and the Soul," in *Spiritus. IV. Colloquio internazionale del Lessico Intelletuale Europeo*, ed. Marta Fattori and Massimo Luigi Bianchi (Rome, 1984), 233–244.

Walker, Daniel Pickering: "Medical Spirits in Philosophy and Theology from Ficino to Newton," in *Arts du spectacle et histoire des idées. Recueil offert en hommage à Jean Jacquot* (Tours, 1984), 287–300.

Warntz, William: "Newton, the Newtonians, and the *Geographia Generalis Varenii*," *Annals of the Association of American Geographers*, 79 (1989), 165–191.

Weeks, Andrew: *Paracelsus, Theophrastus Bombastus von Hohenheim, 1493–1541: Essential Theoretical Writings* (Leiden – Boston, MA, 2008).

Westfall, Richard S.: *The Construction of Modern Science: Mechanisms and Mechanics* (Cambridge, 1977).

Wilson, Catherine: *Epicureanism at the Origins of Modernity* (Oxford, 2008).

Wilson, Malcolm: *Structure and Method in Aristotle's* Meteorologica: *A More Disorderly Nature* (Cambridge, 2013).

Wohlwill, Emil: "Joachim Jungius und die Erneuerung atomistischer Lehren im 17. Jahrhundert," *Abhandlungen aus dem Gebiete der Naturwissenschaften, herausgegeben vom Naturwissenschaftlichen Verein in Hamburg* (Hamburg, 1887), 3–66.

Woods, Rega and Weisberg, Michael: "Interpreting Aristotle on Mixture: Problems about Elemental Composition from Philiponus to Cooper," *Studies in History and Philosophy of Science*, 35 (2004), 681–706.

Yates, Frances A.: *Giordano Bruno and the Hermetic Tradition* (London, 1964).

Yates, Frances A.: *Theatre of the World* (London, 1969).

Zedler, Johann Heinrich: *Grosses vollständiges Universal-Lexikon aller Wissenschaften und Künste*, 68 vols. (Halle und Leipzig, 1732–1754).

Zilsel, Edgar: "The Origins of Gilbert's Scientific Method," in *Roots of Scientific Thought. A Cultural Perspective*, ed. Philip P. Wiener and Aaron Noland (New York, 1960), 219–250.

Index of Names

Achillini, Alessandro 19, 42
Acsubofen 269
Adami, Tobias 199
Aetius 183
Albertus Magnus 96–97
Alcionio, Pietro 98
Alcmeon 184
Alexander of Aphrodisias 19, 36–38, 70, 183
Alva, Duke of 115
Amalarius, Bishop of Metz 165
Amato, Barbara 176*n*2, 186*n*27
Anaxagoras 39–40, 141, 179, 184–185, 195
Anaximenes 126
Anglicus, Ricardus. *See* Ricardus
Antonius, Joannes 34
Aquinas, Thomas 37–38, 59, 98, 233, 234*n*23, 262*n*64
Aretino, Rinuccio 25
Aristotle 1, 3, 6, 10–12, 18–19, 21–22, 26–27, 30–31, 35–37, 40, 44–47, 49–55, 57, 59–62, 64–71, 75, 76*n*83, 84*n*122, 85, 87–94, 96–105, 108–112, 123–124, 126–127, 130, 132, 135, 139, 142, 148, 163, 176–187, 190–192, 198–199, 218, 219*n*100, 220, 233–236, 241, 262, 273, 278
Arminius, Jacob(us) 27, 232
Arnald of Villanova 72–73
Arndt, Johan 172*n*70
Aslakssøn, Kort 154
Astanus 269
Atanasijević, Ksenjia 293
Aubrey, John 194–195, 198*n*19
Augustine 8, 154, 165*n*47
Averroes 11, 36–38, 52, 59, 62, 83, 95–96, 132
Avicenna 11, 57–59, 62–63, 67–68, 80, 84–86, 93, 107–108

Bacon, Francis 1, 31, 147, 221, 261
Bakker, Frederik 32, 113*n*1
Barbaro, Ermolao 37
Bartholomew of Salerno 20, 61
Basson, Sébastien 6, 11–12, 18, 20, 27, 114*n*3, 135
Batho, Gordon R. 218*n*96
Bayle, Pierre 232
Becanus, Martin 239, 245
Becke, Alexander von der. *See* von der Becke
Beeckman, Isaac 18, 114*n*3, 135
Beguin, Jean 276
Bellis, Delphine 32
Benzi, Ugo 107
Beretta, Marco 200*n*28
Bernardus Silvestris 124
Bernardus Trevisanus 75
Beroaldo, Filippo, the Elder 42
Berthelot, Marcelin 4, 195*n*7
Biesius, Nicolaus 6, 18, 22–23, 27, 30, 113–145
Blair, Ann 27*n*75, 164, 239*n*41
Blank, Andreas 64*n*34, 70*n*61, 106, 228
Boccadiferro, Lodovico 19, 90, 101–105, 109
Bodin, Jean 239
Boethius, Anicius Manlius Torquatus Severinus 124, 190
Bonaventure 166
Borelli, Giovanni Alfonso 16
Borsic, Luka 178
Boyle, Robert 9, 147, 247, 260, 268
Bracciolini, Poggio 8, 23
Brahe, Tycho 110, 147–148, 150*n*4, 158, 256
Brakelants, Louis 124
Brown, Alison 36*n*10, 37
Bruno, Giordano (1548–1600) 2, 6, 12, 14–18, 27, 55, 114, 122–125, 128, 143, 173, 176–177, 185–193, 195–197, 199, 204, 206, 211–218, 226, 249
Bugenhagen, Johan 163
Buridan, John 186
Burton, Robert 198

Cabeo, Niccolò 89, 109
Calcidius 61
Campanella, Tommaso 199
Canone, Eugenio 204*n*44
Cardano, Girolamo 20
Cassirer, Ernst 26
Castellanus, Petrus 143
Cavalcanti, Guido 48
Cellamare, Davide 32
Charleton, Walter 3, 15, 268
Chiaramonti, Scipione 110–111
Cicero, Marcus Tullius 21, 115, 199–200, 243
Clarke, Brian 32

Claves, Etienne Des 6
Clericuzio, Antonio 24*n*69, 56*n*1, 196*n*9, 198*n*16, 210, 247*n*2, 250*n*17, 265*n*76
Clidemus 184
Clucas, Stephen 29, 110*n*117, 195*n*8, 196*n*9, 203*n*39, 221*n*111, 247*n*1, 248*n*7, 249*n*14, 260*n*54, 265*n*75, 266*n*82, 267*n*84
Coimbrãos (Coimbra Commentators) 111
Constantinus Africanus 61
Copernicus, Nicolaus 7, 17, 115, 117, 173, 176, 187, 198, 212
Cremonini, Cesare 111
Cudworth, Ralph 9
Cusanus, Nicholas 13–17, 27, 190

Dalton, John 10, 227
Da Monte, Giambattista 106–108
Damascius 176
Daneau, Lambert 244*n*61
Dante Alighieri 35, 48
Davidson, William 62*n*27, 151*n*8
De Benedictis, Hieronymus Baptista 34
Deitz, Luc 59*n*13, 178*n*5
Democritus 1–3, 7–9, 11–12, 15–17, 21, 23, 25–27, 30–31, 41, 49, 54*n*54, 56, 59, 84–85, 88, 99–100, 104, 108, 110, 126, 141–142, 145, 176–177, 179–193, 195–196, 197*n*14, 198–199, 207, 216, 227, 275
de Monte Snyders, Johann 269
Del Nero, Francesco 36, 42*n*22
Del Prete, Antonella 187*n*31
Descartes, René 3, 6, 9, 13, 17–18, 29–31, 119, 141, 148, 197, 268, 273–275, 284
D'Este, Luigi 47–48
Dick, Steven J. 187*n*31, 194*n*5
Digby, Kenelm 268
Dijksterhuis, Eduard Jan 25, 146, 172–173
Diogenes Laertius 18, 21, 186*n*29–30, 195
Diogenes of Sinope 184
Dionysius Areopagita 132
Donato, Girolamo 37
Drebbel, Cornelis 31
Du Chesne, Joseph 81
DuClo, Gaston 249, 250*n*17
Duhem, Pierre 5
Dumas, Jean-Baptiste 4
Duns Scotus. *See* Scotus
Dürer, Albrecht 132
Düring, Ingemar 19*n*50, 88

Emerton, Norma 59*n*13, 63, 66
Empedocles 27, 88, 110, 126, 179, 181–182, 184–185, 191*n*51
Epicurus 3, 8–9, 11–12, 15–16, 21, 23, 27, 35–36, 39, 41, 49, 54, 56, 59, 85, 141–142, 145, 186–187, 191*n*51, 193, 195–196, 197*n*14, 198–200, 203, 206–208, 216, 243
Erastus, Thomas 154

Fabbri, Natacha 187*n*31
Falloppio, Gabriele 132
Fernel, Jean 24, 57–59, 63–71, 76, 78–79, 82–84, 86
Ficino, Marsilio 21, 23, 33*n*1, 40–41, 44–45, 54, 70, 162, 219
Fienus, Thomas 143
Fracastoro, Girolamo 16, 24, 28, 66, 109, 207–208
Frachetta, Girolamo 33, 40, 46–55
Francus, Raphael 23, 33–43, 52
Frege, Gottlob 17
Frisius, Renierus Gemma 117
Fromondus (Froidmont), Libertus 115, 128, 142, 143
Furley, David 88, 89*n*14, 90

Galen 20, 24, 26, 44, 56, 57*n*3, 57*n*5, 58–62, 65, 67, 69–71, 76, 79, 85, 104–106, 108, 132, 158, 168–169
Galilei, Galileo 17, 22, 173
Ganzarini, Tito Giovanni 46
Gassendi, Pierre 6, 9, 16, 18, 21
Gatti, Hilary 176*n*2, 218*n*94, 249*n*12
Gazes, Theodore 8
Gemma, Cornelius 117
Gemma Frisius, Reinierus. *See* Frisius
Gerard of Cremona 62, 107
Gilbert, William 87*n*2, 103*n*82, 199, 212*n*72, 218
Gonzaga, Scipione 48
Gorlaeus, David 6, 20, 28, 114*n*3, 132, 143, 227–229, 240–246
Grant, Edward 187*n*31, 233*n*21, 241*n*47, 243*n*60

Grasseus, Johann 271
Greenblatt, Stephen 2
Grosseteste, Robert 221*n*111
Gulizia, Stefano 178*n*6, 179*n*7

Haly Abbas 61
Hammer-Jensen, Ingeborg 88–89
Hannequin, Arthur 5
Harriot, Thomas 114*n*3, 247, 250*n*17, 256–258, 260
Healey, John 8
Henry, John 57*n*7, 176*n*1
Henry Aristippus 97
Henry of Harclay 12
Heraclitus 126, 184, 187
Hermes Trismegistus 132, 162, 219
Hershbell, Jackson P. 195*n*7
Heywood, Thomas 170
Highmore, Nathaniel 24, 31
Hill, Nicholas 15–16, 18, 25, 27, 114, 193–195, 198–226
Hippocrates 25–27, 59–60
Hirai, Hiro 24*n*68, 32, 45*n*33, 58*n*9, 64*n*34, 69–70, 209*n*61, 234*n*28
Hobbes, Thomas 256*n*43, 267*n*85, 268
Hohenheim, Philippus Aureolus Theophrastus Bombastus von. *See* Paracelsus
Homer 125, 224
Honorius of Autun 165
Hooke, Robert 31, 268
Horst, Gregor 154
Hugh of St. Victor 165–166

Isidore of Seville 167

Jacobo da Forlì 107
Jacquart, Danielle 20, 61, 74*n*74
Jacquot, Jean 218*n*97, 247*n*1
Jaeger, Werner 88
Johannitius 61, 64
John Buridan. *See* Buridan
John Duns Scotus. *See* Scotus
John of Rupescissa 77
John Wyclif. *See* Wyclif
Jonson, Ben 16, 198
Jungius, Joachim 3, 29–30, 110–111, 273–274, 275*n*20, 276, 284

Kepler, Johannes 17, 196*n*13, 256
Khunrath, Heinrich 164
Knox, Dilwyn 113*n*1, 124*n*38
Køning, Maurits 151*n*11, 154

Lasswitz, Kurd 5, 67, 195, 196
Leibniz, Gottfried Wilhelm 27
Leinkauf, Thomas 178*n*4
Leone, Giuliana 203*n*41
Leucippus 1–2, 8–9, 21, 30, 49, 54, 88, 108, 142, 181, 186, 192, 196, 198–199, 275
Lewis, Clive Staples (C.S.) 148*n*2, 151*n*10, 173
Libavius, Andreas 13, 24, 58, 62, 71–79, 81, 84, 86, 112, 154, 174, 252
Locke, John 22
Lucretius 1, 2, 8, 15, 18, 21, 23–24, 33–56, 85, 124*n*38, 185–186, 193, 195, 196, 199–203, 205–207, 227, 266*n*81
Lull, Ramon 72, 77
Lüthy, Christoph 6, 8*n*19, 15*n*44, 17*n*48, 18, 19*n*52, 20*n*60, 22, 26*n*71, 32, 56*n*1–2, 57*n*6, 67*n*46, 80*n*99, 84*n*119–120, 100*n*69, 112*n*124, 113, 132*n*64, 143*n*98, 176*n*2, 190, 193*n*1, 195*n*8, 196*n*9, 196*n*13, 227*n*1, 228, 236*n*32, 241*n*49, 242*n*52, 245–246, 249*n*12
Lyly, John 171*n*69

Mabilleau, Léopold 1, 5
Machiavelli, Niccolò 218*n*96
Madruzzo, Ludovico 91
McPherson, David 198*n*21
Magnen, Jean-Chrysostôme 6, 8, 9, 11, 28
Maier, Anneliese 11*n*33, 59*n*12, 63, 93*n*35, 95*n*42, 130
Maimonides 11
Mallet, Charles 194*n*4
Manning, Gideon 7, 57*n*6, 113
Mansfeld, Jaap 288
Marchetti, Alessandro 46
Marso, Leonardo 42–43
Martianus Capella 124
Martin, Craig 20, 24, 76, 87, 89*n*18, 101*n*76, 105*n*94, 239
Mathesius, Johann 279
Matteoli, Marco 176*n*2
Maximilian II (Emperor) 115, 132

Meinel, Christoph 28, 56*n*1, 111*n*118, 197*n*15, 274
Melanchthon, Philipp 153, 163–164
Melissus 126, 187
Melsen, Andreas (Andrew) van 1, 123*n*33, 253*n*31
Merchant, Carolyn 173, 194
Mersenne, Marin 256
Michael, Emily 58*n*9, 78*n*93, 80*n*101, 83*n*116, 139, 250*n*17, 261*n*60, 271
Michel, Paul 176*n*2
Micheli, Giuseppe 71, 74*n*76
More, Henry 268
Moreau, Elisabeth 19, 20, 24, 32, 56, 58*n*11, 64*n*35, 71*n*62, 79*n*94
Moschus 8, 18
Moses 18, 27, 224
Mucillo, Maria 176*n*2, 178*n*6
Murdoch, John E. 12*n*35–36, 52*n*50, 66, 123*n*33, 136*n*76
Muscettola, Giovan Francesco 46

Newman, William R. 12, 29–30, 56*n*1, 57*n*6, 58*n*8–10, 61*n*23, 62, 76*n*82, 84, 89, 110, 112*n*123, 130*n*57, 139*n*86, 166*n*53, 195*n*7, 247*n*2, 249*n*16, 250*n*17, 268, 270*n*7, 273*n*14, 279*n*31
Newton, Isaac 29–30, 137, 173*n*72, 227, 268–272, 273*n*15, 275*n*21, 276*n*22, 277*n*25, 279*n*31, 281–284
Nicholas of Autrecourt 13
Nicholas of Cusa. *See* Cusanus
Nicoli, Elena 1, 2, 8, 23–24, 32–33, 41*n*19, 42*n*21, 44*n*29, 45*n*33, 113
Nifo, Agostino 19, 37, 66, 101

Ockham, William of 186
Orpheus 177, 178*n*5, 179

Pagel, Walter 149, 151*n*8
Palmerino, Carla Rita 11*n*29, 32, 113*n*1, 128*n*49, 137*n*79, 142, 143*n*96
Paracelsus 15, 24–25, 58, 72–74, 78, 82*n*111, 147–149, 150*n*4, 155–162, 169–172, 174, 209, 211, 225
Parmenides 55, 126, 179, 184, 187
Pasnau, Robert 6, 242*n*51
Patrizi, Francesco 12, 16, 19, 26–27, 176–185, 192, 195–196, 211, 218–226
Paul of Taranto 12, 29
Pelotti, Giovambattista 36
Pennuto, Concetta 207*n*58
Percy, Henry 194–195, 210, 218, 247
Pereira, Benedict 54, 77
Perna, Pietro 178
Philip (King) 115
Philo of Alexandria 164, 219
Piccolomini, Francesco 20, 47
Pico della Mirandola, Francesco 26
Pico della Mirandola, Giovanni 219*n*99
Pietro d'Abano 106
Pinelli, Gian Vincenzo 196
Pius, Johannes Baptista 33–34, 42–45
Plastina, Sandra 15–16, 18, 25, 27, 114*n*3, 193, 194*n*2
Plato 2, 8, 18, 21–23, 26–27, 36, 53–54, 61, 85, 104, 121, 125–126, 128–129, 132, 136–139, 162–163, 177, 178*n*5, 179, 184, 192, 218, 236
Plautus 42, 163, 168, 171*n*69
Plotinus 132, 162*n*42, 164, 169
Plutarch 92, 183
Poliziano, Angelo 166
Pomponazzi, Pietro 19, 37, 90–105, 107–109, 111–112
Popkin, Richard 119
Posidonius 27
Power, Henry 31
Pratensis, Johannes 164, 170
Prins, Johannes 211*n*69, 218*n*95, 247*n*1, 262*n*62
Proclus 125, 138, 169, 176, 219*n*99
Pseudo-Geber. *See* Paul of Taranto
Puliafito Bleuel, Anna Laura 219*n*102
Pullman, Bernard 1
Purnell, Fred 178*n*4
Pyle, Andrew 1*n*1
Pythagoras 177–179, 187

Raleigh, Walter 218
Redondi, Pietro 7
Regius, Henricus 6, 227
Reisch, Gregor 166
Rhodius, Ambrosius 154, 174–175
Ricardus Anglicus 74, 75

Ricci, Saverio 218*n*94
Riccoboni, Antonio 47
Robert Grosseteste. *See* Grosseteste
Robin, Léon 88, 89*n*14
Ross, David 51*n*46, 54*n*56, 88, 89*n*14

Sakamoto, Kuni 27, 28, 67*n*48, 80*n*99, 132*n*64, 227, 236*n*33–*n*34, 243*n*57
Santorio, Santorio 108, 111
Scaliger, Julius Caesar 20, 22, 66–67, 79, 80*n*99, 81–84, 236–237, 242–246
Scotus, John Duns 59, 83, 139
Sedley, David 43*n*25, 200*n*28
Sendivogius, Michael 271, 282
Sennert, Daniel 13, 17, 20, 24, 29–30, 55, 58, 62, 78–86, 89, 112, 139, 144, 175, 249, 250*n*17, 261, 264, 267, 273*n*14, 274, 284
Severino, Marco Aurelio 26
Severinus, Petrus 16, 24–25, 72, 73*n*71, 78, 81–82, 146–155, 157–159, 161–172, 174–175, 194, 208–211
Shackelford, Jole 16, 24, 25, 72*n*66, 146, 147*n*1, 150*n*4–*n*7, 151*n*8, 151*n*12, 152*n*14, 154*n*18, 154*n*20, 158*n*30, 161*n*40, 164*n*45, 172*n*70, 174*n*75, 209*n*61
Shakespeare, William 164, 168, 170*n*66, 171
Simplicius 50, 51, 118, 132
Snel van Royen, Willebrord 256
Socrates 119
Solea, Nicolaus 279
Sorel, Charles 196
Sozzini, Fausto 229, 230, 231, 232, 234, 235, 240, 245
Spruit, Leen 2, 14, 15, 16, 26, 32, 37*n*11, 47*n*39, 113*n*1, 176, 188*n*35, 192*n*54, 196*n*10, 228*n*5, 245*n*62
Steuco, Agostino 219*n*102
Strabo 27

Taurellus, Nicolaus 8, 119, 228, 235*n*31, 236
Terence 155, 163, 167, 168, 171, 172
Tertullian 165*n*47
Thales 126, 224
Themistius 36–38
Theophrastus 182–184, 209
Theophrastus von Hohenheim. *See* Paracelsus
Thomas Aquinas. *See* Aquinas
Thorndike, Lynn 118, 124
Thurneisser zum Thurm, Leonhart 277
Torporley, Nathaniel 256–257
Trapolino, Antonio 94–95
Traversari, Ambrogio 21, 195*n*6
Trevisanus, Bernardus. *See* Bernardus Trevisanus
Trevor Roper, Hugh 194*n*4, 203*n*40
Trismegistus, Hermes. *See* Hermes Trismegistus

Vallés, Francisco 19, 89*n*20, 90, 101, 105–107, 109, 111
Valori, Baccio 219*n*100
Van Helmont, Johann Baptista 269
Van Melsen, Andreas. *See* Melsen
Varenius, Bernhard 3, 29, 30, 268, 269, 270, 271, 272, 273, 274, 275, 276, 277, 279, 280, 281, 282, 283, 284
Varro, Michel 260–261
Vasoli, Cesare 218*n*96
Vesalius, Andreas 132, 168–169
Vimercato, Francesco 19, 90, 101
Vincent of Beauvais 166
Vitruvius 167–168
Voetius, Gisbertus 227–228, 244
von der Becke, Alexander 269, 284
Vorstius, Conrad 27–28, 227–229, 232–240, 244, 245–246

Warner, Walter 29, 30, 211*n*69, 218*n*95, 221*n*111
William of Conches 10, 23, 74
William of Moerbeke 97, 98*n*60
Wollaston, William Hyde 10
Wood, Anthony 59*n*12, 198
Wyclif, John 13

Zabarella, Jacopo 76, 80–81, 83, 111
Zanchi, Girolamo 233–234
Zedler, Johann Heinrich 115
Zeno 179
Zetzner, Lazarus 164
Zilsel, Edgar 165, 218*n*98
Zoroaster 219
Zwinger, Theodor 164